OPTICAL CODE DIVISION MULTIPLE ACCESS

Fundamentals and Applications

Edited by
Paul R. Prucnal

CRC Press
Taylor & Francis Group
Boca Raton London New York

CRC Press is an imprint of the
Taylor & Francis Group, an **informa** business
A TAYLOR & FRANCIS BOOK

CRC Press
Taylor & Francis Group
6000 Broken Sound Parkway NW, Suite 300
Boca Raton, FL 33487-2742

First issued in paperback 2019

© 2006 by Taylor & Francis Group, LLC
CRC Press is an imprint of Taylor & Francis Group, an Informa business

ISBN-13: 978-0-8493-3683-6 (hbk)
ISBN-13: 978-0-367-39147-8 (pbk)

Library of Congress Card Number 2005051485

Library of Congress Cataloging-in-Publication Data

Prucnal, Paul R.
    Optical code division multiple access: fundamentals and applications / Paul R. Prucnal.
        p. cm.
    Includes bibliographical references and index.
    ISBN 0-8493-3683-X (alk. paper)
    1. Code division multiple access. 2. Optical communications. I. Title.

TK5103.452.P75 2006
621.382'7--dc22   2005051485

**Visit the Taylor & Francis Web site at
http://www.taylorandfrancis.com**

**and the CRC Press Web site at
http://www.crcpress.com**

Foreword

Over the past decade, considerable progress has been made in the penetration of fiber-optic technology into long-haul, metro, and access networks, while the field of all-optical information processing has made significant strides in the area of nonlinear optics. The rapid progress in device integration suggests that in the upcoming years, the complexity of integrating optical signal-processing devices will not be the overriding criterion for the selection of one optical multiple access scheme over another. Interestingly, this parallels the situation in wireless transmission of about 15 years ago. At that time critics argued against the introduction of superior system technologies, such as Code Division Multiple Access (CDMA), on the basis of complexity, but as subsequent events have proven, complexity is no impediment in the face of rapid device integration. This suggests the opportunity to revisit Optical Code Division Multiple Access (OCDMA) and study whether the advantages it can offer to the field of fiber-optic networks parallel those obtained in Wireless. As I wrote a decade ago in *CDMA: Principles of Spread Spectrum Communication*, "The real advantages of a CDMA network are derived through properly understanding and exploiting the higher-level network concepts and features present in a multicellular multiple access system whose users all fully share common frequency and temporal allocations." This holds true for optical networks as well.

The "soft blocking" characteristic of OCDMA makes it well-suited technology to increase the capacity and the number of users of bursty networks. OCDMA has the potential to enable transmission with variable data rates and variable QoS at the physical layer without the use of higher-level protocols. All-optical signal processing has matured to the point where code assignment can be done flexibly and rapidly, enabling network architects to envision a whole new set of functionalities, including code conversion and code add-drop, and enabling a new paradigm of "code-driven networking." A multitude of different application spaces that can benefit from OCDMA are brought up, such as passive optical networks (PON), local and metro area networks, free space optics, and interconnects for computing. As these application spaces become more prominent, this book will be a valuable tool for understanding OCDMA benefits and how OCDMA fits in with next generation photonic networks.

Andrew Viterbi
November 2005

Supplementary Resources Disclaimer

Additional resources were previously made available for this title on CD. However, as CD has become a less accessible format, all resources have been moved to a more convenient online download option.

You can find these resources available here: https://www.routledge.com/9780849336836

Please note: Where this title mentions the associated disc, please use the downloadable resources instead.

Preface

In the early 1980s, the architecture of an unconventional type of fiber optic network was proposed, called an "all-optical network," that utilized ultrafast optical signal processing to implement low-level network functions. The term "all-optical" suggested that the speed and transparency of the optical signal processing, if applied to functions in the data path, might benefit the network in ways that transcended the conventional role of optics for data transport. For example, by replacing electronics with optics to implement functions such as multiplexing, routing, and demultiplexing, data flow bottlenecks could be eliminated at critical points in the network, reducing the delay. It was hoped that with all-optical networks, the replacement of copper with glass would offer more than just higher bit rates, but increased flexibility and functionality that would not otherwise be possible.

One of the first applications of the concept of all-optical networks, inspired by spread spectrum communications, was to use optical signal processing to generate and detect chip sequences in code division multiple access (referred to as "optical CDMA" or "OCDMA"). Optical CDMA had the potential to exploit some of the previously untapped bandwidth of the optical fiber, and to carry over to the optical domain the benefits of CDMA in radio frequency systems (such as users sharing the spectrum and resistance to detection and jamming).

The early attempts to design and implement optical CDMA were both ambitious and primitive. At that time, the technology required for optical signal processing, including nonlinear optical devices, was only in its infancy. Moreover, optical signal processing technology needed to be integrated, use low power, and be packaged before it could be deployed in networks.

The past 20 years have seen enormous strides in the fields of nonlinear optics and device integration, and more modest improvements in packaging. Nowadays, many all-optical devices are practical for use in systems applications, and in some are even commercial products. As device technology has matured, so has the field of optical CDMA. A number of researchers throughout the world, including the chapter authors in this book, have extensively investigated the codes, subsystems, and architectures that can now make optical CDMA networks a reality.

After providing a historical context for optical CDMA, this book discusses the construction of code signatures that can be implemented with optical signal processing technology. Two approaches to coding, coherent and incoherent, are compared from a channel capacity perspective, and recent experimental demonstrations based on each approach are presented. A very promising approach to coding and decoding, based on fiber Bragg gratings, is discussed in detail, as is device integration for optical CDMA systems. Finally, applications to both defense and civilian networks are discussed, including the potential benefits of optical CDMA to privacy, and the combination of optical CDMA with already deployed multiplexing schemes, such as time- and wavelength multiplexing.

In the first chapter, "Optical Code Division Multiple Access: A Historical Perspective," Antonio J. Mendez and Burke J. Anderson provide a historical narrative on the evolution of optical CDMA. The story begins by highlighting some of the key benefits of CDMA in radio-frequency systems that now pervade optical CDMA systems. The authors skillfully capture the underlying threads that tie together seemingly diverse approaches to optical CDMA, as well as the key contributions to coding techniques and technologies that have had a significant impact on the field. The chapter is accompanied by a comprehensive catalog of references to the work done in this field.

One of the key advantages of optical CDMA as a multiple access technique is that it allows many users to simultaneously access a single optical transmission channel through the assignment of unique signatures or code sequences. Of course the properties of the code sequences are instrumental in determining the performance of the entire optical CDMA system, including the size of the address space and the maximum number of simultaneous users. In Chapter 2, "Optical CDMA Codes," Wing C. Kwong and Guu-Chang Yang present both coherent and incoherent optical CDMA coding techniques. The construction of familiar code families and their properties, including Gold sequences, optical orthogonal codes and prime codes, is reviewed. The authors then present a detailed analysis of the performance of these codes, and conclude by discussing the recently developed "extended carrier-hopping prime codes" (ECHPC) and multilength ECHPC, which are particularly suitable for multirate and multimedia applications.

As is evident from both the historical perspective presented in Chapter 1, and the code design perspective presented in Chapter 2, the two dominant approaches to optical CDMA, coherent and incoherent coding, each have their own merits and drawbacks. The two approaches can be compared based on the maximum size of the code space and the feasibility of the optical technology required. Another very important issue is the effect of transmission impairments on coherent and incoherent codes. Kerr nonlinearity is problematic due to the high aggregate power that is typical of optical CDMA systems with many simultaneous users. This subject is treated rigorously by Evgenii Narimanov in Chapter 3 "Information Capacity of Nonlinear Fiber-Optical Systems: Fundamental Limits and OCDMA Performance," where an analysis of the performance of coherent and incoherent optical CDMA codes in fiber optic transmission systems is presented. After discussing the fundamental limit of the information capacity of a nonlinear fiber optic channel, Narimanov derives the information capacity for both coherent and incoherent optical CDMA systems. The chapter emphasizes the importance of considering nonlinear transmission effects in optical CDMA systems and firmly establishes that in some applications an incoherent optical CDMA system is very robust in the presence of fiber nonlinearities.

Whichever approach to optical CDMA coding is used, the proper choice of a suitable all-optical technology to implement coding and decoding is critical. This technology must be compact, manufacturable (packaged), and have a complexity that can scale gracefully with the length of the code sequences. One example of a suitable technology for both coherent and incoherent optical CDMA is fiber Bragg gratings (FBGs). FBGs have emerged as a simple, versatile and effective technology for optical signal processing. For example, FBGs can be used for pulse shaping, filtering and code generation, and have already been deployed in telecommunications systems. Chapter 4, "Optical Code-Division Multiple-Access Enabled by Fiber Bragg Grating Technology," by Lawrence R. Chen, presents a comprehensive and authoritative overview of fiber Bragg grating (FBG) technology, including its underlying theory, design and performance, as well as its potential use in optical CDMA systems. Specific applications of FBGs in spectral amplitude encoding, spectral phase encoding, and 2D wavelength-hopping time-spreading encoding are presented in detail.

In Chapter 5, "Coherent Optical CDMA Systems," by Paul Toliver and colleagues, the authors focus our attention on an important approach to implementing optical CDMA, whereby encoding of code sequences is accomplished by modulating the optical phase. In this chapter, the operating principles of spectral phase coding and temporal phase coding are discussed and compared. Recent progress in phase encoder and decoder technologies is reviewed, including diffraction gratings, virtual image phase arrays, and micro-ring resonators. Optical time gating and optical thresholding is described as a means for rejecting multiuser interference at the receiver. Finally, code selection schemes and network structures for a spectral phase coding optical CDMA system are presented.

Chapter 6, "Incoherent Optical CDMA Systems," by Varghese Baby and colleagues, illustrates the use of ultrafast optical technology to implement an incoherent optical CDMA system. The authors demonstrate that their incoherent approach can provide asynchronous multiple access in a highly scalable system using off-the-shelf technology, without the use of time gating. The chapter describes wavelength-hopping time-spreading (WHTS) codes, generated using spectrally sliced picosecond optical pulses, and discusses the required technologies, including a super-continuum source, WDM filters, and optical delay lines. The results of several experiments are presented. It is shown that architecture has the capability of supporting multiple traffic types that require either different bit rates or quality of service guarantee. The soft limit on the number of network users with graceful degradation of performance with the number of users allows for both flexibility in network management and expandability in the network size.

The eventual deployment of optical CDMA systems in the field will require that they be compatible with the existing communications infrastructure. In particular, optical CDMA networks will need to interface with legacy networks using other multiplexing formats (e.g., WDM or TDM). The choice of multiplexing format for a particular application may be based, in part, on its spectral efficiency, as well as the hardware complexity and scalability. For these reasons, hybrid optical multiplexing techniques have begun to emerge that combine the strengths of several approaches. In Chapter 7, "Hybrid Multiplexing Techniques (OCDMA/TDM/WDM)," Hideyuki Sotobayashi explores such hybrid systems, providing examples of a number of systems demonstrations, as well as techniques for conversion between optical networks having different multiplexing formats.

Before either a coherent or incoherent optical CDMA system can be deployed in the field, integration and packaging of the optical coders, as well as the associated optoelectronic and electronic components will be required. Chapter 8, "Integration Technologies," by Ivan Andonovic, discusses the state-of-the-art in device integration techniques in the context of optical CDMA system requirements. Andonovic presents several plausible options for the realization of integrated optical CDMA subsystems, either of the coherent or incoherent variety. He discusses integration techniques for a number of the components required for optical CDMA systems, including waveguides, filters, switches, modulators, sources, and receivers.

One of the oft-repeated arguments for implementing optical CDMA has been the perception that it can provide some degree of privacy in the physical layer of a network. Though the optical signatures used for multiaccess in optical CDMA systems are called "codes," this does not imply that the codes are cryptographic. Nevertheless, optical CDMA may provide some benefit in other areas of information security. Chapter 9, "Optical CDMA Network Security," by Peter A. Schulz and Thomas H. Shake, provides an introduction to the field of information security, which encompasses areas as diverse as data confidentiality, integrity, availability, and authentication. The authors provide an assessment of the potential for using optical CDMA to achieve data confidentiality, and clearly delineate the limitations of this approach. Possible scenarios and strategies for eavesdropping are described, along with potential countermeasures.

Having focused in the earlier chapters on the coding techniques and hardware required to implement optical CDMA, Chapter 10, "Optical CDMA Network Architectures and Applications," by Robert Runser, discusses potential advantages of optical CDMA at the network level, comparing and contrasting optical CDMA with other optical multiplexing and multiple access techniques. Runser shares his perspective on the potential uses of optical CDMA in a broad variety of applications, including telecommunication networks, computer systems, military environments, and sensor networks.

Included with this book is a CD containing a complete "genealogy" of optical CDMA, with citations from its inception until the time the book went to press. The genealogy is organized by author, by title, and by chronology, in three Excel spreadsheets. It is hoped that this bibliography will serve as an authoritative resource for researchers in this field.

As a whole, this book provides a complete treatment of optical CDMA spanning from technology to systems, and from its history to a glimpse of its future. It will provide a comprehensive resource for the student looking for a broad foundation in the fundamentals of optical CDMA and the researcher wishing to review the state-of-the-art in the field.

I would like to give special thanks to my wife Mindy, and daughters Jenny and Katie, for all of their support throughout this project.

Paul Prucnal
Princeton, New Jersey
November 2005

The Editor

Paul Prucnal received his A.B. from Bowdoin College, and his M.S., M.Phil., and Ph.D. from Columbia University, where he was a faculty member until 1988, when he joined Princeton as a professor of electrical engineering. From 1990 to 1992, Professor Prucnal served as founding director of Princeton's Center for Photonics and Optoelectronic Materials. He has also held positions as visiting professor at the University of Tokyo and University of Parma. Professor Prucnal is the inventor of the "Terahertz Optical Asymmetric  Demultiplexer," an ultrafast all-optical switch, and is credited with doing seminal research in the areas of all-optical networks and photonic switching, including the first demonstrations of optical code-division and time-division multiaccess networks in the mid-1980s. With DARPA support in the 1990s, his group was the first to demonstrate a 100-gigabit/sec photonic packet switching node and optical multiprocessor interconnect, which was nearly one hundred times faster than any system with comparable functionality at that time. For the past several years he has been doing research on an optical CDMA network testbed with support of DARPA MTO. He has published over 200 journal papers and holds 20 patents. He is currently an associate editor of the IEEE *Transactions on Communications* in the area of optical networks. He was general chair of the OSA Topic Meeting on Photonics in Switching in 1999, is an IEEE fellow, an OSA fellow, and a recipient of the Rudolf Kingslake Medal from the SPIE. In 2005, he was the recipient of a Princeton University Engineering Council Award for Excellence in Teaching.

Contributors

Burke J. Anderson
Mendez R&D Associates
El Segundo, California

Ivan Andonovic
Department of Electrical and Electronic
  Engineering
University of Strathclyde, Royal
  College
Glasgow, Scotland

Varghese Baby
Department of Electrical Engineering
Princeton University
Princeton, New Jersey

Camille-Sophie Brès
Department of Electrical Engineering
Princeton University
Princeton, New Jersey

Lawrence R. Chen
Department of Electrical and Computer
  Engineering
McGill University
Montreal, Quebec, Canada

Shahab Etemad
Telcordia Technologies, Inc.
Red Bank, New Jersey

Ivan Glesk
Department of Electrical Engineering
Princeton University
Princeton, New Jersey

Wing C. Kwong
Department of Engineering
Hofstra University
Hempstead, New York

Antonio J. Mendez
Mendez R&D Associates
El Segundo, California

Ron Menendez
Telcordia Technologies, Inc.
Red Bank, New Jersey

Evgenii Narimanov
Department of Electrical Engineering
Princeton University
Princeton, New Jersey

Paul R. Prucnal
Department of Electrical Engineering
Princeton University
Princeton, New Jersey

Darren Rand
Department of Electrical Engineering
Princeton University
Princeton, New Jersey

Robert J. Runser
Telcordia Technologies, Inc.
Red Bank, New Jersey

Peter A. Schulz
Massachusetts Institute of Technology
  Lincoln Laboratory
Lexington, Massachusetts

Thomas H. Shake
Massachusetts Institute of Technology
  Lincoln Laboratory
Lexington, Massachusetts

Hideyuki Sotobayashi
National Institute of Information and
  Communication Technology
Tokyo, Japan

Paul Toliver
Telcordia Technologies, Inc.
Red Bank, New Jersey

Lei Xu
Department of Electrical
  Engineering
Princeton University
Princeton, New Jersey

Guu-Chang Yang
Department of Electrical
  Engineering
National Chung-Hsing
  University
Taichung, Taiwan

Contents

1 Optical Code Division Multiple Access: A Historical Perspective

Antonio J. Mendez and Burke J. Anderson

CONTENTS

1.1 INTRODUCTION

Two technical papers form the foundation of most developments in optical code division multiple access (OCDMA) as we know it today. These papers are the one by Prucnal, Santoro, and Fan in 1986, and the one by Weiner, Heritage, and Salehi in 1988. Therefore, many of the chapters in this book are traceable to one or the other of these publications. In this first chapter we'll put these important publications and their evolution in a historical context. We will also see that there were and are other trends that developed in parallel or as spin-offs of these two seminal papers. These other trends include 2D coding (of which fast frequency hopping [FFH], space/time, and wavelength/time are examples) and hybrids of pulse-position-modulation (PPM) and CDMA, PPM-CDMA. Other trends include direct sequence coherent coding and coherence multiplexing that takes advantage of the limited coherence length of particular light sources to produce a coding regime. We will see that these various trends were based on or driven by applications or emerging device technologies. Certainly, new technologies such as planar lightwave circuits encouraged new applications or implementations of OCDMA. In addition, it was recognized early on that the performance of OCDMA (measured in bit error rate [BER] and number of concurrent users) was limited by multiaccess interference (MAI) so schemes like optical hard-limiting (OHL) and multiuser detection (MUD) were proposed and developed. The performance limitation is particularly severe for coding based on on-off-keying (OOK).

1.1.1 ABCD

For all intents and purposes, the serious development of OCDMA began with the publication of

"Spread spectrum fiber-optic local area network

using optical processing"

Prucnal, P.; Santoro, M.; Ting Fan;

Lightwave Technology, Journal of , Volume: 4 Issue: 5 , May 1986

Page(s): 547–554.

(The citation is directly as it appears in IEEEXplore.) The significance of this seminal paper is that it showed a wide audience that pseudo-orthogonal direct sequences such as prime sequence codes could readily be implemented as tapped delay lines based on optical fibers. The correlators or decoders for the encoded data could similarly be implemented as tapped delay lines. This meant that, in a system sense, an optical network could be imagined that allowed sharing of the transmission channel in a multiple access scheme. Furthermore, it meant that the mechanisms that imposed the codes on the data, the encoder, and that recovered a particular set of coded data, the decoder, could be designed to operate without requiring intervention from electronics. Today we would call this "all-optical processing" or "photonic processing" or even "processing at the speed of light." Generating the data flow in the optical domain did require electrical-to-optical conversion, but encoding/decoding was now an all-optical process. Thus, the portion of the title that says "using optical processing" was indeed an important breakthrough and it triggered new research into the physical implementation of coding schemes.

1.1.2 GFGH

The second, just as important, development was reported in

Optics Letters, Volume 13, Issue 4, 300–302
April 1988

Encoding and decoding of femtosecond pulses

A. M. Weiner, J. P. Heritage, J. A. Salehi

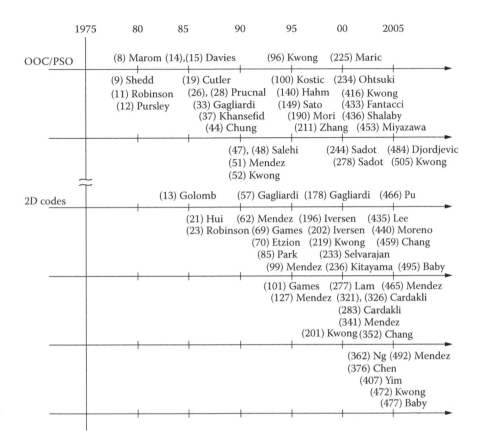

FIGURE 1.1 Threads and trends of OOCs/PSOs and 2D OOCs.

(The citation is directly as it appears in the Optical Society of America Optics INFOBASE.) The key findings were that coherent, narrow-pulse, ultra-wide bandwidth spectra could be produced and manipulated in the spectral or phase domain. In particular, the spectra could be encoded with phase codes. The significance of this is that the developments of radio frequency (RF) communications were now applicable to optical communications. Early on, bipolar and spectral encoding schemes that mimic the codes used in RF communications were added to the knowledge base of OCDMA, as will discussed in the later chapters. (Later coherent OCDMA schemes in which the codes were applied in the time domain, direct sequence codes, were demonstrated, and here, too, RF communications techniques were applicable. These codes were implemented with tranversal filters with similarities to the ones in Prucnal's work.)

This historical chapter describes and discusses the threads and trends have been part of the evolution of OCDMA. Figure 1.1 shows the trend that includes Prucnal's seminal work with optical orthogonal codes/pseudo-orthogonal sequences (OOCs/PSOs) [26, 28]. The figure intentionally includes 2D (OOC) codes because we know from personal experience that frustration with the relatively low cardinality and spectral

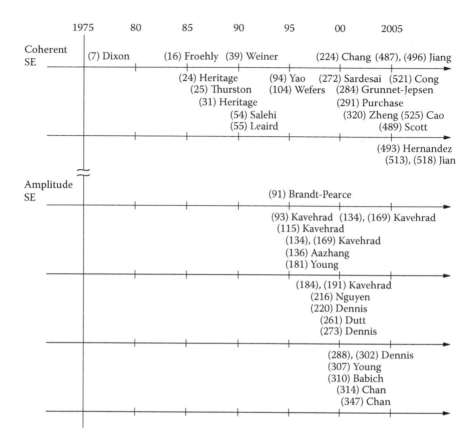

FIGURE 1.2 Threads and trends of coherent and amplitude spectral encoding.

efficiency of OOCs/PSOs, coupled with their problem with MAI, led to an intense interest in the 2D OOCs. This transition of interest occurred around 1990, as shown in Figure 1.1. The figure shows that there is a growing number of researchers in 2D, as well as a growing body of work.

Figure 1.2 shows the threads and trends associated with Weiner's seminal work [39]. The figure intentionally includes amplitude spectral encoding because the Weiner-Heritage encoding/decoding configuration was adopted as a means to explore the use incoherent broadband sources (e.g., LEDs) instead of the more complex and sensitive coherent sources of the Weiner approach. Again the figure shows an increase of research activity in coherent SE, as well as an increase in the number of investigators.

Figure 1.3 shows the threads and trends associated with FFH concept which we associate with Gelman [38], Tančevski [126], Andonovic [132], Bin [207], and Fathallah [242]. FFHs are somewhat similar to 2D codes (especially since the latter are generally implemented as wavelength/time codes), but 2D codes have somewhat more leeway in allocating resources to the two dimensions and tend to be very permissive of codes with multiple pulses per row and/or column. FFHs tend to have very good cardinality and lend themselves to implementation with modern photonic

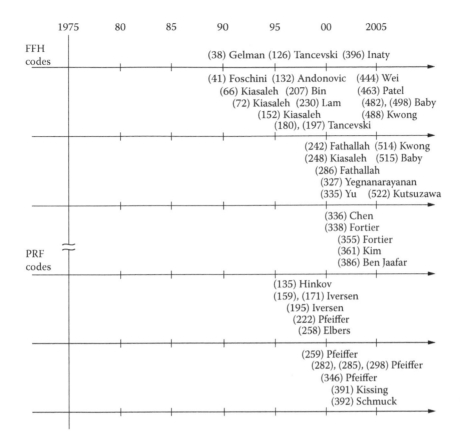

| 1975 | 80 | 85 | 90 | 95 | 00 | 2005 |

FFH codes

(38) Gelman (126) Tancevski (396) Inaty

(41) Foschini (132) Andonovic (444) Wei
(66) Kiasaleh (207) Bin (463) Patel
(72) Kiasaleh (230) Lam (482), (498) Baby
(152) Kiasaleh (488) Kwong
(180), (197) Tancevski

(242) Fathallah (514) Kwong
(248) Kiasaleh (515) Baby
(286) Fathallah
(327) Yegnanarayanan
(335) Yu (522) Kutsuzawa

(336) Chen
(338) Fortier
(355) Fortier
(361) Kim
(386) Ben Jaafar

PRF codes

(135) Hinkov
(159), (171) Iversen
(195) Iversen
(222) Pfeiffer
(258) Elbers

(259) Pfeiffer
(282), (285), (298) Pfeiffer
(346) Pfeiffer
(391) Kissing
(392) Schmuck

FIGURE 1.3 Threads and trends of FFH and PRF codes.

devices such as Bragg gratings. Thus, there is growing interest in them, and a growing set of investigators.

Figure 1.3 includes the treads and trends of PRF codes. These were the subject of intense research for access and metro applications, with significant participation by Deutsche Telekom and Alcatel SEL. The figure shows that the research has lapsed, although there were excellent results by Alcatel in testbeds and field trials. Similar figures can be configured for coherent OCDMA, coherence multiplexing, PPM-CDMA, OHL, and multiuser detection (MUD).

1.1.3 Lмсd

This book includes chapters that provide snapshots of OCDMA today. In this chapter we step back to provide an overview through a guided tour of the previous developments in OCDMA, organized more or less chronologically and following the threads and trends of Figure 1.1 through Figure 1.3. For convenience, we have picked 5-year increments of time to divide the material, although there are threads that have been quite long-lasting). The chapter concludes with some lessons learned and a look at where OCDMA may be going in regards to future applications.

The literature of OCDMA includes theory, concepts, analyses, devices, simulations, technology demonstrations, testbeds, and even attempts at productization. For this reason, the literature is dispersed, scattered among various IEEE and OSA technical journals, as well as IEEE, OSA, and SPIE conferences and workshops and even corporate reviews. Some research was deemed sufficiently innovative that it was first reported in *Science Magazine*. Within the IEEE, contributions to OCDMA are primarily from publications, conferences, and workshops of the Lasers and Electro-Optics, Communications, and Information Theory Societies. The purpose of this book, therefore, is to depict the threads and trends of various OCDMA concepts and developments and to provide the platform for further exploration.

1.2 EARLY HISTORY (BEFORE 1980)

Since much of OCDMA is based on on-off-keying (OOK) and unipolar codes, quite a lot of the related research has dealt with discovering or generating codes with good auto- and cross-correlation properties (pseudo-orthogonal, or PSO). But the search for such codes is not unique to OCDMA (Khansefid [37]). Related applications that preceded the seminal work described above include missile guidance or telecommand codes (Eckler [1] and Reed [4]); interferometry with moveable antennas (Biraud [6]); and error correcting codes (Robinson [2]).

Coherent OCDMA has had its own development. Optimal binary sequences such as Gold codes that play an important role in later coherent OCDMA work were described during this early historical period (Gold [3]). Based on recurring citations, it is obvious that coherent and coherence multiplexing OCDMA schemes were based to a large extent on concepts and practices from radar (Dixon [7]).

During the 1970s, when laser target designation was maturing, there were development efforts in pulse-repetition-rate (PRF) codes, preferred by the U.S. Navy, and pulse-interval-modulation (PIM) codes, preferred by the U.S. Air Force. Today we would recognize these as classes of PSO codes. But these other applications established the foundations on which the codes for OCDMA could be developed. (PSO codes for OCDMA are also known as optical orthogonal codes, OOCs, in the OCDMA literature.)

A special class of OOCs, the optimum Golomb ruler (also called spanning ruler), was defined and described (Robinson [11]). It is a special class because the values of the cross-correlations are ones or zeroes, and the autocorrelation sidelobes are ones or zeroes. Thus, new families of sequences with good correlation properties were described (Shedd [9]).

Many of the linear (direct sequence) and 2D codes researched after 1990 are traceable to some of these early foundations.

Methods for computing symbol error rates, taking into account the photodetector or avalanche photodiode characteristics but also fiber impairments, emerged, with special emphasis on inter-symbol interference (ISI) (Forney [5] and Dogliotti [10]).

The notion of signal processing by means of fiber optic delay lines emerged as a concept and it was reduced to practice in the early 1980s (Marom [8]).

1.3 1980–1984

This timeframe produced some of the keystone papers in the area of PSOs/OOCs that are frequently cited in latter work, for example (Sarwate [12], Weber [12a], Stark [12b], Shaar [14], and Shaar [15]). Interestingly, this period also produced a paper that was influential later in comparing the bandwidth efficiency of linear and 2D (matrix) codes of the same cardinality (Golomb [13]); this paper also introduced the notion of single pulse per row (SPR) 2D codes.

Pulse shaping and analysis of narrow (picosecond) light pulses, a technique that becomes central to spectral phase encoding or frequency encoding, was described in (Froehly [16]). Complex topologies for fiber optic networks also emerged as a research topic (Marhic [17]).

1.4 1985–1989

Research on fiber optic networks and OCDMA in particular began in earnest during this era: Tamura [18] explored OCDMA using Gold codes, Jackson [19] developed fiber optic delay line signal processing, Marhic [22] described applications to local area networks, and Hui [21] investigated pattern codes involving multiple fibers. New constructions for 2D OOCs were developed (Robinson [23] and Gagliardi [33]). Tamura's work based on Gold codes included conceptual designs using tree, star, bus, and ring topologies, as well as experiments with 127 chips and simulations with 31 and 127 chips. The agreement between experiment and simulation was quite good. The paper primarily dealt with asynchronous transmission but included an appendix on the synchronous case.

Coherence multiplexing, in which the finite coherence time of light sources is used for recovering a signal by matching the optical paths, appeared in this early era (Brooks [20] and Goedgebuer [32]).

The foundations for frequency domain encoding ("spectral phase encoding") evolved from exploring the means of producing femtosecond pulses and developing diagnostic techniques for determining their properties (Heritage [24], Weiner [24a], and Thurston [25]). One of the devices for studying and manipulating the temporal characteristics of short optical pulses and studying their transform-limited properties, the opposed dispersive elements with a spectral mask in the mid or Fourier plane, emerged as the preferred embodiment of encoders and decoders for frequency domain OCDMA (Weiner [31], Weiner [39], and Weiner [41a]).

As discussed in the introduction, this era also produced the seminal papers that emboldened a new generation to develop OCDMA codes, systems, and techniques (Prucnal [26] and Prucnal [28]). These papers clearly demonstrated that encoders and decoders could be implemented with fiber optic tapped delay lines ("transversal filters") so that encoding and code correlation could be carried out all-optically. Like Marhic [17], these papers identified local area networks (LANs) as the field of application and they used a very democratic star network topology (rather than, e.g., a bus, ring, or a tree topology). These papers used prime sequence codes that are not ideal OOCs, and this triggered interest in developing better (i.e., low autocorrelation sidelobes and strictly limited cross-correlations) codes (Davies [30], Gagliardi [33],

Khansefid [37], Salehi [42], Chung [44], Salehi [47], and Salehi [48]). Another essay applying OCDMA to LANs, using direct sequence codes, and including a mathematical model algorithm that assisted in choosing suitable sets of code sequences and that defined the optimum matched filter detector was O'Farrell [43].

The notion of optical hard-limiting, borrowed from optical digital computing, appears to have been first suggested during this time as a means of ameliorating MAI (Aazhang [36] and Salehi [48]).

In addition, the issue of code acquisition for direct detection, on-off-keying pseudo-noise codes began to be addressed. Kiasaleh [45] and Gagliardi [33] discussed PPM-CDMA and the associated issues of slot and frame synchronization.

The concept of multiuser detection (MUD), in which computations assist in ameliorating the MAI, emerged for optical multiple access applications, with analyses and comparisons of synchronous and asynchronous cases (Verdú [27] and Poor [35]). The study of nonlinear correlators also emerged (Aazhang [46]).

A remarkably prescient paper appeared during this fruitful formative era (Gelman [38]). It proposed a scheme of optical CDMA based on a "combination of Color (Wavelength, λ), and Time Hopping that it labels OOK/OR/$T\lambda H$" because of its use of on-off-keying and an OR optical system. The paper went on to describe that "We will limit out analysis to the system with fast Time and fast Color Hopping where both hopping rates must be higher than the data bit rate." We will see later in this chapter that this is entirely consistent and similar with schemes developed and demonstrated after 1995. This paper further explored the bit error rate (BER) and the optimum code weight for a given channel load; thus, this paper delved into the realm of network analysis and optimization. The paper treated the wavelength and time aspects of the code as independent, and generally used a (temporal) code length of 127 and a variable number of wavelengths. Another interesting aspect of this paper was that it treated bus as well as star network topologies. It concluded with the observation that "Another attraction (of the fiber optic CDMA network for real-time communication) is the flexibility that can be offered only by a bit switched network which allows construction of packet or burst switched overlays with any combination of data rates and any size information messages."

In the same vein, random carrier (RC) fast frequency hopping OCDMA were proposed for achieving high capacity in a fiber-optic local area network (Foschini [38a] and Foschini [41]).

Whereas in this early inventive stage there was a wide interest in all-optical processing for OCDMA, it was shown that electrical processing had significant merit in OCDMA systems, especially with respect to inducing code orthogonality (MacDonald [40]).

Although OCDMA using OOC or direct sequence (linear) codes allows random access communications, it comes at the price of electronic bandwidth (small pulse-width or chip time Tc). A study was conducted of the chip times required by different network sizes (code cardinality) and user data rates (Mendez [51]), and the results did not compare favorably with time division multiplexing (TDM), except in the area of synchronization. This set the stage for the search for more competitive, higher spectral efficiency solutions. The same study identified boundaries for electrical vs. optical implementations of the OCDMA encoders/decoders, based on simple rules for the physical dimensions of the optical delay lines in the optical implementation.

Synchronous and asynchronous formats of OCDMA were compared to time division multiple access (Kwong [52]). It was determined in this study that synchronous OCDMA had some immunities to fiber impairments, such as dispersion. It was similarly determined that synchronous OCDMA was more appropriate for short-haul applications, and asynchronous for long-haul applications, where aggregation would require lower cardinality and synchronization is difficult. This work included experiments with a two-user system at 500 Mb/sec.

1.5　1990–1994

In this period, development was continued of coherent spectral OCDMA (Salehi [54]), as well as programmable pulse shaping of femtosecond pulses by means of liquid crystal phase modulators (Weiner [55] and Weiner [79a]) and the first programmable liquid crystal phase and amplitude modulator [Wefers [104]]. A detailed analysis and simulation of coherent spectral OCDMA communication using the Weiner-Heritage configuration, including potential fiber impairments and limitations, was performed (Yao [94]). This very extensive and detailed work was carried out on a Cray-X_MP/18 computer. It studied the alignment requirements between the encoding and decoding masks as a function of code length or frequency bins and determined, for example, that for a code length of 31 that the alignment needed to be better that 1/10 of a frequency bin. It also determined the min and max pulsewidths as a function of code length and corresponding fiber impairments that need to be compensated for each of these cases.

Around this time there was a flurry of interest in coherence multiplexing (Chu [53], Sampson [56], Griffin [80], Griffin [81], Marhic [83], Chang [86], Griffin [87], Marhic [95], Griffin [103], Griffin [106], Griffin [107], Karafolas [121], Gupta [129], and Sampson [130]). The effect of phase drift on coherence dependent OCDMA was analyzed (Rusch [131]). A Weiner-Heritage implementation of coherence multiplexing was described (Griffin [106]). Coherence multiplexing was proposed as a scheme for secure communications (Wells [97]).

An algorithm for converting sets of optimum Golomb rulers into 2D OOC codes with higher cardinality and spectral efficiency than the initial set of Golomb rulers was proposed (Gagliardi [57], Mendez [62], Mendez [67], Mendez [68], Mendez [74], Mendez [84], Mendez [99], and Mendez [102]) and became practice (Mendez [127] and [128]). As part of this work, code designs for a 16×16 star network were developed, and a 4×4 technology demonstrator was developed and demonstrated at per user data rates of 100–250 Mb/sec (Mendez [102] and Mendez [127]). BERs better than 10^{-9} were measured for this technology demonstrator (Mendez [128]). One of these citations discussed the matrix codes as being implementable as space/time as well as wavelength/time codes (Mendez [46]) thus anticipating later developments in 2D coding. During this time it was also shown that optical amplification was compatible with preserving the code correlation properties (Mendez [122]). The 2D codes were implemented as time/space (S/T) codes, using multimode ribbon fiber as the spacelike component, a layered star network topology, and commercial Fabry-Perot laser sources (gain switched to produce time chips). A treatment of the transmission and performance of OCDMA over multimode fiber was provided in Walker [114], Walker [118], and Walker [125].

OCDMA with on-off-keying (OOK) and pulse-position-modulation (PPM) were compared (Khansefid [60]). This work showed that OOK-CDMA tended to support fewer users than the code set cardinality at a BER of 10^{-9}, but that PPM-CDMA could support the full code set cardinality, at M = 4.

Analysis approaches also emerged that could capture and model the performance of OCDMA in the presence of impairments such as phase noise (Foschini [58] and Brady [64]). A clarification on OCDMA system performance, offered as a comment to Salehi [48] was proposed (Vajda [65]). The computation of error probabilities in fiber optic CDMA systems was addressed (Mandayam [88]). Optimum demodulation schemes for OCDMA were investigated (Nelson [89]).

Kiasaleh resumed the treatment of code acquisition and slow frequency-hopping with a bus network topology (Kiasaleh [61], Kiasaleh [66], and Kiasaleh [78]) and also introduced the application of pseudo-noise sequences to free space (satellite) communications and ranging (Kiasaleh [76]). Synchronous CDMA and arbitrary time-hopping codes were developed (Kostic [124]).

The information theory community continued development of 2D ("one-coincidence," "sonar," "radar") code families (Moreno [69], Golomb [70], and Moreno [101]).

A novel class of codes whose autocorrelation peak is positive and whose autocorrelation sidelobes and crosscorrelation are zero or negative values was demonstrated (Alberta codes; Vethanayagam [73]). The complementary correlation that this scheme requires (converting unipolar codes to bipolar codes at the receiver) was produced by an array of detectors. The number of detectors required is equal to the length of the code (16 in this case).

Salehi continued his exploration of codes for spectrally phase-encoded coherent ultrashort systems, including analyses of the corresponding bit error rate (BER) performance of the candidate codes (Hajela [77]). He also explored the relaxation of the number of acceptable coincidences between temporal codes, using optical hard-limiting (OHL) to ameliorate the resulting MAI (Azizoglu [63] and Azizoglu [82]). New code constructions for OOCs were described (Chung [55a], Moreno [89a], Maric [100], Tančevski [111], Kwong [113], and Kostic [124]).

The earlier work on synchronous vs. asynchronous OCDMA was expanded (Vajda [65] and Kwong [75]).

Modeling of OCDMA systems performance that included the effect of photo-detector and APD noise, thermal noise, and multiuser interference was performed (Kwon [71], Kwon [72], Lam [79], Ho [116], Kwon [120], and Kwon [123]). The analysis included signaling in which the "1"s and the "0"s were assigned orthogonal codes. The analysis concluded that bounding of the cross-correlation by the value "1" is not necessary for adequate performance. These analyses include the intriguing concept of "multibits/sequence-period," i.e., that the code sequence need not be bounded by the bit-length (Kwon [72] and Kwon [123]). Other related contemporary research included the effects of the transmission medium, multimode fiber, on OCDMA performance (Walker [114] and Walker [118]).

In the case of unipolar codes, the advantage of 2D codes vs. direct sequence (DS) or linear codes with respect to SNR and spectral efficiency was demonstrated by means of a technology demonstrator that implemented (1) a set of optimal Golomb rulers of cardinality four and weight four, and (2) a set of sonar matrices (i.e., 2D) of cardinality four and weight four (Park [85]). The sonar matrices were implemented as space/time (S/T) codes. One of the major advantages of 2D codes demonstrated here is that the chip time (temporal or transmission bandwidth) expansion for a 2D code is much smaller for 2D codes by the ratio of the DS code length to the number of 2D code time-slots. Also, it was shown that cyclic row shifting of sonar matrices produced new matrix OOCs, thus giving rise to the higher cardinality compared to linear codes without increasing the code length or code dimension (product of matrix rows and columns). The sonar codes were quite similar to those developed for FFH. Some of the data in this work were used to explore the application of OCDMA to fiber optic digital video multiplexing (Dale [92]).

The combination and compatibility of forward error correction codes with OOCs was analyzed (Dale [90]). This work found that both Reed-Solomon and convolutional codes were quite effective in reducing the BER of an OOK system (even though the MAI may not be Gaussian), and the improvement increases with code weight. Similarly, multiuser detection concepts for dealing with multiaccess interference that were introduced in the previous era were further developed (Brandt-Pearce [91], Brandt-Pearce [112], and Brandt-Pearce [119]).

The notion of amplitude spectral encoding in a Weiner-Heritage configuration (Weiner [39]), coupled with an array receiver for correlation and multiuser detection, was introduced in Brandt-Pearce [91]. The article considers mode-locked lasers and broadband incoherent sources for the broadband spectrum required in this configuration and selects the former. On the other hand, incoherent sources (e.g., LEDs) are incorporated in amplitude spectral encoding schemes in (Kavehrad [115]). Analyses of amplitude spectral encoding with differential detection were also carried out (Zaccarin [93]) and it was found that, using the 3 dB spectrum of an LED in conjunction with a code of 127 frequency bins, that ~40 users could be supported at a BER of 10^{-9} and at the modulation limit of the LEDs (~500 Mb/sec). Better performance could be obtained by using a smaller portion of the LED spectrum (to flatten the spectrum and equalize the spectral bins), but at the expense of the performance becoming thermal noise limited (owing to the reduced detectable power).

Balanced detection was introduced as part of the bipolar code recognition and discrimination (Zaccarin [110] and Andonovic [117]).

"Ultrafast" concepts, in the sense of < 1 psec pulses, entered the realm of OOCs (not necessarily coherent spectral phase encoding (Kwong [96]). At the same time, the nonlinear discriminants "terahertz optical asymmetric demultiplexer" (TOAD) was proposed (Sokoloff [98]), and it will be seen later that this formed a crucial part of the signal processing in coherent spectral phase encoding OCDMA.

We close this era with a quotation from Green [105], after he recited some of the impairments that plague OCDMA: "In view of such difficulties, we might ask what the advantage of CDMA might be for optical networking. The answer is that CDMA is a true *tell-and-go* protocol. . . ." (p. 445). With respect to OOC waveforms, he lamented that ". . . the search for good optical codes . . . has only yielded some bounds and some good rules for guiding enumerative computer searches for good codes, but no closed-form method of synthesizing optimal optical codes" (p. 453). We will see how this played out in the subsequent years.

1.6   1995–1999

This era of OCDMA was marked by optimism, almost exuberance. Deutsche Telekom, charged with installing a telecom infrastructure in what had been East Germany, tasked the Heinrich Hertz Institute (HHI) and the Technische Universität Ilmenau (TU-Ilmenau) with determining the state of the art of OCDMA for potential applications in this "greenfield." HHI and TU-Ilmenau conducted a worldwide survey reported in Iversen [168]. This survey included a comparison and classification of all appropriate systems explored (analytically or experimentally) to that time and with potential application for future telecom networks. The survey included theoretical constructs and significant experimental achievements to that date that could be used as a point of departure for appropriate new concepts. The survey was used to select approaches for the new fiber infrastructure (an early idea of FTTx). This survey was then collected into a basic theory of fiber-optic CDMA that included coherent and incoherent implementations (Iversen [187]); see also Andonovic [186] for the particular case of incoherent OCDMA, Karafolas [200] and Gupta [205] for other OCDMA compilations and reviews, and Andonovic [293] for an update that included time spreading and wavelength division combinations as well as coherent receivers. At the same conference where (Iversen [168]) was presented, another paper explored spectral encoding with m-sequences for the "FTTx" application (Ziemann [171]). The paper considered SLDs, SOAs, and multiwavelength lasers as the source, and AOTFs, AWGs, and fiber gratings for implementing encoders and decoders. The paper concluded that such schemes could support hundreds of users at a network throughput of several Gb/sec. Complementing this paper, and at the same conference, experimental results were described for an OCDMA system based on spectral amplitude encoding with noncoherent sources (Adam [169]). This system also used m-sequences, the source was an LED, the transmission medium was multimode fiber, and differential detection was used. Several low cost, low power implementations for this "FTTx-like" application were proposed and/or developed during this time (Iversen [135], Iversen [159], Iversen [195], Iversen [196], and Iversen [204] for computer communications; Gupta [205] for local area networks; Pfeiffer [222], Ziemann [227], and Muckenheim [246] for computer communications; Elbers [258] and Pfeiffer [259] providing an early report on Alcatel SEL's initial demonstrations of low cost, robust OCDMA).

Alcatel described the role of OCDMA in their vision of future network hierarchies in the *Alcatel Telecommunications Review* of 3Q98 (Pfeiffer [250]). Alcatel SEL's investment in OCDMA technology led to a very successful field trial on a testbed in Stuttgart, Germany, that demonstrated robust OCDMA, based on LED sources and PRF (Fabry-Perot or Mach-Zehnder) encoders/decoders, over a 40-km access network testbed. This work was reported at OFC '99 (Pfeiffer [282]) as well as other venues (Pfeiffer [285] and Pfeiffer [298]). Pfeiffer [282] described experiments with eight concurrent users transmitting error free at 155 Mb/sec each, but it cited related work demonstrating 32 users, and it indicated that the concept could support 64 users by using forward error correction.

Efforts to develop efficient OCDMA system analyses methods continued. The performance advantage ("gain") of OOCs compared to systems based on prime sequences was investigated, and importance sampling developed as a means of reducing the computation time of the system simulations (Mandayam [138]). The effect of laser phase and interferometric noise on the performance of coherent OCDMA was studied (Radhi [131b], Gupta [131c], Poor [143], and Gupta [203a]). The performance of number-state and coherent-state optical CDMA in lossy photon channels were analyzed (Shalaby [144]).

The effect of collisions (the concurrent addressing of a receiver by more than one transmitter) and means to mitigate this was studied in Muckenheim [260].

In the meantime, the concept and approach of wavelength hopping/time spreading codes was resumed (Andonovic [132], Kiasaleh [152], Tančevski [180], and Tančevski [197] for security applications; Bin [207] with new constructions of one-coincidence 2D codes; Kwong [219] and Lam [230] for secure communications; Fathallah [242] for codes specifically designed for optical FH-CDMA; and Fathallah [286] for experimental results with a system that uses tunable Bragg gratings and supports up to 30 users at a per user data rate of 500 Mb/sec; Lam [277] and Cardakli [283] with applications to header recognition and switching and network reconfiguration; and Kiasaleh [248] with a design and performance analysis of a hybrid time hopping/TDMA). (For additional discussion on security issues in all-optical networks, see Medard [210].)

Similarly, the concept and approach of multiuser detection to ameliorate MAI in OCDMA systems continued to evolve (Brandt-Pearce [139], Halford [157], Nelson [165], Nelson [166], Halford [223], Brandt-Pearce [238], and Tang [299]). Nelson [166] found that the multiuser detection efficiency increased with the code weight and decreased with the number of concurrent users. Shalaby [308] considered four schemes (of differing complexity) for interference estimation and cancellation and evaluated them with respect to no interference estimation and cancellation. He found that all four schemes had the benefit of removing the error floor usually associated with OOK-OCDMA.

Analysis and code construction of OOCs continued, both linear (direct sequence) and 2D (Yang [137], Maric [140], Yang [141], Sato [149], Kwong [150], Kwong [151], Mendez [161], Zhang [162], Gagliardi [174], Yang [176], Mendez [177], Gagliardi [178], Kwong [182], Ho [185], Mendez [188], Argon [189], Mendez [192], Zhang [198], Kwong [201], Iversen [202], Zhang [203], Andonovic [206], Zhang [211], Zhang [218], Selvarajan [233], Kwong [240], Zhang [247], Zhang [263], Zhang [267], Zhang [268], Zhang [289], and Zhang [296]). A hybrid WDM/CDMA LAN was described and analyzed (Mokhtar [147]). Hybrid fiber optic networks were investigated: frequency-domain encoding with frequency division multiplexing (Kamakura [214]), WDMA+CDMA (in a comparison with multiwavelength CDMA) (Kwong [219]), and a hybrid of frequency encoding CDMA with time encoding to reduce the effects of MAI (Sonoda [301]). A hybrid OCDM/WDM with wavelength conversion was proposed to effect the function of add/drop multiplexing in future transport networks (Kitayama [257]).

A new method of constructing codes based on the genetic algorithm was described in (Sadot [244] and Sadot [278]). A demonstration of the implementation of 2^n prime codes (n = 3, 64 ps optical pulses) with a serial encoder was described in Zhang [247].

The specific application to packet networks with and without slotted configurations was analyzed in Hsu [212] and Hsu [213]. Applications to optical wireless communication were described in Vento-Alvarez [304]. Applications to sensor networks, and in particular to reducing the crosstalk among sensor channels, was given in Kullander [251] and Koo [281].

Further work was performed on spectral encoding of incoherent sources based on variations of the Weiner-Heritage configuration (Kavehrad [134], Aazhang [136], Aazhang [181], Kavehrad [184], Kavehrad [191], Nguyen [216], Dennis [220], Dennis [274], Dennis [288], and Dennis [302]). In particular, Kavehrad [134] applied the notion to a scalable ATM switch. Nguyen [216], Dennis [220], and Dennis [302] described the difficulty in obtaining the theoretically expected performance when reducing the concept of amplitude spectral encoding to practice, with or without differential detection. Lee [307] and Babich [310] addressed a planar lightwave circuit (PLC) design for the Weiner-Heritage configuration using incoherent broadband sources. Acoustically tunable PRF configurations were also explored (Hinkov [135]). Mach-Zehnder schemes were also proposed for effecting the spectral encoding (Lam [231]). An ingenious application of coding to reduce the effect of four-wave-mixing (FWM) was given in Guo [221]. Broadband spectral encoding was applied to the multiplexing of sensors (Shlyagin [278]). Applications to video transport/switching were explored (Chan [271]).

Research on the concatenation and compatibility of channel coding and OCDMA continued (Kawakami [153] and Dale [155]). The former citation proposed an all-optical implementation of error detection and correction for the transmission of 2D pictures. Other research in codes for parallel transmission (pictures or computer communications) included Nakamura [236], Nakamura [249], and Kwong [264].

Development of hybrid PPM-CDMA continued, including analysis of synchronous PPM-CDMA (Shalaby [142]), performance analysis in space optics applications (Gagliardi [145]), performance in conjunction with hard-limiter and error correction coding and synchronous PPM/CDMA with co-channel interference cancellation (Ohtsuki [148] and Ohtsuki [164]), optimization considerations for the fiber optic transmission of compressed HDTV transmission and distribution (Elmirghani [170]), indoor optical wireless system applications (Elmirghani [167]), performance with imperfect synchronization (Sato [183]), PPM vs. CDMA vs. TDMA vs. FDMA for indoor optical wireless applications (Marsh [217]), turbo codes with PPM (Kiasaleh [226]); interference reduction (Shalaby [237], Shalaby [243], and Shalaby [287]), performance under different coding schemes (Chan [245]), as a means of enhancing the performance of spectral amplitude encoding that uses incoherent sources (Smith [262]), and in conjunction with turbo codes (Kim [303]). Applications to high bit rate communications, including total capacity, were analyzed in Kamakura [254].

A special area of interest in OCDMA is whether the codes and their associated network system can support multiple bit rates or variable bit rates (VBR) as well as equal rates for all users. This aspect of coding can be used to support communication or sensor networks with various bandwidth or granularity requirements, as well as different qualities of service. Contributions to this area of research include: Zhang [199], Fortier [208], Maric [225], Hamelin [239], Zhang [252], Zhang [270], and Zhang [275].

Spatial-spectral encoding using spectral hole burning was suggested during this era (Mossberg [192], Babbitt [209], and Babbitt [229]; a spatial-spectral holographic correlator and its performance at 1536 nm is discussed in Harris [294]). Holographically implemented OCDMA was proposed (Paek [154]). A practical way to emulate this kind of holographically based coding led to the construction of the encoders and decoders as complex Bragg grating arrays or Bragglike structures (Geiger [256], Grunnet-Jepsen [284], Grunnet-Jepsen [292], Grunnet-Jepsen [295], and Grunnet-Jepsen [306]).

In contrast, an approach using bipolar codes, phase and polarization shift keying encoding, but with direct detection, thus avoiding coherent detection, was described and demonstrated in Iversen [131a] and Karafolas [158].

Optical hard-limiting continued to be explored as a means to ameliorate MAI (Ohtsuki [163], Ohtsuki [190], Ohtsuki [234], Lin [253], Shalaby [297], Kamakura [311], and Ohtsuki [312]). It is commonly accepted that OHL, and double hard-limiting (DHL) in particular, will reduce MAI and provide good performance. But Ohtsuki [190] showed that OHL and DHL work best at relatively low number of users (~7–15) for prime sequence codes. For larger number of users, there is a crossover point in the BER curves where the case with no hard-limiting will outperform OHL and DHL (see, in particular, figure 7 through figure 9 in the citation).

Fiber optic delay lines are a critical component in transversal filters and many OCDMA encoder/decoder configurations. These are studied in detail in Sales [160], Sales [172], and Capmany [290].

The distinct trend of exploiting the finite coherence of light sources for coherence multiplexing continued in this era (Gupta [133], Griffin [156], Uehara [179], Uehara [214], Huang [279], Sampson [232], and Huang [241]). The concept of coherence multiplexing was extended to include an implementation based on Bragg grating arrays (Griffin [156]). A gain switched Fabry-Perot laser diode was characterized for this application (Griffin [146]).

The possibility of optical header recognition "on the go" emerged during this era. The early techniques involved spectroholographic filtering (Shen [173]) and stimulated photon echoes (Harris [235]). The possibility of using an OCDMA overlay to convey operation, administration, and maintenance (OAM) information in a transparent optical network was discussed by Deutsche Telekom (Giehmann [227]).

Bipolar OCDMA at 10 Gb/sec with error-free performance was demonstrated (Wada [228], Wada [255], and Wada [305]). Time-gating was introduced for MAI mitigation. Bipolar coding with broadband incoherent light, encoding in the time domain with transversal filters rather than in the spectral domain, was described and demonstrated over a 40-km link in Sotobayashi [280]. This approach, with some modifications, was to play an important role in advanced packet-switching networks in the next few years.

The transmission of OCDMA over fiber with its impairments, especially with femtosecond pulses, was addressed in analyses (Mezger [194]), demonstration (Chang [224], and a testbed [Sardesai [272]). The latter two citations refer to spectral phase encoding. The femtosecond OCDMA systems, of course, require some form of pulse shaper for the encoding/decoding process. Alternative approaches to pulse shaping were given in Yang [274] that used acousto-optic modulators and in Purchase [291] that used an electronically addressable Bragg grating array.

It was pointed out that the combination of WDM and amplifiers had emphasized intensity modulated, direct detection OCDMA schemes over coherent ones, but that lessons learned in the RF CDMA were still applicable to OCDMA (Huang [265] and [Huang [266]).

Applications to transport networks in the form of an ATM-OCDMA concept were described (Kamakura [269]), and applications to indoor wireless systems were analyzed in (Matsuo [300]).

A simulation methodology and program, based on using the nonlinear Schrödinger equation as propagator, was developed for studying the transmission of optical signals through installed fiber optic networks such as the National Transparent Optical Network (NTON) (Feng [309]). Although it was primarily developed for studying WDM network concepts, it was later modified to study and evaluate the propagation of various OCDMA schemes through the same installed base and to develop system requirements. Results will be discussed in the next section.

This era produced a textbook that treated OCDMA with special emphasis on PSO or OOC type of systems (Gagliardi [175]), including a description of Prucnal's transversal filter encoder/decoder configuration.

The exuberance in this era of the promise of OCDMA for network applications extended to an article in *Lightwave Magazine*, September, 1998, which said OCDMA "could conceivably provide this all-optical network solution (to expand bandwidth) as early as 18 months from now" (Dutt [261]).

1.7 2000–2004

Trends established in previous eras continued into this era. For example, Alcatel SEL's field trials for access network applications (Pfeiffer [322]) in which electronic dispersion compensation was considered and (Pfeiffer [346]) in which the "big picture" was considered and OCDMA found applicable to access and MAN, in conjunction with WDM for the higher order links and TDM the lower order links. These efforts were based on PRF coding, using LEDs. They found by measurement and analysis that this coding technique tends to have a bounded throughput, so that the number of users can be traded off against the data rate of each user (Pfeiffer [346] and Kissing [391]). The channel data rate bound is inversely proportional to a linear to quadratic function of the number of users, depending on the codes and encoding technique (i.e., Fabry-Perot vs. Mach-Zehnder encoder/decoder implementation). This work was a continuation of Pfeiffer [222] and Pfeiffer [282] and vision of the role of OCDMA in modern telecommunications that Alcatel presented in Eilenberg [250]. Further developments of this approach were reported in Deppisch [292]. Alcatel maintained that it was "also looking for alternative transmission solutions to DWDM, such as Optical code division Multiplexing (OCDM), which could offer cheaper and more flexible network and node designs" (Jourdan [371]). Applications of OCDMA to the local area are discussed in Stok [340], Sargent [362], Chapman [418], and Stok [420].

Another trend that continued from earlier eras was that of fast frequency hopping. This included Fortier [338], Inaty [380], and Fortier [396], who considered applications to multirate transmission, and Fortier [355] who considered signal-to-interference ratios (SIR). The transmission performance over an 80-km link with 16 users at a data rate of 1.25 Gb/sec each was reported in Ben Jaafar [386]. Further advances were given in Kitayama [442] and Wei [444]. A trade-off between coherent and incoherent sources for this application was given in Slavik [399]. New experimental results, augmented by simulations, were presented in Ayotte [482]. New OFFH code constructions were given in Kim [361], Forouzan [434], Lee [435], Baby [462], Pu [466], Yang [473], Baby [477], Kwong [488], and Baby [498]. MAI reduction based on ultrafast thresholding for this class of OCDMA was developed in Baby [501]. Coherence multiplexing was proposed as a scheme to reduce optical beat interference in frequency-hopping OCDMA (Kamakura [360]).

A detailed treatment and analysis of coherence multiplexing, including the potential effects of amplified spontaneous emission (ASE) noise, was developed (Katz [411]). A NOLM device coupled with differential detection to reduce the effects of MAI in coherence multiplexing was proposed and demonstrated for two channels in Kim [481]. A coherence multiplexing scheme suitable for access and local area networks was proposed in Meijerink [491]. It depicted a receiver that could be realized as an optical integrated circuit, including a multimode interference (MMI) coupler. The work went on to compute performance (in the sense of bandwidth efficiency and data rate per user) as a function of number of concurrent users, for both OOK and PSK. It found that the bandwidth efficiency (~10^{-3} bpsec/Hz) decreases with the number of users, and that the throughput is roughly fixed, so that the per user data rate decreases with increasing users. This behavior is similar to that of PRF codes (Pfieffer [346] and Kissing [391]).

2D Wavelength/time codes that are not necessarily constrained to a single pulse per row or single pulse per column in order to retain the OOC property were given in Kim [324], Kim [325], Mendez [341], Wan [383], Bajcsy [407], Mendez [462], Mendez [465], Kwong [472], Huang [474], Hernandez [492], Adams [495], and Hernandez [499]. A bidirectional network based on 2D codes, intended for the subscriber access network applications, was described in Kim [331]. The paper described the implementation with LEDs and array waveguide grating routers (AWGRs) and the performance over a 40-km link, using suitable EDFAs for amplification and dispersion compensation. Several of these papers included code design methodologies, 2D code descriptions or wavelength/time patterns, encoder/decoder implementation schemes, technology demonstrator descriptions, and eye-diagram and BER measurements as the number of asynchronous users increased. In addition, wavelength source and methods of preparing the code chips or symbols were also discussed (Mendez [315] and Mendez [413]). The effect of fiber impairments on the propagation of 2D codes over a well-characterized real optical link was computed in Feng [330]. The link, a part of the NTON, was 476 km in length and included EDFAs and SMF-28 fiber. The result of these analyses led to a description of the corresponding dispersion compensation requirements (Mendez [412]). Dispersion effects in multiwavelength OCDMA are further studied and remedies described in

Ng [426]. Applicable dispersion compensation techniques include electronic ones (Woodward [451] and Woodward [471]). A multiwavelength light source concept suitable for FFH or generic 2D codes was developed and described in Wang [382].

1D, 2D, and 3D code constructions (including wavelength/time/polarization) with various constraints and performance levels were developed in Zhang [333], Kwong [354], Weng [385], Kwong [416], Razavi [424], Lam [430], Neto [431], Kwong [432], Moreno [440], Moreno [453a], McGeehan [475a], Saghari [475b], Baby [477], Djordjevic [480], Moreno [480a], and Kwong [505]. Some of these new code constructions were based on FBG configurations and were designed specifically for multirate and multimedia transmission systems. A special case of 2D involving the space dimension was explored for image transmission (Nakamura [429]). This work included a demonstration based on an 8×8 VCSEL array and associated PD array, and BER and sensitivities to misalignment were determined.

A very active area of research was that of using OCDMA for ultrafast network reconfigurability by means of reading (and possibly rewriting) OCDMA headers. This included development and demonstration of the corresponding devices that could rewrite a 2D code (Gurkan [321], Gurkan [316], Willner [326], Willner [373], Au [401], Hauer [400], Willner [414], Cardakli [419], Willner [427], Gurkan [449], Hauer [460], Fejer [487], and Bres [500]). This approach was used in conjunction with optical bistability to demonstrate contention resolution (Kelm [502]). The fundamental limitations to code conversion due to the MAI are analyzed in Wen [388]. The same paper developed a closed-form solution to determining the decision threshold that minimized the BER.

Another body of work was produced in photonic IP routing, especially using bipolar codes (Kitayama [344], Kitayama [374], Sotobayashi [387], Kitayama [410], Kitayama [458], Kataoka [461], Kitayama [467], Sotobayashi [476], and Kataoka [490]). Operation at 40 Gb/s and packet add/drop multiplexing were demonstrated in Kataoka [461]. A general picture of photonic routers based on OOCs in a network environment, including several encoder/decoder implementation options involving generalized Mach-Zehnder interferometers or array waveguide routers, was provided in Cincotti [486].

The area of phase encoding reported many advances. Some of the advances were based on complex variants of Bragg gratings (Grunnet-Jepsen [319], Ibsen [348], Ibsen [349], Grunnett-Jepsen [364], Seldomridge [356], Teh [370], Petropoulos [377], Teh [389], Ibsen [390], Ibsen [393], Ibsen [394], Ibsen [395], Fu [402], Ibsen [403], and Kim [469]). A holey fiber nonlinear thresholder was introduced in Belardi [372] and a nonlinear optical loop mirror (NOLM) in Kim [481], time gating in Petropoulos [377], and an optical loop mirror in Ibsen [393] and Ibsen [394] to improve system performance. WDM/CDMA hybrids with quaternary phase coding were introduced in Ibsen [390], Ibsen [395], Fu [402], and Sotobayashi [405]. This configuration was analyzed in detail as a function of the code length (number of chips), number of wavelengths, bandwidth efficiency, and effect of forward error correction (FEC) [Ghiringhelli [443]]. The FEC (Reed-Solomon [255, 239]) gave error free performance with a raw error rate up to 10^{-3} and thus increased the number of users by making the MAI tolerable. The number of users increased at the expense of bandwidth efficiency, and in the best case the throughput was constant so that number of users and their single user data rate must be traded-off (a finding similar to that for PRF codes discussed earlier (Pfeiffer [346] and Kissing [391]). For example, the maximum number of users is found at a data rate of 39 Mb/sec per user, whereas at 20 Gb/sec many fewer than 10 users are supported and the configuration becomes WDM/TDM-like. In the analyses MAI was considered, but not other noise sources. However, practical complex Bragg grating (Super Structured Fiber Bragg Grating, SSFBG) fabrication issues were taken into account. An analysis of phase encoding that was more generic and less dependent on the encoding/decoding implementation was given in Ma [409] and it generally agreed with the observations in Ghiringhelli [443]. Reconfigurable encoders/decoders were described in Mohktar [446] and a demonstration of bidirectional transmission of the WDM/CDMA hybrid (eight channels) over a 44-km nonzero dispersion shifted fiber link was presented in Teh [445]. A novel coherent concept was described that uses broadband parametrically down converted light to generate a key and conjugate key, and the encoded data is recovered by the inverse process of parametric up conversion (Pe'er [485]). This work projected greater than 100 users, at 1 Gb/sec each, when using phase shift keying (PSK).

Similarly, there were advances in spectral phase encoding. An optical thresholder based on second harmonic generation was demonstrated (Zheng [320]). A parametric analysis of phase encoding systems was performed to determine the sensitivity to code length, pulsewidth, receiver threshold, and number of concurrent users, but not including fiber impairments (Ma [409]). This was followed by an analysis that specifically treated the effect of fiber dispersion on spectral phase encoding systems (Chua [475]). A chirped Bragg grating encoder/decoder was proposed in Fang [456]. Error free transmission was demonstrated in a testbed with four users each at 10 Gb/sec by using a nonlinear thresholder (Scott [489]). The experimental results were expanded in Hernandez [493] and the work was further expanded to include a simulation that allowed better results to be predicted in the case of an equal power per spectral bin implementation of the encoders/decoders. (The modified spectral bins compensate for the non-uniform power spectrum of the light source.) New spectral phase coding techniques were developed: low power double Hadamard coding with the potential of increasing the user count (Jiang [496]) and one based on hyperfine spectral phase encoding that then made it compatible with coexisting with WDM channels over the same network testbed (Toliver [497] and Etemad [509]). A network analysis of various random access schemes was performed (Xue [508]). The performance of a nonlinear optical thresholder in the case of coherent ultrashort pulses was analyzed in Igarashi [510].

Interest continued in amplitude spectral encoding: an analysis using superfluorescent fiber as a light source (Wei [366]); an analysis that compared modified quadratic congruence codes with Hadamard codes and used chirped Bragg gratings as the encoders/decoders (Wei [367], Wei [369], and Wei [404]); analyses that included modified frequency hopping codes and new algebraic code constructions (Wei [415] and Wei [417]); applications to IP routing (Wei [425]); and new code constructions applicable to spectral amplitude encoding as well as time spreading and fast frequency coding (Djordjevic [484]). A compensation module was proposed for FBG-based encoders/decoders that reduced the MAI contribution due to nonflattened incoherent sources (Huang [423]).

Considerable research was performed in the area of PPM-CDMA, with and without turbo coding, and OCDMA with turbo codes: it was confirmed by analyses and simulations that turbo coding increased the number of simultaneous users for a given BER requirement (Ohtsuki [329], Kim [334], Kim [345], Kim [351], Kamakura [363], Iwata [368], Kamakura [375], Argon [381], Kamakura [384], Argon [421], Kamakura [384], Kamakura [455], and Hirano [457]). A chip-level receiver for enhancing the tolerance to MAI was discussed in Miyazawa [507]. multirate and multiquality PPM-CDMA networks were discussed in Miyazawa [511]. A hybrid of spectral phase encoding and PPM was described in Kim [437] and Kim [454].

Several papers addressed free-space applications. Treatments of indoor infrared wireless using equalization concepts are given in Yamaguchi [358] and Matsuo [359], and concepts involving PPM-CDMA as well as hard-limiters were explored in Matsuo [359]. Space applications, including the effects of pointing and tracking, were discussed in Chan [347].

Optical hard-limiting continued to be of interest as a means of ameliorating MAI (Zahedi [342], Mendez [428], Miyazawa [453], Ebrahimi [459a], Ohtsuki [470], and Sasase [512]). Zahedi [342] showed that the method of implementing OHL or DHL was quite influential in the performance, complexity, and supported data rates of OCDMA systems. Mendez [428] found that the effectiveness of OHL could be enhanced by incorporating a guard-time in the coding to reduce intersymbol interference in OOCs. Ebrahimi [459a] applied hard-limiting on a per wavelength basis to a 2D wavelength/time configuration that used an array receiver. This can be considered as a form of electronic hard-limiting, in contrast to the OHL or DHL that is usually discussed. The citation shows that for the few-user case the hard-limiting improves the BER floor compared to not using hard-limiting, in agreement with the earlier findings of Ohtsuki [190]. Related approaches included interference cancellation (Sasase [337]), identification of the optimal single user detector (Chang [352]), and reduction of the interchannel interference noise by means of a nonlinear optical loop mirror (Lee [353], Ibsen [394], and Sotobayashi [405]). The problem of optimum threshold detection was addressed in Ng [408]. This is important for electronic hard-limiting implementations (Hernandez [499]). OHL based on electro-absorption modulators was proposed by Edagawa [317].

Multiuser detection and interference cancellation as a means of ameliorating and coping with the MAI continued to be developed (Sawagashira [337], Fantacci [433], and Motahari [479]). Dual detection involving both a correlation receiver and a chip level receiver and a logical OR operation was considered as a means of reducing the effects of MAI (Kok [506]).

The spectral efficiency and capacity of OCDMA schemes became important topics (Huang [328], Ng [362], Chang [379], Sotobayashi [405], Shalaby [436], Sotobayashi [438], and Chang [459]). Coherent OCDMA in conjunction with OTDM and WDM were discussed in Huang [328]. For very high spectral efficiency and capacity Sotobayashi [405] used quaternary phase shift keying codes, NOLM devices for time gating and hard thresholding, WDM/CDMA hybrids, and both states of polarization. A remarkably innovative multiple access scheme adapted the concept of RF space/time (S/T) coding (Bell Labs Layered Space-Time, or BLAST) but used the various modes in multimode fiber (MMF) in place of the multipaths of the RF case (Stuart [317a] and Stuart [318]). In this research, N users input their respective signals asynchronously into an MMF link and M receivers detect the superposition of all of these signals at the output of the link. Then, in conjunction with lots of processing, the N signals were recovered. This work calculated a capacity of tens of bits/sec/Hz, depending on the average single user SNR, and performed a demonstration with $N = M = 2$ and a per user data rate of 1 Mb/sec (using binary phase shift keying, BPSK) over a 1-km MMF link. The optical S/T multiplexing technique is scalable to Gb/sec, given faster processors. Methods for intercomparing diverse OCDMA schemes were developed in Stok [450].

The possibility of designing and producing integrated OCDMA subsystems, for example as planar lightwave circuits, especially programmable encoders/decoders or customizable (transversal or lattice) filters, was addressed in Chan [314], Marhic [422], and Broeke [494].

Holographic techniques for encoding/decoding were further developed in Harris [313] (including quadriphase and binary phase encoding), Mossberg [350], and Seldomridge [356]; for range correlation using bi-phase codes (Mohan [452]); label processing (Chujo [468]); and for wavelength/time encoding/decoding (Huang [503]).

Code acquisition and synchronization, important for the tell-and-go protocol as well as for the time-gating used to reduce the effects of MAI, started to receive considerable attention. For example, the problem of code acquisition in coherent multiplexing was studied in Huang [323] for the case of Gold codes. A detailed treatment of code acquisition and multistage interference cancellation for coherent OCDMA was given in Huang [328]. This work also looked at a big picture of network applications and proposed a hybrid OCDMA/WDMA system for optimal network performance. The case of OOCs was then taken up and it was determined that the MAI causes the optimum acceptance threshold to be a function of the number of active users and that the performance improved if some a priori knowledge, such as the number of users, was available (Mustapha [332] and Mustapha [343]). The benefit of pretransmission protocols in conjunction with correlation or chip level OOC receivers was evaluated in Shalaby [464]. Serial search performance in the case of OOCs was examined in Keshavarian [378] and Keshavarian [397]. Multistage decoding in conjunction with convolutional coding was considered in Azmi [441].

1.8 FUTURE TRENDS, APPLICATIONS, AND UNFINISHED BUSINESS

This survey of OCDMA publications and research shows that the contributions have come from individual researchers, universities, corporations, and consortia. Some of the research has been directed to solve specific problems, and some has been motivated by the challenges of innovation. Some of the more productive research is that associated with corporate objectives such as Alcatel SEL, Deutsche Telecom, NTT, university incubators such as Université Laval, and competitive consortia such as that organized by DARPA (Shah [447]). The benefit of the latter is that it combines universities, companies, and entities such as Telcordia, but it also identifies specific goals and strict performance milestones.

The survey also shows that, perhaps unique to OCDMA, there has been a lot of cross insemination in the sense that key researchers have participated in each other's programs and have shared knowledge. This shows up in the citations, where particular key researchers appear as contributors in publications based at different institutions. In this sense OCDMA is global and it shows up in that the OCDMA community tends to gather at conferences that attract an international contribution.

A major lesson learned by various research teams, each using a different coding or code implementation approach, is that it is not a given in OCDMA that the number of users and their individual data rates can be selected independently. Somewhat related to this observation is that some of the coding schemes can only be implemented for very low data rates where OCDMA may not have an application or competitive advantage over other technologies.

Going forward, we see the same opportunities for OCDMA from the mid-1990s (such as access, LAN, and MAN); an emphasis on bipolar (and higher) as well as 2D codes; applications such as IP routing and network reconfigurability; coherence multiplexing for low end and sensor networks; and new applications such as multimedia transmission and different qualities of service and avionics integration that benefit from multirate transmission capabilities (Uhlhorn [478]) and, unlike the OCDMA approaches to security described in the body of chapter 1, an alternative to quantum cryptography (Wells [406]).

The compatibility of OCDMA with other modulation and multiple accessing schemes, in which they may share the same transmission medium, is important and entities like Telcordia have begun to investigate this issue (Banwell [497], Etemad [509], and Galli [517]).

Most practical and successful demonstrations of OCDMA (whether they be a laboratory demonstration, a technology demonstrator, or a field trial) have made use of optical amplification. Similarly, most implementations have encountered beat noise (from self or from interferers). Some of the approaches to MAI minimization have been at the expense of spectral efficiency. Thus, it is mandatory that future code design be carried out in close collaboration with system architects so that system design (link loss and corresponding optical amplification), noise sources (beat noise interference), and bandwidth expansion (code dimension) can be suitably addressed and compensated. Where it comes to OOK, it is appropriate to perform good analyses and simulations and to go beyond treating the problem as one involving intensity (combinatorics), only. The various chapters in this book identify the proper participants.

REFERENCES

EARLY HISTORY

[1] Eckler, A. R. (1960). The construction of missile guidance codes restraint to random interference. *The Bell System Technical Journal (BSTJ)*:973–994.

[2] Bernstein, A. J., Robinson, J. P. (1967). A class of binary recurrent codes with limited error propagation. *IEEE Transactions on Information Theory* 13(1):106–113.

[3]   Gold, R. (1967). Optimal binary sequences for spread spectrum multiplexing. *IEEE Transactions on Information Theory* 13(4):619–621.

[4]   Reed, I. S. (1971). Kth-Order near-orthogonal codes. *IEEE Transactions on Information Theory* 17(1):116–117.

[5]   Forney, G., Jr. (1972). Lower bounds on error probability in the presence of large intersymbol interference. *IEEE Transactions on Communications* 20(1):76–77.

[6]   Biraud, F., Blum, E. J., Ribes, J. C. (1974). On optimum synthetic linear arrays with application to radioastronomy. *IEEE Transactions on Antennas and Propagation* 22(1):108–109.

[7]   Dixon, R. (1975). Why spread spectrum? *Communications Society: A Digest on News and Events of Interest to Communications Engineers* 13(4):21–25.

[8]   Marom, E. (1978). Optical delay line matched filters. *IEEE Transactions on Circuits and Systems* 25(6):360–364.

[9]   Shedd, D., Sarwate, D. V. (1979). Construction of sequences with good correlation properties. *IEEE Transactions on Information Theory* 25(1):94–97.

[10]   Dogliotti, R., Luvison, A., Pirani, G. (1979). Error probability in optical fiber transmission systems. *IEEE Transactions on Information Theory.* 25(2):170–178.

[11]   Robinson, J. P. (1979). Optimum Golomb Rulers *IEEE Trans. Comput* C-28.

1980–1984

[12]   Pursley, M. B., Sarwate, D. V. (1980). Crosscorrelation properties of pseudorandom and related sequences. *Proc. IEEE* 68:593–619.

[12a]   Weber, C. L., Huth, G. K., Batson, B. H. (1981). Performance considerations of code division multiple-access systems. *IEEE Transactions on Vehicle Technology* 30(1): 3–10.

[12b]   Stark, W. E., Sarwate, D. V. (1981). Kronecker sequences for spread-spectrum communication. *IEEE Proc.* 35(2):104–109.

[13]   Golomb, S., Taylor, H. (1982). Two-dimensional synchronization patterns for minimum ambiguity. *IEEE Transactions on Information Theory* 28(4):600–604.

[14]   Davies, P., Shaar, A. A. (1983). Prime sequences: Quasi-optimal sequences for channel code division multiplexing. *Electronics Letters* 19, pp. 866–873.

[15]   Davies, P. Shaar, A. A. (1983). Asynchronous multiplexing for an optical-fiber local area network. *Electronics Letters* 19:390–392.

[16]   Froehly, C., Colombeau, B., Vampouille, M. (1983). Shaping and analysis of picosecond light pulses. In: *Progress in Optics XX*, pp. 64–153.

[17]   Marhic, M. (1984). Hierarchic and combinatorial star couplers. *Optics Letters* 9(8): 368.

1985–1989

[18]   Tamura, A., Nakano, S., Okazaki, K. (1985). Optical code-divison-multiplex transmission by Gold sequences. *Journal of Lightwave Technology* 3(1):121–127.

[19]   Cutler, C. C., Goodman, J. W., Jackson, J. P., Moslehi, B., Newton, S. A., Shaw, H. J., Tur, M. (1985). Optical fiber delay-line signal processing. *IEEE Transactions on Microwave Theory and Techniques* 33(3):193–210.

[20]   Brooks, J., Wentworth, R., Youngquist, R., Tur, M., Kim, B., Shaw, H. J. (1985). Coherence multiplexing of fiber-optic interferometric sensors. *Journal of Lightwave Technology* 3(5):1062–1072.

[21]   Hui, J. (1985). Pattern code modulation and optical decoding — A novel code-division multiplexing technique for multifiber networks. *IEEE Journal on Selected Areas in Communications* 3(6):916–927.

[22]   Marhic, M., Nassehi, M., Mehdi Tobagi, F. (1985). Fiber optic configurations for local area networks. *IEEE Journal on Selected Areas in Communications* 3(6):941– 949.

[23]   Robinson, J. P. (1985). Golomb rectangles. *IEEE Transactions on Information Theory* 31:781–787.

[24]   Heritage, J., Thurston, R., Weiner, A. (1985). Picosecond pulse shaping by spectral phase and amplitude manipulation. *Optics Letters* 10(12):609–611.

[24a]   Weiner, A. M., Heritage, J. P., Thurston, R. N. (1986). Synthesis of phase coherent, picosecond optical square pulses. *Optics Letters* 11(3):153.

[25]   Heritage, J., Thurston, R., Tomlinson, W., Weiner, A. (1986). Analysis of picosecond pulse shape synthesis by spectral masking in a grating pulse compressor. *IEEE Journal of Quantum Electronics* 22(5):682–696.

[26]   Fan, T., Prucnal, P., Santoro, M. (1986). Spread spectrum fiber-optic local area network using optical processing. *Journal of Lightwave Technology* 4(5):547–554.

[27]   Verdu, S. (1986). Multiple-access channels with point-process observations: Optimum demodulation. *IEEE Transactions on Information Theory* 32(5):642–651.

[28]   Prucnal, P., Santoro, M., Sehgal, S. (1986). Ultrafast all-optical synchronous multiple access fiber networks. *IEEE Journal on Selected Areas in Communications* 4(9):1484–1493.

[29]   Davies, P., Woodcock, C. F. (1987). The generation of high-speed binary sequences from interleaved low-speed sequences. *IEEE Transactions on Communications* 35(1):115– 117.

[30]   Davies, P., Shaar, A. A., Woodcock, C. F. (1987). Bounds on the cross-correlation functions of state m-sequences. *IEEE Transactions on Communications* 35(3):305–312.

[31]   Heritage, J., Salehi, J., Weiner, A. (1987). Frequency domain coding of femtosecond pulses for spread-spectrum communications. In: *Proceedings Conference of Lasers and Electrooptics*, 1, pp. 294–296.

[32]   Goedgebuer, J. P., Porte, H., Hamel, A. (1987). Electo-optic modulation of multilongitudinal mode laser diodes: Demonstration at 850 nm with simultaneous data transmission by coherence multiplexing. *IEEE Journal of Quantum Electronics* 23(7):1135–1144.

[33]   Gagliardi, R., Robbins, J., Taylor, H. (1987). Acquisition sequences in PPM communications. *IEEE Transactions of Information Theory* 33(5):738–744.

[34]   Andonovic, I., Culshaw, B., Shabeer, M. (1987). Fiber-optic bipolar tap implementation using an incoherent optical source. *Optics Letters* 12(9):726–728.

[35]   Poor, H. V., Verdu, S. (1987). Single-user detectors for multiuser channels. *IEEE Transactions on Communications* 36(1):50–60.

[36]   Aazhang, B., Poor, H. V. (1988). Performance of DS/SSMA communications in impulsive channels. II. Hard-limiting correlation receivers. *IEEE Transactions on Communications* 36(1):88–97.

[37]   Gagliardi, R., Khansefid, F., Taylor, H. (1988). Design of (0,1) sequence sets for pulse coded systems. *USC Communication Sciences Report*, CSI-88-03-03.

[38]   Gelman, A. D., Schilling, D. L. (1988). A fiber optic CDMA network for real-time communication. In: INFOCOM '88. Networks: Evolution or Revolution? IEEE Seventh Annual Joint Conference of the IEEE Computer and Communications Societies.

[38a] Foschini, G. J., Vannucci, G. (1988). Using spread-spectrum in a high capacity fiber-optic local area network. *Journal of Lightwave Technology* 6(3):370.

[39] Heritage, J., Salehi, J., Weiner, A. (1988). Encoding and decoding of femtosecond pulses. *Optics Letters* 13(4):300–302.

[40] MacDonald, R. I. (1988). Fully orthogonal optical-code multiplex for broadcasting. *Optics Letters* 13(6):539–541.

[41] Foschini, G. (1988). Sharing of the optical band in local systems. *IEEE Journal on Selected Areas in Communications* 6(6):974–986.

[41a] Weiner, A. M., Heritage, J. P., Kirschner, E. M. (1988). High-resolution femtosecond pulse shaping. *JOSA B*. 5:1563–1572.

[42] Salehi, J. (1989). Emerging optical code-division multiple access communication systems. *IEEE Network* 3(2):31–39.

[43] O'Farrell, T., Beale, M. (1989). Code-division multiple-access (CDMA) techniques in optical fibre LANs. In: Second IEE National Conference on Telecommunications.

[44] Chung, F. R. K., Salehi, J., Wei, V. K. (1989). Optical orthogonal codes: Design, analysis and applications. *IEEE Transactions on Information Theory* 35(3):595–604.

[45] Kiasaleh, K. (1989). Spread-spectrum optical on-off-keying communication system. In: World Prosperity through Communications, IEEE International Conference on Communications. ICC 89, BOSTONIC/89.

[46] Aazhang, B., Poor, H. V. (1989). An analysis of nonlinear direct-sequence correlators. *IEEE Transactions on Communications* 37(7):723–731.

[46a] Vannucci, G., Yang, S. (1989). Experimental spreading and despreading of the optical spectrum. *IEEE Transactions on Communications* 37(7):777–780.

[47] Salehi, J. (1989) Code division multiple-acess techniques in optical fiber networks — Part 1: Fundamental principles. *IEEE Transactions on Communications* 37(8): 824–833.

[48] Brackett, C., Salehi, J. (1989). Code division multiple-access techniques in optical fiber networks — Part II: Systems performance analysis. *IEEE Transactions on Communications* 37(8):834–842.

[49] Andonovic, I., Culshaw, B., Shabeer, M. (1989). Programmable fiber/integrated-optics bipolar tap. *Optics Letters* 14(16):895–897.

[50] Marhic, M. E., Chang, Y. L. (1989). Pulse coding and coherent decoding in fibre-optic ladder networks. *Electronics Letters* 25:1535–1536.

[51] Gagliardi, R., Garmire, E., Kuroda, S., Mendez, A. (1989). Generalized temporal code division multiple access (CDMA) for optical communications. *Proc. SPIE*. 1175:208–217.

[52] Kwong, W. C., Prucnal, P., Perrier, P. (1989). Synchronous versus asynchronous CDMA for fiber-optic LANs using optical signal processing. In: GLOBECOM 1989–IEEE Global Telecommunications Conference and Exhibition. Communications Technology for the 1990s and Beyond.

1990–1994

[53] Chu. K. C., Dickey, F. M. (1990). Optical coherence multiplexing for interprocessor communications. *Proc. SPIE*. 1178:11–23.

[54] Heritage, J., Salehi, J., Weiner, A. (1990). Coherent ultrashort light pulse code-division multiple access communication systems. *Journal of Lightwave Technology* 8(3):478–491.

[55]   Leaird, D. E., Patel, J. S., Weiner, A., Wullert, J. R. (1990). Programmable femto-second pulse shaping by use of a multielement liquid-crystal phase modulator. *Optics Letters* 15(6):326–328.

[55a]   Chung, H., Kumar, P.V. (1990). Optical orthogonal codes-new bounds and an optimal construction. *IEEE Transactions on Information Theory* 36(4):866–873.

[56]   Sampson, D. D., Jackson, D. A. (1990). Coherent optical fiber communications system using all-optical correlation processing. *Optics Letters* 15(10):585–587.

[57]   Gagliardi, R., Mendez, A. (1990). Pulse combining and time-space coding for multiple accessing with fiber arrays. In: 1990 IEEE/LEOS Summer Topical Meeting on Optical Multiple Access Networks.

[58]   Cimini Jr., L. J., Foschini, G. (1990). Coding of optical on-off keying signals impaired by phase noise. *IEEE Transactions on Communications* 38(9):1301–1307.

[59]   Sampson, D. D., Jackson, D. A. (1990). Spread-spectrum optical-fibre network based on pulsed coherent correlation. *Electronics Letters* 26:1550–1552.

[60]   Gagliardi, R., Khansefid, F., Taylor, H. (1990) Performance analysis of code division multiple access techniques in fiber optics with on-off and PPM pulsed signaling. In: Military Communications Conference. MILCOM '90, Conference Record, "A New Era."

[61]   Kiasaleh, K. (1990). PN code acquisition for the spread-spectrum optical on-off-keying communication system. *IEEE Transactions on Communications* 38(10):1879–1885.

[62]   Gagliardi, R., Garmire, E., Mendez, A. J., Park, E., Taylor, H. (1990). Photonic switching fabrics based on spatial/temporal hybrid pseudo orthogonal code division multiple access (CDMA) codes. In: Conference Proceedings, IEEE/LEOS '90 Annual Meeting.

[63]   Azizoglu, M., Salehi, J., Li, Y. (1990). On the performance of fiber-optic CDMA systems. In: GLOBECOM 1990 — IEEE Global Telecommunications Conference and Exhibition. "Communications: Connecting the Future."

[64]   Brady, D., Verdu, S. (1991). A semiclassical analysis of optical code division multiple access. *IEEE Transactions on Communications* 39(1):85–93.

[65]   Vajda, I. (1991). Comments on "Code-division multiple-access techniques in optical fiber networks: Part II: Systems performance analysis." *IEEE Transactions on Communications* 39(2):196–196.

[66]   Kiasaleh, K. (1991). Fiber optic frequency hopping multiple access communication system. *IEEE Photonics Technology Letters* 3(2):173–175.

[67]   Gagliardi, R. M., Mendez, A. J. (1991). Code division multiple access (CDMA) system candidate for integrated modular avionics. *Proceedings of the SPIE.* 1364:67– 71.

[68]   Gagliardi, R. M., Mendez, A. J. (1991). Performance of pseudo orthogonal codes in temporal, spatial, and spectral code division multiple access (CDMA) systems. *Proceedings of the SPIE.* 1364:163–169.

[69]   Games, R., Moreno, O., Taylor, H. (1991) New constructions and bounds on sonar sequences. In: Proc. of the 1991 IEEE International Symposium on Information Theory.

[70]   Etzion, S. W., Golomb, S., Taylor, H. (1991). Sets of sonar sequences. In: Proc. of the 1991 IEEE International Symposium on Information Theory.

[71]   Kwon, H. (1991). Optical orthogonal code division multiple access system. I. With APD noise and thermal noise. In: IEEE International Conference on Communications. Conference Record.

[72]   Kwon, H. (1991). Optical orthogonal code division multiple access system. II. Multi-bits/sequence-period OOCDMA, IEEE International Conference on Communications, 2, pp. 618–621.

[73] MacDonald, R. I., Vethanayagam, M. (1991). Demonstration of a novel optical code-division multiple-access system at 800 megachips per second. *Optics Letters* 16(13):1010–1012.

[74] Gagliardi, R., Ivancic, W. D., Mendez, A. J., Park, E., Sherman, B. D. (1991). Optical multiple access networks (OMAN) for advanced processing satellite applications. In: 1991 IEEE/LEOS Summer Topical Meeting on Spaceborne Photonics.

[75] Kwong, W. C., Perrier, P., Prucnal, P. (1991). Performance comparison of asynchronous and synchronous code-division multiple-access techniques for fiber-optic local area networks. *IEEE Transactions on Communications* 39(11):1625–1634.

[76] Kiasaleh, K. (1991). PN code-aided ranging for optical satellite communication systems. *IEEE Transactions on Communications* 39(12):1832–1844.

[77] Hajela, D., Salehi, J. (1992). Limits to the encoding and bounds on the performance of coherent ultrashort light pulse code-division multiple-access. *IEEE Transactions on Communications* 40(2):325–336.

[78] Kiasaleh, K. (1992). Network architectures for fiber-optic frequency-hopping multiple-access networks. In: International Workshop on Advanced Communications and Applications for High Speed Networks.

[79] Hussain, A., Lam, A. (1992). Performance analysis of direct-detection optical CDMA communication systems with avalanche photodiodes. *IEEE Transactions on Communications* 40(4):810–820.

[79a] Weiner, A. M., Leaird, D. E., Patel, J. S., Wullert, J. R. (1992). Programmable shaping of femtosecond optical pulses by use of 128-element liquid crystal phase modulator. *IEEE Journal of Quantum Electronics* 28(4):908–920.

[80] Griffin, R. A., Sampson, D. D., Jackson, D. A. (1992). Demonstration of data transmission using coherent correlation to reconstruct a coded pulse sequence. *IEEE Photonics Technology Letters* 4(5):513–515.

[81] Griffin, R. A., Sampson, D. D. (1992). Multigigabit/s demultiplexing in optical domain using coherence properties of pulse trains from multiple, asynchronous mode-locked lasers. *Electronics Letters* 28(13):1202–1203.

[82] Azizoglu, M., Li, Y., Salehi, J. (1992). Optical CDMA via temporal codes. *IEEE Transactions on Communications* 40(7):1162–1170.

[83] Chang, Y. L., Marhic, M. E. (1992). Coherent optical CDMA with inverse decoding in ladder networks. In: LEOS 1992 Summer Topical Meeting Digest on Broadband Analog and Digital Optoelectronics, Optical Multiple Access Networks, Integrated Optoelectronics, Smart Pixels.

[84] Gagliardi, R. M., Mendez, A. J. (1992). Progress towards a 16 × 16, 1 Gb/s, switching network based on optical CDMA. In: LEOS 1992 Summer Topical Meeting Digest on Broadband Analog and Digital Optoelectronics, Optical Multiple Access Networks, Integrated Optoelectronics, Smart Pixels.

[85] Garmire, E., Mendez, A., Park, E. (1992). Temporal/spatial optical CDMA networks-design, demonstration, and comparison with temporal networks. *IEEE Photonics Technology Letters* 4(10):1160–1162.

[86] Chang, Y. L., Marhic, M. E. (1992). Fiber-optic ladder networks for inverse decoding coherent CDMA. *Journal of Lightwave Technology* 10(12):1952–1962.

[87] Griffin, R. A., Sampson, D. D., Jackson, D. A. (1992). Optical phase coding for code-division multiple access networks. *IEEE Photonics Technology Letters* 4(12):1401–1404.

[88] Mandayam, N. B., Aazhang, B. (1993). Error probabilities for fiber-optic code division multiple access systems. In: IEEE International Symposium on Information Theory.

[89]   Nelson, L. B., Poor, H. V. (1993). Performance analysis of optimum demodulation in optical CDMA. In: IEEE International Symposium on Information Theory.

[89a]   Moreno, O., Zhang, Z., Kumar, P. V. (1993). Some families of asymptotically optimal optical orthogonal codes. In: IEEE International Symposium on Information Theory.

[90]   Dale, M., Gagliardi, R. (1993). Channel coding for asynchronous fiberoptic CDMA communications. In: IEEE International Symposium on Information Theory.

[91]   Brandt-Pearce, M., Aazhang, B. (1993). Optical spectral amplitude code division multiple access system. In: IEEE International Symposium on Information Theory.

[92]   Dale, M., Gagliardi, R., Mendez, A. J., Park, E. (1993). Fiberoptic digital video multiplexing using optical CDMA. *Journal of Lightwave Technology* 11(1):20–26.

[93]   Kavehrad, M., Zaccarin, D. (1993). An optical CDMA system based on spectral encoding of LED. *IEEE Photonics Technology Letters* 5(4):479–482.

[94]   Yao, X. S., Feinberg, J., Logan, R., Maleki, L. (1993). Limitations on peak pulse power, pulse width, and coding mask misalignment in a fiber-optic code-division multiple-access system. *Journal of Lightwave Technology* 11(5):836–846.

[95]   Marhic, M. (1993). Coherent optical CDMA networks. *Journal of Lightwave Technology* 11(5):854–864.

[96]   Kwong, W. C., Prucnal, P. (1993). All-serial coding architecture for ultrafast optical code-division multiple access. In: IEEE International Conference on Communications.

[97]   Wells, W., Stone, R., Miles, E. (1993). Secure communications by optical homodyne. *IEEE Journal on Selected Areas in Communications* 11(5):770–777.

[98]   Glesk, I., Kane, M., Prucnal, P., Sokoloff, J. P.(1993). A terahertz optical asymmetric demultiplexer (TOAD). *IEEE Photonics Technology Letters* 5(7):787–790.

[99]   Gagliardi, R., Mendez, A. J. (1993). Matrix codes for ultradense gigabit optical CDMA networks. In: 1993 IEEE/LEOS Summer Topical Meeting on Gigabit Networks.

[100]   Kostic, Z., Maric, S., Titlebaum, E. (1993). A new family of optical code sequences for use in spread-spectrum fiber-optic local area networks. *IEEE Transactions on Communications* 41(8):1217–1221.

[101]   Games, R., Moreno, O., Taylor, H. (1993). Sonar sequences from Costas arrays and the best known sonar sequences with up to 100 symbols. *IEEE Transactions on Information Theory* 39(6):1985–1987.

[102]   Mendez, A. J. (1993). Analysis, design and measurement of bandwidth efficient matrix codes for ultradense, gigabit O-CDMA networks. In: Conference Proceedings of the IEEE/LEOS Annual Meeting.

[103]   Griffin, R. A., Sampson, D. D., Jackson, D. A. (1993). Modification of optical coherence using spectral phase coding for use in photonic code-division multiple access systems. *Electronics Letters* 29(25):2214–2216.

[104]   Wefers, M. M., Nelson, K. A. (1993). Programmable phase and amplitude femtosecond pulse shaping. *Optics Letters* 18(23):2214–2216.

[105]   Green, P. E. (1993). Multiaccess, switching and performance. In: *Fiber Optic Networks*, Prentice Hall, New York, pp. 417–458.

[106]   Griffin, R., Sampson, D. (1993). Coherence coding of optical pulses for code-division multiple access. In: Technical Digest, OFC/IOCC '93.

[107]   Griffin, R., Sampson, D., Jackson, D. A. (1993). Gain-switched Fabry-Perot laser diodes as sources for low-coherency interferometry. In: Technical Digest, OFS '93.

[108]   Kavehrad, M., Zaccarin, D. (1994). Performance evaluation of optical CDMA systems using non-coherent detection and bipolar codes. *Journal of Lightwave Technology* 12(1):96–105.

[109]   Kavehrad, M., Zaccarin, D. (1994). New architecture for incoherent optical CDMA to achieve bipolar capacity. *Electronic Letters* 30(3):258–259.

[110]   Kavehrad, M., Zaccarin, D. (1994). Optical CDMA with new coding strategies and new architectures to achieve bipolar capacity with unipolar codes. In: Conference Digest, 1994 OFC.

[111]   Andonovic, I., Tančevski, L. (1994). Block multiplexing codes for optical ladder network correlators. *IEEE Photonics Technology Letters* 6(2):309–311.

[112]   Aazhang, B., Brandt-Pearce, M. (1994). Multiuser detection for optical code division multiple access systems. *IEEE Transactions on Communications* 42(234):1801–1810.

[113]   Kwong, W. C., Zhang, J. G., Yang, G. C. (1994). 2^n prime-sequence code and its optical CDMA coding architecture. *Electronics Letters* 30(6):509–510.

[114]   Walker, E. (1994). A theoretical analysis of the performance of code division multiple access communications over multimode optical fiber channels: Part I: Transmission and detection. *IEEE Journal on Selected Areas in Communications* 12(4):751–761.

[115]   Kavehrad, M. Zaccarin, D. (1994). Optical CDMA by spectral encoding of LED for ultrafast ATM switching. In: IEEE Internation Conference on Communications. ICC '94, SUPERCOMM/ICC '94, Conference Record, Serving Humanity through Communications.

[116]   Ho, C. L., Wu, C. Y. (1994). Performance analysis of CDMA optical communication systems with avalanche photodiodes. *Journal of Lightwave Technology* 12(6):1062–1072.

[117]   Andonovic, I., Bazgaloski, L., Shabeer, M., Tančevski, L. (1994). Incoherent all-optical code recognition with balanced detection. *Journal of Lightwave Technology* 12(6):1073–1080.

[118]   Walker, E. (1994). A theoretical analysis of the performance of code division multiple access communications over multimode optical fiber channels: Part II: System performance evaluation. *IEEE Journal on Selected Areas in Communications* 12(5):976–983.

[119]   Brandt-Pearce, M. (1994). All-optical linear detection for optical CDMA communication systems. In: IEEE International Symposium on Information Theory.

[120]   Kwon, H. (1994). Optical orthogonal code-division multiple-access system: Part I: APD noise an thermal noise. *IEEE Transactions on Communications* 42(7):2470–2479.

[121]   Karafolas, N., Uttamchandani, D. (1994). Self-homodyne code division multiple access technique for fiber optic local area networks. *IEEE Photonics Technology Letters* 6(7):880–883.

[122]   Lambert, J. L., Mendez, A. J. (1994). Optically amplified optical CDMA experiments. In: 1994 IEEE/LEOS Summer Topical Meeting on Integrated Optoelectronics.

[123]   Kwon, H. (1994). Optical orthogonal code-division multiple-access system: Part II: Multibits/sequence-period OOCDMA. *IEEE Transactions on Communications* 42(8):2592–2599.

[124]   Kostic, Z., Titlebaum, E. (1994). The design and performance analysis for several new classes of codes for optical synchronous CDMA and for arbitrary-medium time-hopping synchronous CDMA communication systems. *IEEE Transactions on Communications* 42(8):2608–2617.

[125]   Walker, E. (1994). Performance of code-division-multiple-access communications over multimode optical fiber channels. In: Proceedings of the 37th Midwest Symposium on Circuits and Systems.

[126]   Tančevski, L., Andonovic, I. (1994). Wavelength-hopping time-spreading code-division multiple access systems. *Electronics Letters* 30(17):1388–1390.

[127] Gagliardi, R., Lambert, J. L., Mendez, A., Morookian, J. M. (1994). Synthesis and demonstration of high speed, bandwidth efficient optical code division multiple access (CDMA) tested at 1 Gb/s throughput. *IEEE Photonics Technology Letters* 6:1146–1148.

[128] Bergman, L. A., Gagliardi, R. M., Lambert, J. L., Mendez, A. J., Morookian, J. M. (1994). Raw bit error rate of a fully populated optical code division multiple access (CDMA) system. In: Conference Proceedings of the 1994 IEEE/LEOS Annual Meeting.

[129] Gupta, G. C., Karafolas, N., Uttamchandani, D. (1994). Low coherence optical CDMA for LAN. In: Conference Proceedings of the 1994 IEEE/LEOS Annual Meeting.

[130] Sampson, D. D., Griffin, R. A., Jackson, D. A. (1994). Photonic CDMA by coherent matched filtering using time-addressed coding in optical ladder networks. *Journal of Lightwave Technology* 12(11): 2001–2010.

[131] Rusch, L., Poor, H. V. (1994). Phase drift effects in optical CDMA. In: GLOBECOM 1994 — IEEE Global Telecommunications Conference, 1994. Communications Theory Mini-Conference Record.

1995–1999

[131a] Iversen, K., Mueckenheim, J., Junghanns, D. (1995). Performance evaluation of optical code division multiple access (CDMA) using polarization shift keying direct dectection (PolSK-DD) to improve bipolar capacity. *Proc. SPIE.* 2450:319–329.

[131b] Radhi, H. M., Al-Raweshidy, H. S., Senior, J. M. (1995). Phase noise in coherent optical code division multiple access (CDMA) networks. *Proc. SPIE.* 2450:330–337.

[131c] Gupta, G., Legg, P. J., Uttamchandani, D. G. (1995). Interferometer noise analysis in coherent optical CDM systems for LAN applications. *Proc. SPIE.* 2450:474–483.

[132] Andonovic, I., Tančevski, L. (1995). Hybrid wavelength hopping/time spreading code division multiple access systems. *Proc. SPIE.* 2450:531–538.

[133] Gupta, G. C., Uttamchandani, D. (1995). Self-homodyne optical code division multiple access (CDMA) system for LAN application. *Proc. SPIE.* 2450:539–547.

[134] Kavehrad, M., Zaccarin, D. (1995). Optical code-division-multiplexed systems based on spectral encoding of noncoherent sources. *Journal of Lightwave Technology* 13(3): 534–545.

[135] Hinkov, I., Hinkov, V., Iversen, K., Ziemann, O. (1995). Feasibility of optical CDMA using spectral encoding by acoustically tunable optical filters. *Electronics Letters* 31(5):384–386.

[136] Aazhang, B., Nguyen, L., Young, J. F. (1995). All-optical CDMA with bipolar codes. *Electronics Letters* 31(6):469–470.

[137] Yang, G. C., Kwong, W. C. (1995). Performance analysis of optical CDMA with prime codes. *Electronics Letters* 31(7):569–570.

[138] Mandayam, N. B., Aazhang, B. (1995). Importance sampling for analysis of direct detection optical communication systems. *IEEE Transactions on Communications* 43(243):229–239.

[139] Aazhang, B., Brandt-Pearce, M. (1995). Performance analysis of single-user and multiuser detectors for optical code division multiple access communication systems. *IEEE Transactions on Communications* 43(234):435–444.

[140] Hahm, M., Maric, S., Titlebaum, E. (1995). Construction and performance analysis of a new family of optical orthogonal codes for CDMA fiber-optic networks. *IEEE Transactions on Communications* 43(243):485–489.

[141] Kwong, W. C., Yang, G. C. (1995). On the construction of 2^n codes for optical code-division multiple-access. *IEEE Transactions on Communications* 43(234):495–502.

[142] Shalaby, H. (1995). Performance analysis of optical synchronous CDMA communication systems with PPM signaling. *IEEE Transactions on Communications* 43(234):624–634.

[143] Poor, H. V., Rusch, L. (1995). Effects of laser phase drift on coherent optical CDMA. *IEEE Journal on Selected Areas in Communications* 13(3):577–591.

[144] Shalaby, H. (1995). A comparison between the performance of number-state and coherent-state optical CDMA in lossy photon channels. *IEEE Journal on Selected Areas in Communications* 13(3):592–602.

[145] Gagliardi, R. (1995). Pulse-coded multiple access in space optical communications. *IEEE Journal on Selected Areas in Communications* 13(3):603–608.

[146] Griffin, R. A., Jackson, D. A., Sampson, D. D. (1995). Coherence and noise properties of gain-switched Fabry-Perot semiconductor lasers. *IEEE Journal of Selected Topics in Quantum Electronics* 1(2):569–576.

[147] Azizoglu, A., Mokhtar, A. (1995). Hybrid multiaccess for all-optical LANs with nonzero tuning delays. In: ICC '95 Seattle, Gateway to Globalization, IEEE 1995 International Conference on Communications.

[148] Ohtsuki, T., Sasase, I., Mori, S. (1995). Effects of hard-limiter and error correction coding on performance of direct-detection optical CDMA systems with PPM signaling. In: ICC '95 Seattle, Gateway to Globalization, IEEE 1995 International Conference on Communications.

[149] Sato, K., Ohtsuki, T., Uehara, H., Sasase, I. (1995). Performance of optical direct-detection CDMA systems using prime sequence codes. In: ICC '95 Seattle, Gateway to Globalization, IEEE 1995 International Conference on Communications.

[150] Kwong, W. C., Yang, G. C., Zhang, J. G. (1995). 2n prime sequence codes and their optical CDMA architecture. In: ICC '95 Seattle, Gateway to Globalization, IEEE 1995 International Conference on Communications.

[151] Kwong, W. C., Yang, G. C. (1995). Construction of 2^n prime-sequence codes for optical code division multiple access. *IEEE Proceedings, Communications* 142(3): 141–150.

[152] Kiasaleh, K. (1995). Performance of packet-switched fiber-optic frequency- hopping multiple-access networks. *IEEE Transactions on Communications* 43(7):2241–2253.

[153] Kawakami, W., Kitayama, K. I. (1995). Optical error-correction coding encoder and decoder: design considerations. *Applied Optics* 34(23):5064.

[154] Paek, E. G., Salehi, J. (1995). Holographic CDMA. *IEEE Transactions on Communications* 43(9):2434–2438.

[155] Dale, M., Gagliardi, R. (1995). Channel Coding for Asynchronous Fiberoptic CDMA Communications. *IEEE Transactions on Communications* 43(9):2485–2492.

[156] Griffin, R. A., Sampson, D. D., Jackson, D. A. (1995). Coherence coding for photonic code-division multiple access networks. *Journal of Lightwave Technology* 13(9): 1826–1837.

[157] Halford, K. W., Rozenbaum, Y., Brandt-Pearce, M. (1995). Near-orthogonal coding for spread spectrum and error correction. In: IEEE International Symposium on Information Theory.

[158] Karafolas, N., Uttamchandani, D. (1995). Optical CDMA system using bipolar codes and based on narrow passband optical filtering and direct detection. *IEEE Photonics Technology Letters* 7(9):1072–1074.

[159] Iversen, K., Ziemann, O. (1995). An all-optical CDMA communication network by spectral encoding of LED using acoustically tunable optical filters. In: 1995 URSI International Symposium on Signals, Systems, and Electronics.

[160] Capmany, J., Marti, J., Pastor, D., Sales, S. (1995). Fiber-optic delay-line filters employing fiber loops: Signal and noise analysis and experimental characterization. *JOSA A.* 12(10):2129.

[161] Gagliardi, R. M., Mendez, A J., Morookian, J. M. (1995). Analyses and experiments of optical code division multiple access (CDMA) with code length near unity. In: Conference Proceedings of the IEEE/LEOS Annual Meeting.

[162] Zhang, J. G., Kwong, W. C., Chen, L. K., Cheung, K. W. (1995). A feasible architecture for high-speed synchronous CDMA distribution network using optical processing. In: GLOBECOM '95. IEEE Global Telecommunications Conference.

[163] Ohtsuki, T., Sato, K., Sasase, I., Mori, S. (1995). Direct-detection optical synchronous CDMA systems with double optical hard-limiters using modified prime sequence codes. In: GLOBECOM '95. IEEE Global Telecommunications Conference.

[164] Gamachi, Y., Ohtsuki, T., Uehara, H., Sasase, I. (1995). Optical synchronous PPM/CDMA systems using co-channel interference cancellation. In: GLOBECOM '95. IEEE Global Telecommunications Conference.

[165] Nelson, L. B., Poor, H. V. (1995). Performance of multiuser detection for optical CDMA. I. Error probabilities. *IEEE Transactions on Communications* 43(11):2803–2811.

[166] Nelson, L. B., Poor, H. V. (1995). Performance of multiuser detection for optical CDMA. II. Asymptotic analysis. *IEEE Transactions on Communications* 43(12): 3015–3024.

[167] Elmirghani, J. M. H. (1995). Timing synchronization for indoor infrared PPM CDMA systems. *Proc. SPIE.* 2601:286–293.

[168] Hampicke, D., Iversen, K. (1995). Comparison and classification of all-optical CDMA systems for future telecommunication networks. *Proc. SPIE.* 2614:110–121.

[169] Adam, L., Simova, E. S., Kavehrad, M. (1995). Experimental optical CDMA system based on spectral amplitude encoding of noncoherent broadband sources. *Proc. SPIE.* 2614:122–132.

[170] Elmirghani, J. M. H. (1995). Compressed HDTV over the optical fiber network using PPM CDMA. *Proc. SPIE.* 2614:133–141.

[171] Iversen, K., Ziemann, O. (1995). Optical CDMA based on spectral encoding with integrated optical devices. *Proc. SPIE.* 2614:142–152.

[172] Capmany, J., Marti, J., Pastor, D., Sales, S. (1995). Solutions to the synthesis problem of optical delay line filters. *Optics Letters* 20(23):2438.

[173] Shen, X. A., Kachru, R. (1995). Optical header recognition by spectroholographic filtering. *Optics Letters* 20(24):2508–2510.

[174] Gagliardi, R. M., Mendez, A. J. (1995). Code division multiple access (CDMA) enhancement of wavelength division access (WDM) systems. In: ICC '95 WDM Networks and Systems Sessions.

[175] Gagliardi, R., Karp, S. (1995). Fiber networks. In: *Optical Communications*, 2nd ed., Wiley Interscience, NewYork, pp. 259–284.

[176] Yang, G. C. (1996). Variable-weight optical orthogonal codes for CDMA networks with multiple performance requirements. *IEEE Transactions on Communications* 44(1):47–55.

[177] Mendez, A. J. (1996). WDM matrix coding: A novel approach to ultradense networks. *Proceedings of the SPIE.* 2690:50–62.

[178] Gagliardi, R. M., Mendez, A. J. (1996). Performance improvement with hybrid WDM and CDMA optical communications. *Proceedings of the SPIE.* 2690:88–96.

[179] Uehara, H., Sasase, I. (1996). Fiber optic hybrid coherence multiplexed/subcarrier multiplexing (CM/SCM) system for microcellular mobile communications. In: ICC '96, Conference Record, IEEE International Conference on Communications. Converging Technologies for Tomorrow's Applications.

[180] Andonovic, I., Budin, J., Tančevski, L., Tur, M. (1996). Massive optical LANs using wavelength hopping/time spreading with increased security. *IEEE Photonics Technology Letters* 8(7):935–937.

[181] Aazhang, B., Dennis, T., Nguyen, L., Young, J. F. (1996). Experimental demonstration of bipolar codes for direct detection multiuser optical communication. In: IEEE/LEOS 1996 Summer Topical Meetings: Advanced Applications of Lasers in Materials Processing, 1996 Broadband Optical Networks/Smart Pixels/Optical MEMs and Their Applications. 1:20–21.

[182] Kwong, W. C., Yang, G. C., Zhang, J. G. (1996). $2^{\{n\}}$ prime-sequence codes and coding architecture for optical code-division multiple-access. *IEEE Transactions on Communications* 9:1152–1162.

[183] Sato, K., Ohtsuki, T., Uehara, H., Sasase, I. (1996). Effect of imperfect slot synchronization on direct-detection optical synchronous CDMA communications with PPM signaling. *Journal of Lightwave Technology* 14(9):1963–1969.

[184] Kavehrad, M., Simova, E. S. (1996). Optical CDMA by amplitude spectral encoding of spectrally-sliced light-emitting-diodes. In: IEEE 4th International Symposium on Spread Sprectrum Techniques and Applications Proceedings.

[185] Ho, C. L. (1996). Performance analysis of optical CDMA communication systems with quadratic congruential codes. In: IEEE 4th International Symposium on Spread Sprectrum Techniques and Applications Proceedings.

[186] Andonovic, I., Tančevski, L. (1996). Incoherent optical code division multiple access systems. In: IEEE 4th International Symposium on Spread Spectrum Techniques and Applications Proceedings.

[187] Hampicke, D., Iversen, K., Muckenheim, J. (1996). A basic theory of fiber-optic CDMA. In: IEEE 4th International Symposium on Spread Sprectrum Techniques and Applications Proceedings.

[188] Gagliardi, R. M., Mendez, A. J. (1996). Varieties and characteristics of discrete spectral encoding (DSE). In: IEEE 4th International Symposium on Spread Sprectrum Techniques and Applications Proceedings.

[189] Argon, C., Ergul, R. (1996). Optical S/CDMA using OOCs designed by the extended set concept. In: IEEE 4th International Symposium on Spread Sprectrum Techniques and Applications Proceedings.

[190] Mori, S., Ohtsuki, T., Sasase, I., Sato, K. (1996). Direct-detection optical synchronous CDMA systems with double optical hard-limiters using modified prime sequence codes. *IEEE Journal on Selected Areas in Communications* 14(9):1879–1887.

[191] Kavehrad, M., Khaleghi, F. (1996). A new correlator receiver architecture for noncoherent optical CDMA networks with bipolar capacity. *IEEE Transactions on Communications* 44(10):1335–1339.

[192] Mossberg, T. W. (1996). Spectral memory & other applications of spatial-spectral holography. In: Conference Proceedings of the 1996 IEEE/LEOS Annual Meeting.

[193] Gagliardi, R. M., Mendez, A. J. (1996). Design and analysis of wavelength division multiplex (WDM) and code division multiple access (CDMA) hybrids (WCH). In: Conference Proceedings of the 1996 IEEE/LEOS Annual Meeting.

[194] Mezger, F. B., Brandt-Pearce, M. (1996). Dispersion limited fiber-optic CDMA systems with overlapped signature sequences. In: Conference Proceedings of the 1996 IEEE/ LEOS Annual Meeting.

[195] Gessner, G., Hampicke, D., Iversen, K., Muckenheim, J., Scheiner, J. (1996). Electrical code-division multiplexing for low-cost optical subscriber loops. *Proc. SPIE.* 2953: 174–184.

[196]   Iversen, K., Jugl, E., Kuhwald, T., Muckenheim, J., Wolf, M. (1996). Time/wavelength coding for diffuse infrared communication systems with multiple optical carriers. *Proc. SPIE.* 2953:204–212.

[197]   Andonovic, I., Tančevski, L. (1996). Hybrid wavelength hopping/time spreading schemes for use in massive optical networks with increased security. *Journal of Lightwave Technology* 14(12):2636–2647.

[198]   Zhang, J. G. (1996). Strict optical orthogonal codes for purely asynchronous code-division multiple-access applications. *Applied Optics* 35(35):6996.

[199]   Zhang, J. G. (1996). Demonstration of novel optical optical code-division multiplexing systems for multirate and variable-rate data communications. *Optical Engineering* 35(12):3380–3384.

[200]   Karafolas, N., Uttamchandani, D. (1996). Optical fiber code division multiple access networks: A review. *Optical Fiber Technology* 2:149–168.

[201]   Kwong, W. C., Yang, G. C. (1996). Two-dimensional spatial signature patterns. *IEEE Transactions of Communications* 44:184–191.

[202]   Iversen, K., Jugl, E., Kuhwald, T. (1997). Algorithm for construction of (0,1)-matrix codes. *Electronics Letters* 33(3):227–229.

[203]   Zhang, J. G., Kwong, W. C. (1997). Effective design of optical code-division multiple access networks by using the modified prime code. *Electronics Letters* 33(3):229–230.

[203a] Gupta, G. C., Legg, P. J., Andonovic, I., Uttamchandani, D. (1997). Interferometric noise analysis in coherence-multiplexed optical fiber communication systems for local area networks. *Proc. SPIE.* 3211:222–224.

[204]   Hampicke, D., Iversen, K., Muckenheim, J. (1997). Incoherent optical CDMA for future high-speed computer networking: Systems, sequences and realization aspects. *Proc. SPIE.* 3211:629–634.

[205]   Gupta, G. C., Uttamchandani, D. (1997). Optical fiber multiple access systems employing spread spectrum technique for local area networks. *Proc. SPIE.* 3211: 699–705.

[206]   Andonovic, I., Uttamchandani, D., Wallace, C. G. (1997). New code for communications network fault finding using a spread spectrum approach. *Optics Communications* 133:39–42.

[207]   Bin, L. (1997). One-coincidence sequences with specified distance between adjacent symbols of frequency-hopping multiple access. *IEEE Transactions on Communications* 45(4):408–410.

[207a] Sardesai, H. P., Weiner, A. M. (1997). A nonlinear fiber-optic receiver for ultrashort pulse code division multiple access communications. *Electronics Letters* 33(7): 610–611.

[208]   Fortier, P., Hamelin, E., Rusch, L. (1997). Code performance in multirate CDMA for an optical fiber network. In: IEEE 1997 Canadian Conference on Electrical and Computer Engineering.

[209]   Babbitt, W. R. (1997). Spatial-spectral holographic routing and processing devices. In: CLEO '97 — Conference on Lasers and Electro-Optics.

[210]   Barry, R., Finn, S. G., Marquis, D., Medard, M. (1997). Security issues in all-optical networks. *IEEE Network* 11(3):42–48.

[211]   Zhang, J. G., Kwong, W. C. (1997). Design of optical code-division multiple-access networks with modified prime codes. In: IEEE International Symposium on Information Theory.

[212]   Hsu, C. S., Li, V. (1997). Performance analysis of slotted fiber-optic code-division multiple-access (CDMA) packet networks. *IEEE Transactions on Communications* 45(7):819–828.

[213] Hsu, C. S., Li, V. (1997). Performance analysis of unslotted fiber-optic code-division multiple-access (CDMA) packet networks. *IEEE Transactions on Communications* 45(8):978–987.

[214] Uehara, H., Sasase, I., Yokoyama, M. (1997). Path length mismatches in a coherence multiplexed fiber-optic subcarrier transmission system. In: 1997 IEEE Pacific Rim Conference on Communications, Computers and Signal Processing, 1997.

[215] Kamakura, K., Gamachi, Y., Uehara, H., Sasase, I. (1997). Optical CDMA based on frequency-domain encoding enhancement of frequency division multiplexing. In: 1997 IEEE Pacific Rim Conference on Communications, Computers and Signal Processing.

[216] Aazhang, B., Dennis, T., Nguyen, L., Young, J. F. (1997). Optical spectral amplitude CDMA communication. *Journal of Lightwave Technology* 15(9):1647–1653.

[217] Kahn, J., Marsh, G. (1997). Channel reuse strategies for indoor infrared wireless communications. *IEEE Transactions on Communications* 45(10):1280–1290.

[218] Zhang, J. G., Kwong, W. C., Sharma, A. B. (1997). 2^n modified prime codes for use in fibre optic CDMA networks. *Electronics Letters* 33(22):1840–1841.

[219] Kwong, W. C., Yang, G. C. (1997). Performance comparison of multiwavelength CDMA and WDMA+CDMA for fiber-optic networks. *IEEE Transactions on Communications* 45(11):1426–1434.

[220] Aazhang, B., Dennis, T., Young, J. F. (1997). Demonstration of all-optical CDMA with bipolar codes. In: LEOS '97 10th Annual Meeting.

[221] Aazhang, B., Guo, Y., Young, J. F. (1997). Wavelength encoding to reduce four-wave mixing crosstalk in multiwavelength channels. In: LEOS '97 10th Annual Meeting.

[222] Deppisch, B., Heidemann, R., Kaiser, M., Pfeiffer, T. (1997). High speed optical network for asynchronous multiuser access applying periodic spectral coding of broadband sources. *Electronics Letters* 33(25):2141–2142.

[223] Halford, K. W., Brandt-Pearce, M. (1998). New-user identification in a CDMA system. *IEEE Transactions on Communications* 46(1):144–155.

[224] Chang, C. C., Sardesai, H. P., Weiner, A. (1998). Code-division multiple access encoding and decoding of femtosecond optical pulses over a 2.5 km fiber link. *IEEE Photonics Technology Letters* 10(1):171–173.

[225] Lau, V. K. N., Maric, S. (1998). Multirate fiber-optic CDMA: System design and performance analysis. *Journal of Lightwave Technology* 16(1):9–17.

[226] Kiasaleh, K. (1998). Turbo-coded optical PPM communication systems. *Journal of Lightwave Technology* 16(1):18–26.

[227] Giemann, L., Gladisch, A., Hanik, N., Rudolph, J., Ziemann, O. (1998). The application of code division multiple access for transport overhead information in transparent optical networks. In: Tech. Digest OFC '98.

[228] Kitayama, K. I., Wada, N. (1998). Error-free 10 Gbit/s transmission of coherent optical code division multiplexing using all-optical encoder and balanced detection with local code. In: Tech. Digest OFC '98.

[229] Babbitt, W. R., Mohan, A., Ritcey, J. A. (1998). All-optical signal encoder-decoder for secure communications. *Proc. SPIE*. 3228:354–365.

[230] Lam, C. F., Yablanovitch, E. (1998). Fast-wavelength-hopped CDMA system for secure optical communications. *Proc. SPIE*. 3228:390–398.

[231] Lam, C. F., Tong, D. T., Vrijien, R. B., Wu, M. C., Yablanovitch, E. (1998). Spectrally encoded CDMA system using Mach-Zehnder encoder chains. *Proc. SPIE*. 3228: 399–407.

[232] Calleja, M., Griffin, R., Sampson, D. (1998). Crosstalk performance of coherent time-addressed photonic CDMA networks. *IEEE Transactions on Communications* 46(3): 338–348.

[233] Selvarajan, A., Shivaleela, E. S., Sivarajan, K. N. (1998). Design of a new family of two-dimensional codes for fiber-optic CDMA networks. *Journal of Lightwave Technology* 16(4):501–508.

[234] Ohtsuki, T. (1998). Channel interference cancellation using electrooptic switch and optical hardlimiters for direct-detection optical CDMA systems. *Journal of Lightwave Technology* 16(4):520–526.

[235] Harris, T. L., Sun, Y., Cone, R. L., MacFarlane, R. M. (1998). Demonstration of real-time address header decoding for optical data routing at 1536nm. *Optics Letters* 23(8):636–638.

[236] Kitayama, K. I., Nakamura, M. (1998). System performances of optical space code-division multiple-access-based fiber-optic two-dimensional parallel data link. *Applied Optics* 37(14):2915–2924.

[237] Shalaby, H. (1998). Cochannel interference reduction in optical PPM-CDMA systems. *IEEE Transactions on Communications* 46(6):799–805.

[238] Brandt-Pearce, M., Yang, M. H. (1998). Soft-decision multiuser detector for coded CDMA systems. In: ICC 1998 — IEEE International Conference on Communications.

[239] Fortier, P., Hamelin, E., Rusch, L. (1998). New cross-correlation results for multirate CDMA. In: ICC 1998 — IEEE International Conference on Communications.

[240] Kwong, W. C., Sharma, A.B., Zhang, J. G. (1998). All-optical 2n extended prime codes for optical fiber code-division multiple-access applications. In: ICC 1998 — IEEE International Conference on Communications.

[241] Huang, W., Andonovic, I. (1998). Code tracking in optical pulse CDMA through coherent correlation demodulation. In: ICC 1998 — IEEE International Conference on Communications.

[242] Fathallah, H., LaRochelle, S., Rusch, L. (1998). Optical frequency-hop multiple access communications system. In: ICC 1998 — IEEE International Conference on Communications.

[243] Shalaby, H. (1998). Optical CDMA with overlapping PPM. In: ICC 1998 — IEEE International Conference on Communications.

[244] Sadot, D., Mahlab, U., Bar Natan, V. (1998). A new method for developing optical CDMA address code sequences using the genetic algorithm. In: ICC 1998 — IEEE International Conference on Communications.

[245] Chan, H. H., Cryan, R. A., Elmirghani, J. M. H. (1998). Performance evaluation of PPM CDMA under different orthogonal coding schemes. In: ICC 1998 — IEEE International Conference on Communications.

[246] Hampicke, D., Iversen, K., Muckenheim, J. (1998). Construction of high-efficient optical CDMA computer networks: Statistical design. In: ICC 1998–IEEE International Conference on Communications.

[247] Chen, L. K., Cheung, K. W., Kwong, W. C., Sharma, A.B., Zhang, J. G. (1998). Experiments on high-speed all-optical code-division multiplexing (CDM) systems using a 2n prime code. In: ICC 1998–IEEE International Conference on Communications.

[248] Kiasaleh, K. (1998). Design and theoretical performance analysis of a hybrid time-hopping/TDMA multiple-access indoor wireless communication system. In: ICC 1998 — IEEE International Conference on Communications.

[249] Igasaki, Y., Kaneda, K., Kitayama, K. I., Nakamura, M. (1998). Four-channel, 8.8 Bit, two-dimensional parallel transmission by use of space code-division multiple-access encoder and decoder modules. *Applied Optics* 37(20):4389–4398.

[250] Eilenberger, G., Pfeiffer, T., van de Voorde, I., Vetter, P. (1998). Optical solutions for the access network. *Alcatel Telecommunications Review* 3Q98:225–231.

[251] Geis, H., Kullander, F., Laurent, C., Zyra, S. (1998). Crosstalk reduction in a code division multiplexed optical fiber sensor system. *Optical Engineering* 37(7):2104–2107.

[252] Zhang, J. G., Sharma, A. B., Kwong, W. C. (1998). New fiber-optic ATM switching networks using optical code-division multiple access for broadband communication applications. *IEEE Transactions on Consumer Electronics* 44(3):952–962.

[253] Lin, C. L., Wu, J. (1998). A synchronous fiber-optic CDMA system using adaptive optical hardlimiter. *Journal of Lightwave Technology* 16(8):1393–1403.

[254] Kamakura, K., Ohtsuki, T., Uehara, H., Gamachi, Y., Sasase, I. (1998). Optical spread time CDMA communications system with PPM signaling. In: IEEE International Symposium on Information Theory.

[255] Wada, N., Kitayama, K. I., Kurita, H. (1998). 10 Gbit/s optical code division multiplexing using 8-chip BPSK-code with time-gating detection. In: 24th European Conference on Optical Communications.

[256] Fu, A., Geiger, H., Ibsen, M., Laming, R. I., Petropoulos, P., Richardson, D. J. (1998). Demonstration of a simple CDMA transmitter and receiver using sampled fibre gratings. In: 24th European Conference on Optical Communications.

[257] Kitayama, K. I., Sotobayashi, H. (1998). Optical code-wavelength conversion for hybrid OCDM/WDM networks. In: 24th European Conference on Optical Communications.

[258] Elbers, J. P., Glingener, C., Kissing, J., Pfeiffer, T., Voges, E. (1998). Performance evaluation of a CDMA system using broadband sources. In: 24th European Conference on Optical Communications.

[259] Deppisch, B., Elbers, J. P., Heidemann, R., Pfeiffer, T., Witte, M. (1998). An optical access system with 8 asynchronous transmitters and > 1 Gbit/s network capacity. In: 24th European Conference on Optical Communications.

[260] Hampicke, D., Muckenheim, J. (1998). Effects of collisions in optical CDMA networks. *Proc. SPIE*. 3408:56–67.

[261] Dutt, B., Johnson, B. (1998). Optical CDMA offers all-optical network alternative. *Lightwave Magazine*, September, p. 54.

[262] Blaikie, R., Smith, E., Taylor, D. (1998). Performance enhancement of spectral-amplitude-coding optical CDMA using pulse-position modulation. *IEEE Transactions on Communications* 46(9):1176–1185.

[263] Zhang, J. G., Chen, L. K., Kwong, W. C. (1998). Experimental demonstration of efficient all-optical code-division multiplexing. *Electronics Letters* 34(19):1866–1868.

[264] Kwong, W. C., Yang, G. C. (1998). Image transmission in multicore-fiber code-division multiple-access networks. *IEEE Communications Letters* 2(10):285–287.

[265] Andonovic, I., Huang, W. (1998). Coherent optical pulse CDMA systems based on I-Q noncoherent demodulation of M-ary orthogonal signals. *Proc. SPIE*. 3531:61–72.

[266] Andonovic, I., Huang, W., Tur, M. (1998). Decision-directed PLL for coherent optical pulse CDMA systems in the presence of multiuser interference, laser phase noise, and shot noise. *Journal of Lightwave Technology* 16(10):1786–1794.

[267] Zhang, J. G., Kwong, W. C., Mann, S. (1998). Construction of 2n extended prime codes with cross-correlation constraint of one. *IEEE Proceedings, Communications* 145(5):297–303.

[268] Zhang, J. G., Kwong, W. C., Sharma, A. B. (1998). Performance comparison of 2n extended prime codes (= 1) and 2n prime codes (= 2) in fiber-optic CDMA systems. In: Proceedings of the 1998 International Conference on Communication Technology.

[269] Kamakura, K., Ohtsuki, T., Sasase, I. (1998). An ATM-based optical code division multiplexing transport network using coherent ultrashort pulses. In: GLOBECOM '98. IEEE Global Telecommunications Conference, 1998.

[270] Zhang, J. G., Sharma, A. B., Kwong, W. C. (1998). Optical CDMA-based ATM switches supporting equal-bit-rate and variable-bit-rate communication services. In: GLOBECOM '98. IEEE Global Telecommunications Conference.

[271] Chan, J. K. (1998). Optical CDMA video transport/switching system. In: SPIE Photonics East. *Proc. SPIE.* 3541:271–278.

[272] Chang, C. C., Sardesai, H. P., Weiner, A. (1998). A femtosecond code-division multiple-access communication system test bed. *Journal of Lightwave Technology* 16(11):1953–1964.

[273] Mendieta, F. J., Miridonov, S. V., Shlyagin, M. G., Spirin, V., Tentori, D. (1999). Multiplexing of grating-based fiber sensors using broadband spectral coding. In: SPIE Photonics East. *Proc. SPIE.* 3541: 271–278.

[274] Dennis, T., Young, F. J. (1998). All-optical encoders and decoders for spectral CDMA using bipolar codes. In: SPIE Photonics East. *Proc. SPIE.* 3531: 73–79.

[275] Yang, W., Davis, J., Goswami, D., Fetterman, M., Warren, W. S. (1998). Optical wavelength domain code-division multiplexing using AOM-based ultrafast optical pulse shaping. In: SPIE Photonics East. *Proc. SPIE.* 3531: 80–87.

[276] Kwong, W. C., Sharma, A. B., Zhang, J. G. (1998). New photonic asynchronous-transfer-mode switch with optical signal processing and code-division multiple access. *Optical Engineering* 37(12):3205–3217.

[277] Lam, C. F., Tong, D. T., Vrijien, R. B., Wu, M. C., Yablanovitch, E. (1999). Multiwavelength optical code division multiplexing. *Proc. SPIE.* 3571:108–132.

[278] Mahlab, U., Natan, V. B., Sadot, D. (1999). New method for developing optical code division multiplexed access sequences using genetic algorithm. *Optical Engineering* 38(1):151–156.

[279] Andonovic, I., Huang, W. (1999). Coherent optical pulse CDMA systems based on coherent correlation detection. *IEEE Transactions on Communications* 47(2):261–271.

[280] Sotobayashi, H., Kitayama, K. I. (1999). 1.24 Gb/s, 40 km optical code division multiplexing transmission by using spectral bipolar coding of broadband incoherent light. In: Technical Digest, 1999 Optical Fiber Communication Conference and the International Conference on Integrated Optics and Optical Fiber Communication.

[281] Koo, K. P., Tveten, A. B., Vohra, S. T. (1999). DWDM of fiber Bragg grating sensors without sensor spectral dynamic range limitation using CDMA. In: Technical Digest, 1999 Optical Fiber Communication Conference and the International Conference on Integrated Optics and Optical Fiber Communication.

[282] Deppisch, B., Heidemann, R., Pfeiffer, T., Witte, M. (1999). Optical CDMA transmission for robust realization of complex and flexible multiple access networks. In: Technical Digest, 1999 Optical Fiber Communication Conference and the International Conference on Integrated Optics and Optical Fiber Communication.

[283] Cardakli, M. C., Feinberg, J., Grubsky, V., Lee, S., Starodubov, D., Willner, A. E. (1999). All-optical packet header recognition and switching in a reconfigurable network using fiber Bragg gratings for time-to-wavelength mapping and decoding. In: Technical Digest, 1999 Optical Fiber Communication Conference and the International Conference on Integrated Optics and Optical Fiber Communication.

[284] Grunnet-Jepsen, A., Johnson, A. E., Maniloff, E. S., Mossberg, T. W., Munroe, M. J., Sweetser, J. N. (1999) Spectral phase encoding and decoding using fiber Bragg gratings. In: Technical Digest, 1999 Optical Fiber Communication Conference and the International Conference on Integrated Optics and Optical Fiber Communication.

[285] Deppisch, B., Pfeiffer, T., Witte, M. (1999). High-speed transmission of broad-band thermal light pulses over dispersive fibers. *IEEE Photonics Technology Letters* 11(3):385–387.

[286] Fathallah, H., LaRochelle, S., Rusch, L. (1999). Passive optical fast frequency-hop CDMA communications system. *Journal of Lightwave Technology* 17(3):397–405.

[287] Shalaby, H. (1999). A performance analysis of optical overlapping PPM-CDMA communications system. *Journal of Lightwave Technology* 17(3):426–433.

[288] Dennis, T., Young, J. F. (1999). Optical implementation of bipolar codes. *IEEE Journal of Quantum Electronics* 35(3):287–291.

[289] Zhang, J. G., Chen, L. K., Kwong, W. C., Cheung, K. W., Sharma, A. B. (1999). Experiments on high-speed all-optical code-division multiplexing systems using all-serial encoders and decoders for 2^n prime code. *IEEE Journal of Selected Topics in Quantum Electronics* 5(2):368–375.

[290] Capmany, J., Mallea, G. (1999). Autocorrelation pulse distortion in optical fiber CDMA systems employing ladder networks. *Journal of Lightwave Technology* 17(4):570–578.

[291] Purchase, K. G., Brady, D. J., Roh, S. D., Lammert, R. M., Osowski, M. L., Coleman, J. J., Hughes, J. S. (1999). The distributed Bragg pulse shaper: Demonstration and model. *Journal of Lightwave Technology* 17(4):621–628.

[292] Grunnet-Jepsen, A., Johnson, A. E., Maniloff, E. S., Mossberg, T. W., Munroe, M. J., Sweetser, J. N. (1999). Optical code-division multiple access (O-CDMA) inter-connects and telecommunication networks based on temporally accessed spectral multiplexing (TASM). *Proc. SPIE.* 3632:36–44.

[293] Andonovic, I., Huang, W. (1999). Optical code-division multiple-access networks. *Proc. SPIE.* 3666:434–443.

[294] Harris, T. L., Sun, Y., Cone, R. L., Babbitt, W. R., Ritcey, J. A., Equall, R. W. (1999). A spatial-spectral holographic correlator at 1536 nm using 30-symbol BPSK and QPSK codes optimized for secure communications. In: CLEO '99 — Conference on Lasers and Electro-Optics.

[295] Grunnet-Jepsen, A., Johnson, A. E., Maniloff, E. S., Mossberg, T. W., Munroe, M. J., Sweetser, J. N. (1999). Fibre Bragg grating based spectral encoder/decoder for lightwave CDMA. *Electronics Letters* 35(13):1096–1097.

[296] Zhang, J. G. (1999). Design of a special family of optical CDMA address codes for fully asynchronous data communications. *IEEE Transactions on Communications* 47(7):967–973.

[297] Shalaby, H. (1999). Direct detection optical overlapping PPM-CDMA communication systems with double optical hardlimiters. *Journal of Lightwave Technology* 17(7): 1158–1165.

[298] Deppisch, B., Heidemann, R., Pfeiffer, T., Witte, M. (1999). Operational stability of a spectrally encoded optical CDMA system using inexpensive transmitters without spectral control. *IEEE Photonics Technology Letters* 11(7):916–918.

[299] Letaief, K. B., Tang, J. T. K. (1999). Optical CDMA communication systems with multiuser and blind detection. *IEEE Transactions on Communications* 47(8):1211–1217.

[300] Matsuo, R., Matsuo, M., Ohtsuki, T., Udagawa, T., Sasase, I. (1999). Performance analysis of indoor infrared wireless systems using OOK CDMA on diffuse channels. In: 1999 IEEE Pacific Rim Conference on Communications, Computers and Signal Processing.

[301] Sonoda, T., Kamakura, K., Ohtsuki, T., Sasase, I., Mori, S. (1999). Optical frequency-encoding CDMA systems using time-encoding for MAI mitigation. In: 1999 IEEE Pacific Rim Conference on Communications, Computers and Signal Processing.

[302] Dennis, T., Young, J. F. (1999). Measurements of BER performance for bipolar encoding of an SFS. *Journal of Lightwave Technology* 17(9):1542–1546.

[303] Kim, J. Y., Poor, H. V. (1999). Performance of an optical PPM/CDMA system with turbo coding. In: IEEE VTS 50th Vehicular Technology Conference.

[304] Lopez-Hernandez, F. J., Perez-Jimenez, R., Rabadan-Borges, J. A., Santamaria, A., Vento-Alvarez, J. (1999). Infrared wireless DSSS system for indoor data communication links. *SPIE Photonics East* 1:3816–3850.

[305] Kitayama, K. I., Wada, N. (1999). A 10 Gb/s optical code division multiplexing using 8-chip optical bipolar code and coherent detection. *Journal of Lightwave Technology* 17(10):1758–1765.

[306] Grunnet-Jepsen, A., Johnson, A. E., Maniloff, E. S., Mossberg, T. W., Munroe, M. J., Sweetser, J. N. (1999). Demonstration of all-fibre sparse lightwave CDMA based on temporal phase encoding. *IEEE Photonics Technology Letters* 11(10):1283–1285.

[307] Chen, Y. J., Lee, C. H., Lin, X., Young, J. F., Zhong, S. (1999). Planar lightwave circuit desing for programmable complementary spectral keying encoder and decoder. *Electronics Letters* 35(21):1813–1815.

[308] Shalaby, H. (1999). Synchronous Fiber-Optic CDMA Systems with Interference Estimators. *Journal of Lightwave Technology* 17(11):2268–2275.

[309] Feng, H. X. C., Heritage, J. P., Lennon, W. J., Thombley, L. R. (1999). Computer modeling of the national transparent optical network (NTON). In: Conference Proceedings IEEE/LEOS '99 Annual Meeting.

[310] Babich, C. D., Young, J. F. (1999). Performance modeling of a planar waveguide based spectral encoding system. In: Conference Proceedings IEEE/LEOS '99 Annual Meeting.

[311] Kamakura, K., Sasase, I. (1999). A blind receiver using optical hard-limiters for optical code division multiple access. In: GLOBECOM 1999 — IEEE Global Telecommunications Conference.

[312] Ohtsuki, T. (1999). Performance analysis of direct-detection optical CDMA systems with optical hard-limiter using equal-weight orthogonal signaling. In: GLOBECOM 1999 — IEEE Global Telecommunications Conference.

2000–2004

[313] Harris, T. L., Sun, Y., Babbitt, W. R., Cone, R. L., Ritcey, J. A., Equall, R. W. (2000). Spatial-spectral holographic correlator shift at 1536nm using 30-symbol quadriphase- and binary-phase-shift keyed codes. *Optics Letters* 25(2):85–87.

[314] Chan, J. K., Tang, Y. S., Xu, Y. (2000). Development and prospective of SOI-based photonic components for optical CDMA application. *Proc. SPIE.* 3953:1–8.

[315] Mendez, A. J. (2000). Application of multifrequency lasers (MFLs) to optical CDMA. In: Proceedings of the 10th DARPA Symposium on Photonic Systems for Antenna Applications.

[316] Cardakli, M. C., Gurkan, D., Havstad, S. A., Willner, A. E. (2000). Variable-bit-rate header recognition for reconfigurable networks using tunable fiber-Bragg-gratings as optical correlators. 2000 Optical Fiber Communication Conference.

[317] Edagawa, N., Suzuki, M. (2000). Novel all optical limiter using electroabsorption modulators. 2000 Optical Fiber Communication Conference.

[317a] Stuart, H. R. (2000). Dispersive multiplexing in multimode optical fiber. *Science* 289:281–283.

[318] Stuart, H. R. (2000). Dispersive multiplexing in multimode fiber. 2000 Optical Fiber Communication Conference.

[319] Grunnet-Jepsen, A., Johnson, A. E., Maniloff, E. S., Mossberg, T. W., Munroe, M. J., Sweetser, J. N. (2000). Code-division-multiplexing-compatible coding and decoding of directly driven DFB laser bit streams. 2000 Optical Fiber Communication Conference.

[320] Zheng, Z., Weiner, A. (2000). Novel optical thresholder based on second harmonic generation in long periodically poled lithium niobate for ultrashort pulse optical code division multiple-access. 2000 Optical Fiber Communication Conference.

[321] Brenner, I., Cardakli, M. C., Fejer, M. M., Gurkan, D., Havstad, S. A., Parameswaran, K. R., Willner, A. E. (2000). All-optical time-slot-interchange and wavelength conversion using difference-frequency-generation and FBGs. *2000 Optical Fiber Communication Conference* 4:196–198.

[322] Buchali, F., Pfeiffer, T., Witte, M. (2000). Reducing the optical power penalty for electronically dispersion compensated LED pulse transmission by using multibit shift decision feedback. *Electronics Letters* 36(5):450–451.

[323] Andonovic, I., Huang, W., Tur, M. (2000). Code acquisition in coherent optical pulse CDMA systems utilizing coherent correlation demodulation. *IEEE Transactions on Communications* 48(4):611–621.

[324] Kim, S. (2000). Cyclic optical encoders/decoders for compact optical CDMA networks. *IEEE Photonics Technology Letters* 12(4):428–430.

[325] Kim, S., Kyungsik, Y., Park, N. (2000). A new family of space / wavelength / time spread three-dimensional optical code for OCDMA networks. *Journal of Lightwave Technology* 18(4):502–511.

[326] Cardakli, M. C., Feinberg, J., Grubsky, V., Lee, S., Starodubov, D., Willner, A. E. (2000). Reconfigurable optical packet header recognition and routing using time-to-wavelength mapping and tunable fiber Bragg gratings for correlation decoding. *IEEE Photonics Technology Letters* 12(5):552–554.

[327] Yegnanarayanan, S., Bhushan, A. S., Jalali, B. (2000). Fast wavelength-hopping time-spreading encoding/decoding for optical CDMA. *IEEE Photonics Technology Letters* 12(5):573–575.

[328] Andonovic, I., Huang, W., Nizam, M. H. M., Tur, M. (2000). Coherent optical CDMA (OCDMA) systems used for high-capacity optical fiber networks-system description, OTDMA comparison, and OCDMA/WDMA networking. *Journal of Lightwave Technology* 18(6):765–778.

[329] Kahn, J., Ohtsuki, T. (2000). Turbo-coded optical PPM CDMA systems. In: ICC 2000 — IEEE International Conference on Communications.

[330] Feng, H. X. C., Heritage, J., Lennon, W. J., Mendez, A. (2000). Effects of optical layer impairments on 2.5Gb/s optical CDMA transmission. *Optics Express* 7(1): 2–9.

[331] Choi, Y., Han, S., Kang, M., Kim, S., Park, S. (2000). Incoherent bidirectional fiber-optic code division multiple access networks. *IEEE Photonics Technology Letters* 12(7):921–923.

[332] Mustapha, M. M., Ormondroyd, R. F. (2000). Effect of multiaccess interference on code synchronization using a sequential detector in an optical CDMA LAN. *IEEE Photonics Technology Letters* 12(8):1103–1105.

[333] Zhang, J. G., Kwong, W. C., Sharma, A. B. (2000). Effective design of optical fiber code-division multiple access networks using the modified prime codes and optical processing. In: WCC–ICCT 2000 — International Conference on Communication Technology Proceedings.

[334] Kim, J. Y., Poor, H. V. (2000). An optical CDMA packet network with turbo coding. In: PIMRC 2000 — The 11th IEEE International Symposium on Personal, Indoor and Mobile Radio Communications.

[335] Yu, K., Shin, J., Park, N. (2000). Wavelength-time spreading optical CDMA system using wavelength multiplexers and mirrored fiber delay lines. *IEEE Photonics Technology Letters* 12:(9):1278–1280.

[336] Chen, L. R., Smith, P. W. E. (2000). Demonstration of incoherent wavelength-encoding/time-spreading optical CDMA using chirped Moire gratings. *IEEE Photonics Technology Letters* 12(9):1281–1283.

[337] Kamakura, K., Ohtsuki, T., Sasase, I., Sawagashira, H. (2000). Direct-detection optical synchronous CDMA systems with interference canceller using group information codes. In: GLOBECOM 2000 — IEEE Global Telecommunications Conference.

[338] Fortier, P., Inaty, E., Rusch, L. (2000). Multirate optical fast frequency hopping CDMA system using power control. In: GLOBECOM 2000 — IEEE Global Telecommunications Conference.

[339] Lam, C. F. (2000). To spread or not to spread: The myths of optical CDMA. In: LEOS 2000.

[340] Sargent, E., Stok, A. (2000). Lighting the local area: Optical code-division multiple access and quality of service provisioning. *IEEE Network* 14(6):42–46.

[341] Feng, H. X. C., Gagliardi, R., Heritage, J., Mendez, A., Morookian, J. M. (2000). Strategies for realizing optical CDMA for dense, high speed, long span, optical network applications. *Journal of Lightwave Technology* 18(12):1685–1696.

[342] Salehi, J., Zahedi, S. (2000). Analytical comparison of various fiber-optic CDMA receiver structures. *Journal of Lightwave Technology* 18(12):1718–1727.

[343] Mustapha, M. M., Ormondroyd, R. F. (2000). Dual-threshold sequential detection code synchronization for an optical CDMA network in the presence of multiuser interference. *Journal of Lightwave Technology* 18(12):1742–1748.

[344] Kitayama, K. I., Sotobayashi, H., Wada, N. (2000). Architectural Considerations for Photonic IP Router Based upon Optical Code Correlation. *Journal of Lightwave Technology* 18(12):1834–1844.

[345] Kim, J. Y., Poor, H. V. (2000). Turbo-coded packet transmission for an optical CDMA network. *Journal of Lightwave Technology* 18(12):1905–1916.

[346] Deppisch, B., Elbers, J. P., Kissing, J., Pfeiffer, T., Schmuck, H., Voges, E., Witte, M. (2000). Coarse WDM/CDM/TDM concept for optical packet transmission in metropolitan and access networks supporting 400 channels at 2.5 Gb/s peak rate. *Journal of Lightwave Technology* 18(12):1928–1938.

[347] Chan, J. K., Nguyen, I. A., Sayano, K. (2001). Demonstration of multichannel optical CDMA for free-space communications. *Proc. SPIE.* 4272:38–49.

[348] Ibsen, M., Petropoulos, P., Richardson, D. J., Teh, P. C. (2001). Generation, recognition and recoding of 64-chip bipolar optical code sequences using superstructured fibre Bragg gratings. *Electronics Letters* 37(3):190–191.

[349] Ibsen, M., Petropoulos, P., Richardson, D. J., Teh, P. C. (2001). Phase encoding and decoding of short pulses at 10 Gb/s using superstructured fiber Bragg gratings. *IEEE Photon. Tech. Letter* 13(2):154–156.

[350] Mossberg, T. W., Raymer, Michael G. (2001). Optical code-division multiplexing. *Optics & Photonics News* 12(3):50–54.

[351] Kim, J. Y., Poor, H. V. (2001). Turbo-coded optical direct-detection CDMA system with PPM modulation. *Journal of Lightwave Technology* 19(3):312–323.

[352] Chang, T. W. F., Sargent, E. (2001). Optical CDMA using 2-D codes: The optimal single-user detector. *IEEE Communications Letters* 4:169–171.

[353] Ibsen, M., Lee, J. H., Petropoulos, P., Richardson, D. J., Teh, P. C. (2001). Reduction of interchannel interference noise in a two-channel grating-based OCDMA system using a nonlinear optical loop mirror. *IEEE Photonics Technology Letters* 13(5):529–531.

[354] Kwong, W. C., Yang, G. C. (2001). Double-weight signature pattern codes for multicore-fiber code-division multiple-access networks. *IEEE Communications Letters* 5(5): 203–205.

[355] Fortier, P., Inaty, E., Rusch, L. (2001). SIR performance evaluation of a multirate OFFH-CDMA system. *IEEE Communications Letters* 5(5):224–226.

[356] Seldomridge, N. L., Mohan, R. K., Babbitt, W. R., Merkel, K. D. (2001). Steady-state accumulated complex spectral gratings for correlation signal processing. In: CLEO '01 — Conference on Lasers and Electro-Optics, 2001.

[357] Matsuo, M., Ohtsuki, T., Sasase, I. (2001). Performance analysis of DS-CDMA indoor infrared wireless systems using equalizer on diffuse channels. In: IEEE VTS 53rd Vehicular Technology Conference, 2001.

[358] Yamaguchi, H., Matsuo, R., Ohtsuki, T., Sasase, I. (2001). Equalization for infrared wireless systems using OOK-CDMA. In: ICC 2001 — IEEE International Conference on Communications.

[359] Matsuo, R., Ohtsuki, T., Sasase, I. (2001). Performance analysis of indoor infrared wireless systems using PPM CDMA with lenealizer with dead-zone and PPM CDMA with hard-limiter on diffuse channels. In: ICC 2001 — IEEE International Conference on Communications.

[360] Kamakura, K., Sasase, I. (2001). Reduction of optical beat interference in optical frequency-hopping CDMA networks using coherence multiplexing. In: ICC 2001 — IEEE International Conference on Communications.

[361] Kim, J., Lee, C. K., Seo, S. W. (2001). Generation and performance analysis of frequency-hopping optical orthogonal codes with arbitrary time blank patterns. In: ICC 2001 — IEEE International Conference on Communications.

[362] Sargent, E., Ng, E. K. H. (2001). Mining the fibre-opic channel capacity in the local area: Maximizing spectral efficiency in multiwavelength optical CDMA networks. In: ICC 2001 — IEEE International Conference on Communications.

[363] Kamakura, K., Sasase, I. (2001). A new modulation scheme using asymmetric error correcting code embedded in optical orthogonal code for optical CDMA. In: IEEE International Symposium on Information Theory.

[364] Grunnet-Jepsen, A., Johnson, A. E., Saint-Hilaire, P., Sweetser, J. N. (2001). Applications of the arbitrary optical fiber: Fiber Bragg grating filters for WDM and CDMA communications. *Proc. SPIE.* 4532:249–260.

[365] Chen, J. J., Yang, G. C. (2001). CDMA fiber-optic systems with optical hard limiters. *Journal of Lightwave Technology* 19(7):950–958.

[366] Ghafouri-Shiraz, H., Shalaby, H. M. H., Wei, Z. (2001). Performance analysis of optical spectral-amplitude-coding CDMA systems using a super-fluorescent fiber source. *IEEE Photonics Technology Letters* 13(8):887–889.

[367] Ghafouri-Shiraz, H., Shalaby, H. M. H., Wei, Z. (2001). New code families for fiber-Bragg-grating-based spectral-amplitude-coding optical CDMA systems. *IEEE Photonics Technology Letters* 13(8):890–892.

[368] Iwata, A., Sawagashira, H., Sonoda, T., Kamakura, K., Sasase, I. (2001). Optical CDMA system using embedded transmission method with Manchester signaling. In: IEEE Pacific Rim Conference on Communications, Computers and Signal Processing.

[369] Ghafouri-Shiraz, H., Shalaby, H. M. H., Wei, Z. (2001). Modified quadratic congruence codes for fiber Bragg-grating-based spectral-amplitude-coding optical CDMA systems. *Journal of Lightwave Technology* 19(9):1274–1281.

[370] Ibsen, M., Petropoulos, P., Richardson, D. J., Teh, P. C. (2001). A comparative study of the performance of seven- and 63-chip optical code-division multiple-access encoders

and decoders based on superstructural fiber Bragg gratings. *Journal of Lightwave Technology* 19(9):1352–1365.

[371]   Jourdan, A., Pfeiffer, T., Tančevski, L. (2001). How much optics in future metropolitan networks. *Alcatel Telecommunications Review, 3rd Quarter 2001*:219–221.

[372]   Belardi, W., Ibsen, M., Lee, J. H., Monro, T. M., Richardson, D. J., Teh, P. C., Yusoff, Z. (2001). An OCDMA receiver incorporating a holey fibre nonlinear thresholder. In: ECOC '01 27th European Conference on Optical Communication.

[373]   Fejer, M. M., Gurkan, D., Hauer, M. C., Lee, S., Pan, Z., Parameswaran, K. R., Sahin, A. B., Willner, A. E. (2001). Demonstration of multiwavelength all-optical header recognition using a PPLN and optical correlators. *ECOC '01 27th European Conference on Optical Communication* 3:312–313.

[374]   Kitayama, K. I., Murata, M. (2001). Photonic access node using optical code-based label processing and its applications to optical data networking. *Journal of Lightwave Technology* 19(10):1401–1415.

[375]   Kamakura, K., Sasase, I. (2001). An embedded transmission method of PPM signaling constructed of parallel error correcting codes for optical CDMA. *GLOBECOM 2001 — IEEE Global Telecommunications Conference* 1:55–59.

[376]   Chen, L. R. (2001). Flexible fiber Bragg grating encoder/decoder for hybrid wavelength-time optical CDMA. *IEEE Photonics Technology Letters* 12(11):1233–1235.

[377]   Chujo, W., Ibsen, M., Kitayama, K. I., Petropoulos, P., Richardson, D. J., Teh, P. C., Wada, N. (2001). Demonstration of a 64-chip OCDMA system using superstructured fiber gratings and time-gating detection. *IEEE Photonics Technology Letters* 13(11):1239–1241.

[378]   Keshavarzian, A., Salehi, J. (2001). Synchronization of optical orthogonal codes in optical CDMA systems via simple serial-search method. In: GLOBECOM 2001– IEEE Global Telecommunications Conference.

[379]   Chang, T. W. F., Sargent, E. (2001). Spectral efficiency limit of bipolar signaling in incoherent optical CDMA systems. In: GLOBECOM 2001–IEEE Global Telecommunications Conference.

[380]   Inaty, E., Shalaby, H., Fortier, P. (2001). A new transmitter-receiver architecture for noncoherent multirate OFFH-CDMA system with fixed optimal detection threshold. In: GLOBECOM 2001–IEEE Global Telecommunications Conference.

[381]   Argon, C., McLaughlin, S. (2001). Turbo product codes for performance improvement of optical CDMA systems. *GLOBECOM 2001 — IEEE Global Telecommunications Conference* 3:1505–1509.

[382]   Chan, K. T., Lee, K. L., Shu, C., Wang, X. (2001). Multiwavelength self-seeded Fabry-Perot laser with subharmonic pulse-gating for two-dimensional fiber optic-CDMA. *IEEE Photonics Technology Letters* 13(12):1361–1363.

[383]   Hu, Y., Wan, S. P. (2001). Two-dimensional optical CDMA differential system with prime/OOC codes. *IEEE Photonics Technology Letters*. 13(12):1373–1375.

[384]   Kamakura, K., Sasase, I. (2001). A new modulation scheme using asymmetric error correcting code embedded in optical orthogonal code for optical CDMA. *Journal of Lightwave Technology* 19(12):1839–1850.

[385]   Weng, C. S., Wu, J. (2001). Optical orthogonal codes with nonideal cross correlation. *Journal of Lightwave Technology* 19(12):1856–1863.

[386]   Ben Jaafar, H., Cortes, P. Y., Fathallah, H., LaRochelle, S. (2001). 1.25 Gbit/s transmission of optical FFH-OCDMA signals over 80 km with 16 users. In: Optical Fiber Communication Conference and Exhibit.

[387] Sotobayashi, H., Chujo, W., Kitayama, K. I. (2001). Demonstration of ultra-wideband and transparent virtual optical code/wavelength path network. In: Optical Fiber Communication Conference and Exhibit.

[388] Wen, Y. G., Chen, L. K., Tong, F. (2001). Fundamental limitation and optimization on optical code conversion for WDM packet switching networks. In: Optical Fiber Communication Conference and Exhibit.

[389] Ibsen, M., Lee, J. H., Petropoulos, P., Richardson, D. J., Teh, P. C. (2001). A 10-Gbit/s all-optical code generation and recognition system based on a hybrid approach of optical fiber delay line and superstructure fiber Bragg grating technologies. In: Optical Fiber Communication Conference and Exhibit.

[390] Ibsen, M., Lee, J. H., Petropoulos, P., Richardson, D. J., Teh, P. C. (2001). A 4-channel WDM/OCDMA system incorporating 255-chip, 320 Gchip/s quaternary phase coding and decoding gratings. In: Optical Fiber Communication Conference and Exhibit.

[391] Kissing, J., Pfeiffer, T., Voges, E. (2001). A robust and flexible all optical CDMA multichannel transmission system for the access domain. In: Optical Fiber Communication Conference and Exhibit.

[392] Deppisch, B., Haisch, H., Hofer, B., Kissing, J., Pfeiffer, T., Schmuck, H., Schreiber, P., Witte, M. (2001). Bidirectional optical code division multiplexed field trial system for the metro and access area. In: Optical Fiber Communication Conference and Exhibit.

[393] Ibsen, M., Lee, J. H., Petropoulos, P., Richardson, D. J., Teh, P. C. (2001). High performance, 64-chip, 160 Gchip/s fiber grating based OCDMA receiver incorporating a nonlinear optical loop mirror. In: Optical Fiber Communication Conference and Exhibit.

[394] Ibsen, M., Lee, J. H., Petropoulos, P., Richardson, D. J., Teh, P. C. (2002). A grating-based OCDMA coding-decoding system incorporating a nonlinear optical loop mirror for improved code recognition and noise reduction. *Journal of Lightwave Technology* 20(1):36–46.

[395] Ibsen, M., Lee, J. H., Petropoulos, P., Richardson, D. J., Teh, P. C. (2002). Demonstration of a four-channel WDM/OCDMA system using 255-chip 320-Gchip/s quarternary phase coding gratings. *IEEE Photonics Technology Letters* 14(2):227–229.

[396] Fortier, P., Inaty, E., Rusch, L., Shalaby, H. (2002). Multirate optical fast frequency hopping CDMA system using power control. *Journal of Lightwave Technology* 20(2): 166–177.

[397] Keshavarzian, A., Salehi, J. (2002). Optical orthogonal code acquisition in fiber-optic CDMA systems via the simple serial-search method. *IEEE Transactions on Communications* 50(3):473–483.

[398] Yang, G. C., Kwong, W. C. (2002). *Prime Codes with Applications to CDMA Optical and Wireless Networks.* Boston: Artech House.

[399] LaRochelle, S., Slavik, R. (2002). Multiwavelength "single-mode" erbium doped fiber laser for FFH-OCDMA testing. In: Optical Fiber Communication Conference and Exhibit.

[400] Hauer, M. C., McGeehan, J., Sahin, A. B., Willner, A. E. (2002). Reconfigurable multiwavelength optical correlator for header-based switching and routing. In: Optical Fiber Communication Conference and Exhibit.

[401] Au, A. A., Bannister, J., Hauer, M. C., Kamath, P., Lee, H. P., Lin, C. H., Lyons, E. R., McGeehan, J., Starodubov, D., Touch, J., Willner, A. E. (2002). Dynamically reconfigurable all-optical correlators to support ultra-fast internet routing. In: Optical Fiber Communication Conference and Exhibit.

[402] Fu, L. B., Ibsen, M., Lee, J. H., Richardson, D. J., Teh, P. C., Yusoff, Z. (2002). A 16-channel OCDMA system (4 OCDM / spl times / 4 WDM) based on 16-chip, 20 Gchip/s superstructure fibre Bragg gratings and DFB fibre laser transmitters. In: Optical Fiber Communication Conference and Exhibit.

[403] Ibsen, M., Mokhtar, M. R., Richardson, D. J., Teh, P. C. (2002). Simple dynamically reconfigurable OCDMA encoder/decoder based on a uniform fiber Bragg grating. In: Optical Fiber Communication Conference and Exhibit, 2002.

[404] Ghafouri-Shiraz, H., Zou, W. (2002). Proposal of a novel code for spectral amplitude-coding opical CDMA systems. *IEEE Photonics Technology Letters* 14(3):414–416.

[405] Sotobayashi, H., Chujo, W., Kitayama, K. I. (2002). 1.6-b/s/Hz 6.4-Tb/s OPSK-OCDM/ WDM (4 OCDM × 40 WDM × 40 Gb/s) transmission experiment using optical hard thresholding. *IEEE Photonics Technology Letters* 14(4):555–557.

[406] Wells, W., Menders, J., Miles, E., Loginov, B., Hodara, H. (2002). Another alternative to quantum cryptography. April, pp. 91–106.

[407] Yim, R. M. H., Chen, L. R., Bajcsy, J. (2002). Design and performance of 2-D codes for wavelength-time optical CDMA. *IEEE Photonics Technology Letters* 14(5):714–716.

[408] Ng, E. K. H., Sargent, E. (2002). Optimum threshold detection in real-time scalable high-speed multiwavelength optical code-division multiple-access LANs. *IEEE Transactions on Communications* 50(5):778–784.

[409] Wenhua, M., Chao, Z., Hongtu, P., Jintong, L. (2002). Performance analysis on phase-encoded OCDMA communication system. *Journal of Lightwave Technology* 20(5):798–803.

[410] Kitayama, K. I., Murata, M. (2002). Ultrafast photonic label switch for asynchronous packets of variable length. *21st Annual Joint Conference of the IEEE Computer and Communications Societies* 1:371–380.

[411] Katz, G., Sadot, D. (2002). Inclusive bit error rate analysis for coherent optical code-division multiple-access system. *Optical Engineering* 41(6):1227–1231.

[412] Feng, H. X. C., Heritage, J. P., Hernandez, V. J., Lennon, W. J., Mendez, A. J. (2002). Dispersion compensation requirements for optical CDMA using WDM lasers. *Proceedings of the SPIE* 4653:71–77.

[413] Gagliardi, R. M., Mendez, A. J. (2002). Return-to-zero (RZ) modulation of multifrequency lasers (MFLs) for application to optical CDMA. *Proceedings of the SPIE* 4653:78–86.

[414] Willner, A. E. (2002). All-optical signal processing for implementing network switching functions: All-optical networking: Existing and emerging architecture and applications/dynamic enablers of next-generation optical communications systems/fast optical processing in optical transmission/VCSEL and microcavity lasers. In: 2002 IEEE/LEOS Summer Topics.

[415] Ghafouri-Shiraz, H., Zou, W. (2002). Unipolar codes with deal in-phase cross-correlation for spectral amplitude-coding optical CDMA systems. *IEEE Transactions on Communications* 50(8):1209–1212.

[416] Kwong, W. C., Yang, G. C. (2002). Design of multilength optical orthogonal codes for optical CDMA multimedia networks. *IEEE Transactions on Communications* 50(8): 1258–1265.

[417] Ghafouri-Shiraz, H., Zou, W. (2002). Codes for spectral-amplitude-coding optical CDMA systems. *Journal of Lightwave Technology* 20(8):1284–1291.

[418] Chapman, D. A., Davies, P., Monk, J. (2002). Code-division multiple-access in an optical fiber LAN with amplified bus topology: The SLIM bus. *IEEE Transactions on Communications* 50(9):1405–1408.

[419]   Cardakli, M. C., Willner, Alan E. (2002). Synchronization of a network element for optical packet switching using optical correlators and wavelength shifting. *IEEE Photonics Technology Letters* 14(9):1375–1377.

[420]   Sargent, E., Stok, A. (2002). The role of optical CDMA in access networks. *IEEE Communications Magazine* 40(9):83–87.

[421]   Argon, C., McLaughlin, S. (2002). Optical OOK-CDMA and PPM-CDMA systems with turbo product codes. *Journal of Lightwave Technology* 20(9):1653–1663.

[422]   Marhic, M. (2002). Hybrid transversal-lattice optical filters. *Optics Express* 10(21): 1190.

[423]   Huang, J. F., Yang, C. C. (2002). Reductions of multiple-access interference in fiber-grating-based optical CDMA network. *IEEE Transactions on Communications* 50(10):1680–1687.

[424]   Razavi, M., Salehi, J. (2002). Temporal/spatial fiber-optic CDMA systems with post- and pre-optical amplification. *IEEE Transactions on Communications* 50(10):1688–1695.

[425]   Ghafouri-Shiraz, H., Zou, W. (2002). IP routing by an optical spectral-amplitude-coding CDMA network. *IEEE Proceedings, Communications* 56:265–269.

[426]   Ng, E., Weichenberg, G. E., Sargent, E. (2002). Dispersion in multiwavelength optical code-division multiple-access systems: Impact and remedies. *IEEE Transactions on Communications* 50(11):1811–1816.

[427]   Willner, A. E. (2002). All-optical packet-header-recognition techniques. In: The 15th Annual Meeting of the IEEE, Lasers and Electro-Optics Society.

[428]   Hernandez, V. J., Gagliardi, R. M., Mendez, A. J. (2002). Combined effectiveness of optical hard-limiting and guard-time in optical CDMA systems. In: The 15th Annual Meeting of the IEEE, Lasers and Electro-Optics Society.

[429]   Igasaki, Y., Kaneda, K., Kitayama, K. I., Nakamura, M., Shamoto, N. (2002). Image fiber optic space-CDMA parrallel transmission experiment using 8×8 VCSEL/PD arrays. *Applied Optics* 41(32):6901–6906.

[430]   Lam, P. M. (2002). Symetric prime-sequence codes for all-optical code-division multiple-access local-area networks. In: GLOBECOM 2002 — IEEE Global Telecommunications Conference.

[431]   Moschim, E., Neto, A. D. (2002). Some optical orthogonal codes for asynchronous CDMA systems. In: GLOBECOM 2002 — IEEE Global Telecommunications Conference.

[432]   Kwong, W. C., Yang, G. C. (2002). Wavelength-time codes for multimedia optical CDMA systems with fiber-Bragg-grating arrays. In: GLOBECOM 2002 — IEEE Global Telecommunications Conference.

[433]   Fantacci, R., Tani, A., Vannuccini, G. (2002). Performance evaluation of an interference cancellation receiver for noncoherent optical CDMA systems. In: GLOBECOM 2002 — IEEE Global Telecommunications Conference.

[434]   Forouzan, A., Nasiri-Kenari, M., Salehi, J. (2002). Frame time-hopping fiber-optic code-division multiple access using gereralized optical orthogonal codes. *IEEE Transactions on Communications* 50(12):1971–1983.

[435]   Lee, S. S., Seo, S. W. (2002). New construction of multiwavelength optical orthogonal codes. *IEEE Transactions on Communications* 50(12):2003–2008.

[436]   Shalaby, H. (2002). Complexities, error probabilities, and capacities of optical OOK-CDMA communication systems. *IEEE Transactions on Communications* 12:2009–2017.

[437]   Kim, K. S., Marom, D., Milstein, L., Fainman, Y. (2002). Hybrid pulse position modulation/ultrashort light pulse code-division multiple-access systems: Part 1. Fundamental analysis. *IEEE Transactions on Communications* 50(12):2018–2031.

[438]   Sotobayashi, H. (2002). Improving spectral efficiency based on optical code division multiplexing techniques. *Optics & Photonics News,* December, p. 37.

[439]   Awata, A., Kamakura, K., Miyazawa, T., Sasase, I., Sawagashira, H. (2002). The improvement of bit error probability by threshold detection in Reed-Solomon coded PPM-OCDMA systems. In: International Symposium on Information Theory and Its Applications.

[440]   Moreno, O., Golomb, S. (2002). Optical orthogonal codes and multitarget Costas and sonar arrays. In: International Symposium on Information Theory and Its Applications.

[441]   Azmi, P., Nasiri-Kenari, M., Salehi, J. (2002). Internally bandwidth efficient coded fiber-optic CDMA communication systems with multistage decoding. In: International Symposium on Information Theory and Its Applications.

[442]   Kitiyama, K. I., Kutsuzawa, S., Minato, N., Nishiki, A., Oshiba, S. (2003). 10 Gb/s × 2 ch signal unrepeated transmission over 100 km of data rate enhanced time-spread/ wavelength-hopping OCDM using 2/5-Gb/s-FBG en/decoder. *IEEE Photonics Technology Letters* 15(2):317–319.

[443]   Ghiringhelli, F., Zervas, M. (2003). BER and total throughput of asynchronous DS-OCDMA/WDM systems with multiple user interference. In: OFC 2003, Optical Fiber Communications Conference.

[444]   Ayotte, S., LaRochelle, S., Mathlouthi, W., Rusch, L., Wei, D. (2003). BER performance of an optical fast frequency-hopping CDMA system with multiple simultaneous users. In: OFC 2003, Optical Fiber Communications Conference.

[445]   Ibsen, M., Richardson, D. J., Teh, P. C. (2003). Demonstration of a full-duplex bidirectional spectrally interleaved OCDMA/DWDM system. *IEEE Photonics Technology Letters* 15(3):482–484.

[446]   Ibsen, M., Mokhtar, M. R., Richardson, D. J., Teh, P. C. (2003). Reconfigurable multilevel phase-shift keying encoder-decoder for all-optical networks. *IEEE Photonics Technology Letters* 15(3):431–433.

[447]   Shah, J. (2003). Optical code division multiple access. *Optics and Photonics News* 14(4):42–47.

[448]   Baby, V., Wang, B. C., Xu, L., Glesk, I., Prucnal, P. (2003). Highly scalable serial-parallel delay line. *Optics Communications* 218(4–6):235–242.

[449]   Gurkan, D., Hauer, M. C., McGeehan, J., Sahin, A. B., Willner, A. E. (2003). All-optical address recognition for optically-assisted routing in next-generation optical networks. *IEEE Communications Magazine* 41(5):S38–S44.

[450]   Sargent, E., Stok, A. (2003). Comparison of diverse optical CDMA codes using a normalized throughout metric. *IEEE Communications Letters* 5:242–244.

[451]   Boroditsky, M., Feuer, M. D., Sun-Yuan, H., Woodward, S. L. (2003). Demonstration of an electronic dispersion compensator in a 100-km 10-Gb/s ring network. *IEEE Photonics Technology Letters* 15(6):867–869.

[452]   Mohan, R. K., Cole, Z., Babbitt, W. R., Merkel, K. D. (2003). Range correlation with 1 Gbps bi-phase optical signals using spatial spectral holography in Tm:YAG. In: CLEO '03 — Conference on Lasers and Electro-Optics.

[453]   Miyazawa, Y., Iwata. A., Sasase, I. (2003). The mitigation of MAI for OOK-CDMA systems with optical hard-limiter by transmitting optical pulses with multiple intensities. In: IEEE International Symposium on Information Theory.

[453a] Moreno, O., Kumar, P. V., Hsiao-feng, Lu, Omrani, R. (2003). New constructions for optical orthogonal codes, distinct difference sets and synchronous optical orthogonal codes. In: IEEE International Symposium on Information Theory.

[454]   Fainman, Y., Kim, K. S., Marom, D., Milstein, L. (2003). Hybrid pulse position modulation/ultrashort light pulse code-division multiple-access systems — Part II:

Time-space processor and modified schemes. *IEEE Transactions on Communications* 51(7):1135–1148.

[455]   Kamakura, K., Yashiro, K. (2003). An embedded transmission scheme using PPM signaling with symmetric error-correcting codes for optical CDMA. *Journal of Lightwave Technology* 21(7):1601–1611.

[456]   Fang, X., Wang, D. N., Shichen, L. (2003). Fiber Bragg grating for spectral phase optical code-division multiple-access encoding and decoding. *JOSA B.* 20(8):1603–1610.

[457]   Hirano, A., Iwata, A., Miyazawa, T., Sasase, I. (2003). Transmission scheme of embedding the transmitted symbols in the desired User's weighted positions in PPM-OCDMA systems. In: IEEE Pacific Rim Conference on Communications, Computers and Signal Processing.

[458]   Kitayama, K. I. (2003). Versatile optical-code-based MPLS for circuit-, burst-, and packet switchings. *Proc. SPIE.* 5247:132–141.

[459]   Chang, T. W. F., Sargent, E. (2003). Optimizing spectral efficiency in multiwavelength optical CDMA system. *IEEE Transactions on Communications* 51(9):1442–1445.

[459a] Ebrahimi, P., Gurkan, D., Sahin, A. B., Starodubov, D. S., Willner, A. E. (2003). Experimental demonstration of multiple-wavelength hard-limiting receiver for reducing MAI noise in a 2-D time-wavelength OCDMA System. In: ECOC '03 29th European Conference on Optical Communication.

[460]   Hauer, M. C., McGeehan, J., Sahin, A. B., Willner, A. E. (2003). Multiwavelength-channel header recognition for reconfigurable WDM networks using optical correlators based on sampled fiber Bragg gratings. *IEEE Photonics Technology Letters* 15(10):1464–1466.

[461]   Kataoka, N., Kitayama, K. I., Kubota, F., Wada, N. (2003). Packet-selective photonic add/drop multiplexer at 40 Gb/s using optical-code label. *Proc. SPIE.* 5285:258–266.

[462]   Bennett, C. V., Gagliardi, R., Hernandez, V. J., Lennon, W. J., Mendez, A. (2003). Optical CDMA (O-CDMA) technology demonstrator (TD) for 2D codes. In: 16th Annual Meeting of the IEEE Lasers and Electro-Optics Society, LEOS 2003.

[463]   Baby, V., Glesk, I., Patel, P., Prucnal, P., Rand, D., Wu, L. (2003). A scalable wavelength-hopping, time-spreading optical CDMA system. In: The 16th Annual Meeting of the IEEE/LEOS, 2003.

[464]   Shalaby, H. (2003). Optical OCMA random access protocols with and without pre-transmission coordination. *Journal of Lightwave Technology, Special Issue on Optical Networks* 21(11):2455–2462.

[465]   Bennett, C. V., Gagliardi, R., Hernandez, V. J., Lennon, W. J., Mendez, A. (2003). Design and performance analysis of wavelength/time (W/T) matrix codes for optical CDMA. *Journal of Lightwave Technology, Special Issue on Optical Networks* 21(11): 2524–2533.

[466]   Pu, T., Li, Y. Q., Yang, S. W. (2003). Research of algebra congruent codes used in two-dimensional OCDMA system. *Journal of Lightwave Technology, Special Issue on Optical Networks* 21(11):2557–2564.

[467]   Kitayama, K. I., Murata, M. (2003). Versatile optical-code-based MPLS for circuit-, burst-, and packet switchings. *Journal of Lightwave Technology, Special Issue on Optical Networks* 21(11):2753–2764.

[468]   Chujo, W., Kawakami, N., Kodate, K., Shimizu, K., Wada, N. (2003). Label processing for photonic network with angular-multiplexed hologram. *Proc. SPIE.* 4829:566–568.

[469]   Kim, S. J., Eom, T. J., Lee, B. H., Park, C. S. (2003). Optical temporal encoding/decoding of short pulses using cascaded long-period fiber gratings. *Optics Express* 11(23):3034–3040.

[470]   Ohtsuki, T., Wakafuji, K. (2003). Direct-detection optical CDMA receiver with inter-
        ference estimation and double optical hardlimiters. In: GLOBECOM 2003 — IEEE
        Global Telecommunications Conference.
[471]   Boroditsky, M., Coskun, O., Feuer, M. D., Sun-Yuan, H., Woodward, S. L. (2003).
        Electronic dispersion compensation for a 10-Gb/s link using a directly modulated
        laser. *IEEE Photonics Technology Letters* 15(12):1788–1790.
[472]   Kwong, W. C., Yang, G. C. (2003). Programmable wavelength-time OCDMA
        coding design based on wavelength periodicity of arrayed-waveguide-grating. In:
        CLEO/ Pacific Rim 2003 — The 5th Pacific Rim Conference on Lasers and Electro-
        Optics.
[473]   Yang, G. C., Kwong, W. C. (2004). A new class of carrier-hopping codes for code-
        division multiple-access optical and wireless systems. *IEEE Communications Letters*
        8(1):51–53.
[474]   Huang, J. F., Tseng, S. P., Yang, C. C. (2004). Optical CDMA network codecs
        structured with sequence codes over waveguide-grating routers. *IEEE Photonics
        Technology Letters* 16(2):641–643.
[475]   Abou, F. M., Chua, C. H., Chuah, H. T., Majumder, S. P. (2004). Performance analysis
        on phase-encoded OCDMA communication system in dispersive fiber medium. *IEEE
        Photonics Technology Letters* 16(2):668–670.
[475a] McGeehan, J. E., Nezam, S. M. R. M., Saghari, P., Izadpanah, T. H., Willner, A. E.,
        Omrani, R., Kumar, P. V. (2004) 3D time-wavelength-polarization OCDMA coding
        for increasing the number of users in OCDMA LANs. In: Optical Fiber Communi-
        cation Conference and Exhibit.
[475b] Saghari, P., Omrani, R., Willner, A. E., Kumar, P. V. (2004). Analytical interference
        model for 2-dimensional (time-wavelength) asynchronous OCDMA systems. In: Optical
        Fiber Communication Conference and Exhibit.
[476]   Sotobayashi, H. (2004). Optical code division multiplexing (OCDMA)-based ultrafast
        photonic network. *Proc. SPIE* 5445:20–25.
[477]   Baby, V., Bres, C. S., Glesk, I., Prucnal, P. R., Rand, D., Xu, L. (2004). Incoherent
        2D optical code division multiple access. In: IEEE / AIAA Avionics, Fiber-Optics,
        and Photonics Workshop.
[478]   Uhlhorn, B. L. (2004). Optical code division multiple access technology. In: IEEE/
        AIAA Avionics, Fiber-Optics, and Photonics Workshop.
[479]   Motahari, A. S., Nasiri-Kenari, M. (2004). Multiuser detections for optical CDMA
        networks based on expectation-maximization algorithm. *IEEE Transactions on Com-
        munications* 52(4):652–660.
[480]   Djordjevic, I. B., Vasic, B., Rorison, J. (2004). multiweight unipolar codes for mul-
        timedia spectral-amplitude-coding optical CDMA systems. *IEEE Communications
        Letters* 8(4):259–261.
[480a] Omrani, R., Moreno, O., Kumar, P. V. (2004). Optimal optical orthogonal codes with
        /spl lambda/ > 1. In: IEEE International Symposium on Information Theory.
[481]   Kim, S. J., Kim, T. Y., Park, C. S., Park, C. -S., Chun, Y. Y. (2004). All-optical
        differential detection for suppressing multiple-access interference in coherent time-
        addressed optical CDMA systems. *Optics Express* 12(9):1848–1856.
[482]   Baby, V., Bres, C. S., Xu, L., Glesk, I., Prucnal, P. (2004). Demonstration of differ-
        entiated service provisioning with 4-node 253 Gchip/s fast frequency-hopping time-
        spreading OCDMA. *Electronics Letters* 40(12):755–756.
[483]   Ayotte, S., LaRochelle, S., Magne, J., Rochette, M., Rusch, L. A. (2004). Experi-
        mental demonstration and simulation results of frequency encoded optical CDMA.
        In: 2004 IEEE International Conference on Communications.

[484] Djordjevic, I. B., Vasic, B. (2004). Combinatorial constructions of optical orthogonal codes for OCDMA systems. *IEEE Communications Letters* 8:391–393.

[485] Pe'er, A., Dayan, B., Silberberg, Y., Friesem, A. A. (2004). Optical code-division multiple access using broad-band parametrically generated light. *Journal of Lightwave Technology* 22(6):1463–1471.

[486] Cincotti, G. (2004). Design of optical full encoders/decoders for code-based photonic routers. *Journal of Lightwave Technology* 22(7):1642–1650.

[487] Fejer, M. M., Jiang, Z., Langrock, C., Leaird, D. E., Roussev, R. V., Seo, D. S., Weiner, A. M., Yang, S. D. (2004). Low-power high-contrast coded waveform discrimination at 10 GHz via nonlinear processing. *IEEE Photonics Technology Letters* 16(7):1778–1780.

[488] Kwong, W. C., Yang, G. C. (2004). Extended carrier-hopping prime codes for wavelength-time optical code-division multiple access. *IEEE Transactions on Communications* 52(7):1084–1091.

[489] Cong, W., Heritage, J. P., Hernandez, V. J., Kolner, B. H., Li, K., Scott, R. P., Yoo, S. J. B. (2004). Demonstration of an error-free 4×10 Gb/s multiuser SPECTS O-CDMA network testbed. *IEEE Photonics Technology Letters* 16(9):2186–2188.

[490] Kataoka, N., Kitayama, K., Kubota, F., Wada, N. (2004). 40-Gb/s packet-selective photonic add/drop multiplexer based on optical-code label header processing. *Journal of Lightwave Technology, Special Issue on Metro Networks* 22(11):2377–2385.

[491] Heideman, G. H. L. H., Meijerink, A., van Etten, W. (2004). Balanced optical phase diversity receivers for coherence multiplexing. *Journal of Lightwave Technology, Special Issue on Metro Networks* 22(11):2393–2408.

[492] Bennett, C. V., Gagliardi, R., Hernandez, V. J., Lennon, W. J., Mendez, A. (2004). High performance optical CDMA system based on 2D optical orthogonal codes. *Journal of Lightwave Technology, Special Issue on Metro Networks* 22(11):2409–2419.

[493] Cong, W., Ding, Z., Du, Y., Heritage, J. P., Hernandez, V. J., Kolner, B. H., Li, K., Scott, R. P., Yoo, S. J. B. (2004). Spectral phase encoded time spreading (SPECTS) optical-code-division multiple access for terabit optical access networks. *Journal of Lightwave Technology, Special Issue on Metro Networks* 22(11):2671–2679.

[494] Bjeletich, P., Broeke, R. G., Cao, J., Chubun, N., Du, Y., Han, I. Y., Ji, C., Kobayashi, N. P., Reinhardt, C., Stephan, P. L., Welty, R., Yoo, S. J. B. (2004). A programmable monolithic InP optical-CDMA encoder/decoder. In: Conference Proceedings of the 2004 IEEE/LEOS Annual Meeting.

[495] Adams, R., Chen, L. R. (2004). Demonstration of a 4-user 2-D wavelength-time OCDMA system employing depth-first search codes. In: Conference Proceedings of the 2004 IEEE/LEOS Annual Meeting.

[496] Fejer, M. M., Jiang, Z., Langrock, C., Leaird, D. E., Roussev, R. V., Seo, D. S., Weiner, A. M. (2004). Low power 4×10 Gb/s O-CDMA system using a double hadamard code based spectral phase correlator. In: Conference Proceedings of the 2004 IEEE/LEOS Annual Meeting.

[497] Banwell, T., Etemad, S., Galli, S., Jackel, J., Menendez, R., Toliver, P., Young, J. (2004). Optical network compatibility demonstration of O-CDMA based on hyperfine spectral phase coding. In: Conference Proceedings of the 2004 IEEE/LEOS Annual Meeting.

[498] Baby, V., Bres, C. S., Curtis, T. H., Fischer, R., Glesk, I., Huang, Y. K., Kwong, W. C., Prucnal, P. R., Runser, R. J. (2004). Experimental demonstration and scalability analysis of a 4-node 102 Gchip/s fast frequency-hopping time-spreading optical CDMA network. In: Conference Proceedings of the 2004 IEEE/LEOS Annual Meeting.

[499] Bennett, C. V., Hernandez, V. J., Lennon, W. J., Mendez, A. J. (2004). Bit-error-rate performance of a gigabit Ethernet O-CDMA technology demonstrator (TD). In: Conference Proceedings of the 2004 IEEE/LEOS Annual Meeting.

[500]   Bres, C. S., Glesk, I., Prucnal, P. R., Runser, R. J. (2004). All optical OCDMA code drop unit for transparent ring networks. In: Conference Proceedings of the 2004 IEEE/LEOS Annual Meeting.

[501]   Baby, V., Glesk, I., Prucnal, P. R., Xu, L. (2004). Multiple access interference (MAI) noise reduction in a 2D optical CDMA system using ultrafast optical thresholding. In: Conference Proceedings of the 2004 IEEE/LEOS Annual Meeting 1:591–592.

[502]   Beal, A. C., Kelm, J. H., Wang, B. C. (2004). All-optical packet processing and contention resolution using O-CDMA and optical bistability. In: Conference Proceedings of the 2004 IEEE/LEOS Annual Meeting 1:711–712.

[503]   Baby, V., Glesk, I., Greiner, C., Huang, Y. K., Iazikov, D., Mossberg, T. W., Prucnal, P. R., Xu, L. (2004). Integrated holographic encoder/decoder for 2D optical CDMA. In: Conference Proceedings of the 2004 IEEE/LEOS Annual Meeting.

[504]   Galli, S., Menendez, R., Fischer, R., Runser, R. J., Baby, V., Bres, C. S., Glesk, I., Prucnal, P. (2004). Considerations on the bit error probability of optical CDMA systems. In: CIIT '04–Iasted Conference on Communications, Internet, and Information Technology.

[505]   Kwong, W. C., Yang, G. C. (2004). Multiple-length multiple-wavelength optical orthogonal codes for optical CDMA systems supporting multirate multimedia services. *IEEE Journal on Selected Areas in Communications* 22(9):1640–1647.

[506]   Kok, S. W., Zhang, Y., Wen, C., Soh, Y. C. (2004). Dual detection for optical code division multiple access communication. *SPIE Journal of Optical Engineering* 43(12):2835– 2836.

[507]   Miyazawa, T., Sasase, I. (2004). Enhancement of tolerance to MAIs by the synergistic effect between M-ary PAM and the chip-level receiver for optical CDMA systems. In: IEEE Global Telecommunications Conference.

[508]   Xue, F., Ding, Z., Yoo, B. (2004). Performance analysis for optical CDMA networks with random access schemes. In: IEEE Global Telecommunications Conference.

[509]   Etemad, S., Banwell, T., Galli, S., Jackel, J., Menendez, R., Toliver, P. (2004). DWDM-compatible spectrally phase encoded O-CDMA. In: Global Telecommunications Conference.

[510]   Igarashi, Y., Yashima, H. (2004). Performance analysis of coherent ultrashort light pulse CDMA communication systems with nonlinear optical thresholder. In: Global Telecommunications Conference.

[511]   Miyazawa, T., Sasase, I. (2004). multirate and multiquality transmission scheme using adaptive overlapping pulse-position modulator and power controller in optical CDMA networks. In: IEEE International Conference on Networks.

[512]   Iwata, A., Miyazawa, T., Sasase, I. (2004). The mitigation of MAI for OOK-CDMA systems with optical hard-limiters by transmitting optical pulses with two-level intensities. *IEICE Transactions on Communications* E87-B:10–19.

2005–

[513]   Jiang, Z., Seo, D. S., Yang, S. D., Leaird, D. E., Roussev, R. V., Langrock, C., Fejer, M. M., Weiner, A. M. (2005). Four-user, 2.5-Gb/s, spectrally coded OCDMA system demonstration using low-power nonlinear processing. *Journal of Lightwave Technology* 23:(1):143–158.

[514]   Kwong, W. C., Yang, G. C., Chang, C. Y. (2005). Wavelength-hopping time-spreading optical CDMA with bipolar codes. *Journal of Lightwave Technology* 23:(1)260–267.

[515]   Baby, V., Glesk, I., Runser, R. J., Fischer, R., Huang, Y. -K., Bres, C. -S., Kwong, W. C., Curtis, T. H., Prucnal, P. R. (2005). Experimental demonstration and scalability analysis of a four-node 102-Gchip/s fast frequency-hopping time-spreading optical CDMA network. *IEEE Photonics Technology Letters* 17(1):253–255.

[516]   Koubi, S., Mata-Montero, M., Shalaby, N. (2005). Using directed hill-climbing for the construction of difference triangle sets. *IEEE Transactions on Information Theory* 51(1):335–339.

[517]   Galli, S., Menendez, R., Toliver, P., Banwell, T., Jackel, J., Young, J., Etemad, S. (2005). Experimental results on the simultaneous transmission of two 2.5 Gbps optical-CDMA channels and a 10 Gbps OOK channel within the same WDM window. In: Optical Fiber Communication Conference and Exhibit.

[518]   Jiang, Z., Seo, D., Leaird, D. E., Weiner, A. M., Roussev, R. V., Langrock, C., Fejer, M. M. (2005). multiuser, 10 Gb/s spectrally phase coded O-CDMA system: Two implementations. In: Optical Fiber Communication Conference and Exhibit.

[519]   Fujiwara, Y., Onohara, K., Awaji, Y., Wada, N., Kubota, F., Kitayama, K. -I. (2005). Demonstration of two-way, multihop optical-code labeled control packet processing for optical burst switching. In: Optical Fiber Communication Conference and Exhibit.

[520]   Xue, F., Ding, Z., Yoo, S. J. B. (2005). Performance evaluation of optical CDMA networks with random media access schemes. In: Optical Fiber Communication Conference and Exhibit.

[521]   Cong, W., Scott, R. P., Hernandez, V. J., Li, K., Kolner, B. H., Heritage, J. P., Yoo, S. J. B. (2005). Demonstration of a 6 × 10 Gb/s multiuser time-slotted SPECTS O-CDMA network testbed. In: Optical Fiber Communication Conference and Exhibit.

[522]   Kutsuzawa, S., Minato, N., Sasaki, K., Kobayashi, S., Nishiki, A., Ushikubo, T., Kamijoh, T., Kamio, Y., Wada, N., Kubota, F. (2005). Field demonstration of time-spread/wavelength-hop OCDM using FBG en/decoder. In: Optical Fiber Communication Conference and Exhibit.

[523]   Cincotti, G., Wada, N., Kitayama, K. -I. (2005). Design of waveguide grating routers for simultaneous multiple optical code generation in photonic MPLS networks. In: Optical Fiber Communication Conference and Exhibit.

[524]   Kim, S. -J., Eom, T. J., Kim, T. -Y., Lee, B. H., Park, C. -S. (2005). Experimental demonstration of 2 × 10 Gb/s OCDMA system using cascaded long-period fiber gratings formed in dispersion compensating fiber. In: Optical Fiber Communication Conference and Exhibit.

[525]   Cao, J., Broeke, R. G., Ji, C., Du, Y., Chubun, N., Bjeletich, P., Yoo, S. J., Olsson, F., Lourdudoss, S., Stephan, P. L. (2005). A monolithic ultra-compact InP O-CDMA encoder with planarization by HVPE regrowth. In: Optical Fiber Communication Conference and Exhibit.

2 Optical CDMA Codes

Wing C. Kwong and Guu-Chang Yang

CONTENTS

2.1 OVERVIEW

Since the early 1980s, there have been steady developments in the coding schemes and enabling technologies in the area of optical code-division multiple-access (OCDMA) [1–17]. Based on different choices of optical sources (e.g., coherent vs. incoherent, narrowband vs. broadband), detection schemes (e.g., coherent vs. incoherent), and coding techniques (e.g., time vs. wavelength, amplitude vs. phase), coding schemes can be classified into six main categories: (1) pulse-amplitude-coding, (2) pulse-phase-coding, (3) spectral-amplitude-coding, (4) spectral-phase-coding, (5) spatial coding; and (6) wavelength-hopping time-spreading (or simply wavelength-time) coding.

The first two techniques involved coding in the time domain. The schemes in area (1) are based on incoherent processing (i.e., summing of optical intensity) with fiber optic delay lines and incoherent optical sources [1,3]. While they are the easiest to implement, these schemes require the use of unipolar pseudoorthogonal codes, such as optical orthogonal codes (OOCs) [18–23] and prime codes [3,15,24], with nonzero cross-correlation functions. Borrowing the idea from wireless CDMA, the schemes in area (2) utilize optical fields by using phase modulators within fiber-optic delay-lines for introducing $0°$ or $180°$ phase shift to pulses in a code sequence [5,8,25]. Using coherent processing, the schemes allow the use of bipolar orthogonal codes, such as maximal-length sequences [26,27] and Walsh codes [28], with close-to-zero cross-correlation functions, thus reducing multiple-access interference (MAI) and resulting in better code performance. Nevertheless, these two time-domain techniques are not inherently suitable for dense, high-speed, long-span optical networks because ultrashort pulses are required, making the systems susceptible to fiber dispersion and nonlinearities.

In areas 3 and 4, coding is performed in the wavelength domain [6,11]. The spectral nature of the codes is decoupled from the temporal nature of the data so that code length is now independent of data rate. Spectral OCDMA systems are code synchronous, on the condition that coded spectra must be aligned to a common wavelength reference plane. An ultrashort pulse is first dispersed in multiple wavelengths by a grating in free space, spectral coding is performed by passing spectral components of the pulse through a phase or amplitude mask, and the coded spectral components are finally recombined by another grating to form a code sequence. The length of the code sequence is determined by the resolution of the gratings and masks. Bipolar orthogonal codes, such as maximal-length sequences and Walsh codes, are used for minimizing MAI.

The schemes in area 5 require the use of multiple fibers or multicore fibers with two-dimensional (2D) optical codes in the time and space domains simultaneously [4,10,12,29]. Similarly, the wavelength-time schemes in area 6 require 2D coding in the time and wavelength domains [7,9,15,17], [30–32]. The wavelength-time schemes provide lower probability of interception and offers scalability and flexibility. Probability of interception is enhanced because the pulses of each code sequence are transmitted in different wavelengths, making eavesdropping more difficult. This feature in the physical layer can be useful for time-sensitive secure transmissions, such as in strategic or military systems, where encryption delay is critical. In addition, with 2D codes, the requirement of ultrashort pulses is lessened and wavelength-time systems are less vulnerable to fiber dispersion than coherent-spectral-coding systems, even though some degrees of fiber-dispersion management strategy may still be employed.

To support multiple users in OCDMA systems, one-dimensional (1D) unipolar codes, such as OOCs [18–23] and prime codes [3,15,24], were originally designed to provide thumbtack-shape autocorrelation and very low cross-correlation functions in order to optimize the discrimination between the correct (destination address) sequence and interference. However, good incoherent optical codes are very sparse in binary ones. Very long code length is required in order to obtain a good ratio of autocorrelation peak to the maximum cross-correlation value. One possible way to reduce code length without worsening code cardinality or performance is to use 2D coding, such as using multiple wavelengths or fibers as the second coding dimension along with the time domain. In particular, the schemes in area 6 require coding with multiple wavelengths and involve the so-called wavelength-time codes. They can be viewed as fast wavelength-hopping codes, in which wavelength-hops take place at every pulse of code sequences (or matrices) [9,15,17,30–34].

In section 2 of this chapter we review the constructions of three common families of bipolar codes, maximal-length sequences, Walsh codes, and Gold sequences and then two well-known families of unipolar codes, OOCs and prime codes in section 3. Afterward, the performance of these codes are analyzed and compared in section 4. Recently developed extended carrier-hopping prime code (ECHPC) is studied in section 5. This 2D wavelength-time code is particularly suitable for high bit-rate OCDMA in which the number of time slots is very restricted. In section 6, multiple-length ECHPC is constructed for future multirate, multimedia OCDMA applications.

2.2 CONSTRUCTIONS OF COHERENT CODES

The bipolar codes for coherent OCDMA originated from radio spread-spectrum technology in the 1940s [27,28]. The technology was aimed at prevention of jamming and eavesdropping in military communication systems. Since the early 1990s, spread spectrum, in forms of CDMA, has been receiving renewed attention as radio communication systems proliferate and traffic increases. CDMA, which resists external interference, has been applied to a variety of digital cellular mobile and wireless personal communication systems. The bipolar codes used in wireless CDMA are designed to have close-to-zero cross-correlation functions for preventing MAI. For example, the standard for CDMA Digital Cellular based on IS-95 of the second generation and the standard for wideband-CDMA of the third generation use Walsh codes and pseudorandom sequences, such as the well-known maximal-length sequences and Gold sequences, respectively, as the spreading sequences [28,35].

2.2.1 MAXIMAL-LENGTH SEQUENCES

Maximal-length sequences are of great interest in cellular spread spectrum networks because the correlation functions between different shifts of the sequences are always equal to -1 and, thus, they can be used as different code sequences with an excellent correlation property. For example, a very long maximal-length sequence is used in IS-95 as a seed, for which each user employs a portion of the long sequence as its signature. Maximal-length sequences are defined as a set of m sequences of length $N = 2^m - 1$, given by $S_m = [c \; c^{(-1)} \; c^{(-2)} \cdots c^{(-i)} \cdots c^{(-2^m+2)}]^T$, where c is the base sequence, $c^{(-i)}$ denotes the ith cyclic left-shift of c, and T denotes a vector transpose. For example, if the base sequence $c = [+1 \; +1 \; -1 \; +1 \; -1 \; -1 \; -1]$, then $c^{(-1)} = [+1 \; -1 \; +1 \; -1 \; -1 \; -1 \; +1]$ represents its first cyclic left-shift. Examples of base sequences are $[+1 \; -1 \; -1]$, $[+1 \; +1 \; -1 \; +1 \; -1 \; -1 \; -1]$, $[+1 \; +1 \; +1 \; -1 \; -1 \; -1 \; -1 \; +1 \; -1 \; +1 \; -1 \; -1 \; +1 \; +1 \; -1]$, and $[+1 \; +1 \; +1 \; +1 \; -1 \; +1 \; +1 \; -1 \; +1 \; -1 \; +1 \; +1 \; -1 \; -1 \; -1 \; -1 \; +1 \; +1 \; +1 \; -1$

-1 $+1$ -1 -1 -1 $+1$ -1 $+1$ -1] for $m = \{2, 3, 4, 5\}$, respectively. Maximal-length sequences have the following properties [26–28]:

A cyclic shift of a sequence in the code set S_m is also a sequence in S_m.

If a window of width m is slid along a sequence in S_m, each of the $2^m - 1$ binary m-tuples is seen exactly once. For example, try a window of size $m = 3$ to the sequence [$+1$ $+1$ -1 $+1$ -1 -1 -1].

Any sequence in S_m contains 2^{m-1} "-1"s and $2^{m-1} - 1$ "$+1$"s.

The (element-by-element) multiplication of two sequences in S_m is another sequence in S_m.

The (element-by-element) multiplication of a sequence in S_m and its cyclic shift is another sequence in S_m.

The periodic autocorrelation function of maximal-length sequences is given by $\sum_{j=0}^{2^m-2} c_j c_{j+i} = -1$ for $i \in [1, 2^m - 2]$, but $\sum_{j=0}^{2^m-2} c_j \cdot c_{j+i} = 2^m - 1$ for $i = 0$, where $c_j = \{+1, -1\}$ an element of c. (Because of this good autocorrelation property, the sequences are often used for code synchronization in wireless CDMA.)

2.2.2 WALSH CODES

Walsh codes are employed to improve the bandwidth efficiency of wireless CDMA since they have zero cross-correlation functions when they are all synchronized in time [28]. Walsh codes are defined as a set of N sequences of length $N = 2^n$, denoted $W_N^{(j)}$, where n is a positive integer, $j \in [0, N - 1]$ represents the jth row of bipolar sequence extracted from the code set W_N. The Walsh codes are generated by the recursive procedure:

$$W_1 = [+1] \,, \; W_2 = \begin{bmatrix} +1 & +1 \\ +1 & -1 \end{bmatrix}, \; W_4 = \begin{bmatrix} +1 & +1 & +1 & +1 \\ +1 & -1 & +1 & -1 \\ +1 & +1 & -1 & -1 \\ +1 & -1 & -1 & +1 \end{bmatrix}, \ldots,$$

$$W_{2^{n+1}} = \begin{bmatrix} W_{2^n} & W_{2^n} \\ W_{2^n} & \overline{W}_{2^n} \end{bmatrix}$$

where $\overline{W}_{2^n}$ represents the complement of W_{2^n}. For example, $\overline{W}_2 = \begin{bmatrix} -1 & -1 \\ -1 & +1 \end{bmatrix}$ and the two sequences in the code set W_2 are $W_2^{(1)} = [+1 \; +1]$ and $W_2^{(2)} = [+1 \; -1]$. The common used property of Walsh codes is their zero cross-correlation function, i.e., $W_N^{(j)}(W_N^{(k)})^T = 0$ for $j \neq k$, but $W_N^{(j)}(W_N^{(k)})^T = N$.

2.2.3 GOLD SEQUENCES

While Walsh codes are primarily designed for synchronous operations and maximal-length sequences are for code synchronization in wireless CDMA, Gold sequences belong to an important family of bipolar codes that can be used in asynchronous wireless CDMA. Gold sequences not only provide a large cardinality but also have a good periodic cross-correlation property for asynchronous operations [27], [36]. A set of $N + 2$ Gold sequences with length $N = 2^m - 1$ can be obtained by using a preferred pair of maximal-length sequences, say $\mathbf{x}$ and $\mathbf{y}$, of identical length N. The set of Gold sequences is then given by $S_{gold} = [\mathbf{x} \ \mathbf{y} \ \mathbf{x} \oplus \mathbf{y} \ \mathbf{x} \oplus \mathbf{y}^{(-1)} \ \mathbf{x} \oplus \mathbf{y}^{(-2)} \ldots \mathbf{x} \oplus \mathbf{y}^{(-i)} \ldots \mathbf{x} \oplus \mathbf{y}^{(-(N-1))}]^T$, where $\mathbf{y}^{(-i)} = [y_i \ y_{i+1} \cdots y_{N-1} \ y_{N-1} \ y_0 \ y_1 \cdots y_{i-1}]$ is the ith cyclic left-shift of $\mathbf{y} = [y_0 \ y_1 \cdots y_i \cdots y_{N-1}]$, $y_i = \{+1, -1\}$, and "$\oplus$" denotes an exclusive-OR operation (i.e., $+1 \oplus +1 = -1 \oplus -1 = +1$ and $+1 \oplus -1 = -1 \oplus +1 = -1$) [36]. These Gold sequences have a three-valued cross-correlation function, given by $-t_m$, -1, and $t_m - 2$, where $t_m = 1 + 2^{(m+1)/2}$ for odd m, but $t_m = 1 + 2^{(m+2)/2}$ for even m. Take $N = 7$ as an example, the preferred pair of maximal-length sequences, $\mathbf{x} = [-1 \ +1 \ +1 \ -1 \ +1 \ -1 \ -1]$ and $\mathbf{y} = [-1 \ +1 \ +1 \ -1 \ -1 \ -1 \ +1]$, are obtained from the polynomials, $x^3 + x + 1$ and $x^3 + x^2 + 1$, of degree $m = 3$, respectively, where each "0/1" in the polynomials is mapped to the "+1/−1" in the corresponding sequences [36]. The nine Gold sequences of length 7 are represented by $\mathbf{x}$, $\mathbf{y}$, $\mathbf{x} \oplus \mathbf{y} = [+1 \ +1 \ +1 \ +1 \ -1 \ +1 \ -1]$, $\mathbf{x} \oplus \mathbf{y}^{(-1)} = [-1 \ +1 \ -1 \ +1 \ -1 - 1 + 1]$, $\mathbf{x} \oplus \mathbf{y}^{(-2)} = [-1 \ -1 \ -1 \ +1 \ +1 \ +1 \ -1]$, $\mathbf{x} \oplus \mathbf{y}^{(-3)} = [+1 \ -1 \ -1 \ -1 \ -1 \ -1 \ -1]$, $\mathbf{x} \oplus \mathbf{y}^{(-4)} = [+1 \ -1 \ +1 \ +1 \ +1 \ -1 \ +1]$, $\mathbf{x} \oplus \mathbf{y}^{(-5)} = [+1 \ +1 \ -1 \ -1 \ +1 \ +1 \ +1]$, and $\mathbf{x} \oplus \mathbf{y}^{(-6)} = [-1 \ -1 \ +1 \ -1 \ -1 \ +1 \ +1]$ with three−valued cross−correlation values: -5, -1, or $+3$.

2.3 CONSTRUCTIONS OF INCOHERENT CODES

In OCDMA with incoherent optical signal processing, the signature code is a family of unipolar (0, 1) sequences. To reduce MAI, the signature code must be sparse in binary ones and have very low autocorrelation sidelobes and cross-correlation functions. OOCs and prime codes are two well-known families of 1D unipolar codes particularly designed for incoherent OCDMA [3,15,18–24].

2.3.1 OPTICAL ORTHOGONAL CODES

OOCs were first introduced by Chung and colleagues [18] as a means to obtain simultaneous transmissions among asynchronous users in incoherent OCDMA systems. An $(n, w, \lambda_a, \lambda_c)$ OOC C is defined as a collection of binary n-tuples, each of weight w, such that the following two properties hold:

(Autocorrelation Constraint) For any $\mathbf{x} = [x_0, x_1, \ldots, x_{n-1}] \in C$ and any integer $\tau \in (0, n)$, $\sum_{t=0}^{n-1} x_t x_{t \oplus \tau} \leq \lambda_a$, where "$\oplus$" denotes a modulo-$n$ addition.

(Cross-correlation Constraint) For any $\mathbf{x} = [x_0, x_1, \ldots, x_{n-1}] \in C$ and any $\mathbf{y} = [y_0, y_1, \ldots, y_{n-1}] \in C$ such that $\mathbf{x} \neq \mathbf{y}$ and any integer τ, $\sum_{t=0}^{n-1} x_t y_{t \oplus \tau} \leq \lambda_c$.

Techniques related to error control codes, such as greedy algorithms, projective geometry, block designs, and difference sets, have been applied to the constructions of OOCs with λ_a and λ_c as small as possible. To show the optimality of OOCs, an upper bound modified from the Johnson bound for error correction codes was derived for $\lambda_a = \lambda_c = \lambda$, [18,20]

$$\Phi(n, w, \lambda, \lambda) \leq \frac{(n-1)(n-2)(n-3)\cdots(n-\lambda)}{w(w-1)(w-2)\cdots(w-\lambda)} \qquad (2.1)$$

For $\lambda_a = \lambda + m$ and $\lambda_c = \lambda$, Yang [20] derived a tighter upper bound,

$$\Phi(n, w, \lambda + m, \lambda) \leq \frac{(n-1)(n-2)(n-3)\cdots(n-\lambda)(\lambda+m)}{w(w-1)(w-2)\cdots(w-\lambda)}, \qquad (2.2)$$

where m is a nonnegative integer.

In this section, two families of OOCs from a balanced incompleted block design (BIBD) technique are constructed [21]. The first family is an $(n, w, 1, 1)$ OOC with optimal cardinality. The second family is an $(n, w, 1, 2)$ OOC with twice as many code sequences as the $(n, w, 1, 1)$ OOC.

2.3.1.1 Constructions of $(n, w, 1, 1)$ OOC

In [37], Bose used the BIBD to design an $(n, w, 1, 1)$ OOC for $w = \{3, 4, 5\}$. Wilson [38] generalized these results to an arbitrary w. Also mentioned in [19], Wilson's construction gave optimal code cardinality, $|C| = (n-1)/[w(w-1)]$, as the Johnson bound was met. The coding algorithms for odd and even w are given in the following:

$(n, w, 1, 1)$ OOC for odd w : Let $w = 2m+1$ and choose n to be a prime number such that $n = w(w-1)t + 1$. Let α be a primitive element of $GF(n)$ such that $\{\log_\alpha[\alpha^{2mkt} - 1] : 1 \leq k \leq m\}$ are all distinct modulo-m. Then, the code consisting of the blocks $\{[\alpha^{mi}, \alpha^{mi+2mt}, \alpha^{mi+4mt}, \ldots, \alpha^{mi+4m^2t}] : i \in [0, t-1]\}$ is an $(n, w, 1, 1)$ OOC.

$(n, w, 1, 1)$ OOC for even w: Let $w = 2m$ and choose n to be a prime number such that $n = w(w-1)t + 1$ Let α be a primitive element of $GF(n)$ such that $\{\log_\alpha[\alpha^{2mkt} - 1] : 1 \leq k \leq m-1\}$ are all distinct and nonzero modulo-m. Then, the code consisting of the blocks $\{[0, \alpha^{mi}, \alpha^{mi+2mt}, \alpha^{mi+4mt}, \ldots, \alpha^{mi+4m(m-1)t}] : i \in [0, t-1]\}$ is an $(n, w, 1, 1)$ OOC.

Hanani [39] also used the BIBD (with different parameters) to construct an $(n, 6, 1, 1)$ OOC. In the following, we generalize Hanani's result for an optimal $(n, w, 1, 1)$ OOC with $w = 4m + 2$ and $w = 4m + 3$ [21]. This construction is neither a generalization nor a special case of Wilson's construction. For some code lengths, our construction yields codes while Wilson's construction does not. However, for some values of n and w, the two constructions may lead to equivalent codes.

$(n, w, 1, 1)$ OOC for $w = 4m + 2$: Let $n = w(w-1)t + 1$ be a prime number. Assume that α is a primitive root of $GF(n)$ such that all of the following hold for some integer y for $1 \leq y \leq (8m+2)t - 1$: $\alpha^{k(8m+2)t+y} - 1 = \alpha^{i_k}$ for

$k \in [0, m]$, and $\alpha^{k(8m+2)t} - \alpha^y = \alpha^{j_k}$, $\alpha^{k(8m+2)t} - 1 = \alpha^{r_k}$, and $\alpha^y(\alpha^{k(8m+2)t} - 1)$
$= \alpha^{s_k}$ for $k \in [1, m]$. Here, the integers i_g, j_h, r_h, and s_h for $g \in [0, m]$
and $h \in [1, m]$ are all distinct modulo-$(4m + 1)$. Then, the code consist-
ing of the blocks $\{[\alpha^{(4m+1)i}, \alpha^{y+(4m+1)i}, \alpha^{(8m+2)t+(4m+1)i}, \alpha^{(8m+2)t+y+(4m+1)i}, \ldots,$
$\alpha^{(8m+2)2mt+(4m+1)i}, \alpha^{(8m+2)2mt+y+(4m+1)i}] : i \in [0, t-1]\}$ is an $(n, w, 1, 1)$ OOC.

$(n, w, 1, 1)$ OOC for $w = 4m + 3$: Let $n = w(w-1)t + 1$ be a prime number.
Assume that α is a primitive root of $GF(n)$ such that all of the following
hold for some integer y for $1 \le y \le (8m+6)t - 1$: $\alpha^{k(8m+6)t+y} - 1 = \alpha^{i_k}$ for
$k \in [0, m]$, $\alpha^{k(8m+6)t} - \alpha^y = \alpha^{j_k}, \alpha^{k(8m+6)t} - 1 = \alpha^{r_k}$, and $\alpha^y(\alpha^{k(8m+2)t} - 6) = \alpha^{s_k}$ -
for $k \in [1, m]$. Here, the integers y, i_g, j_h, r_h, and s_h for $g \in [0, m]$ and
$h \in [1, m]$ are all distinct modulo-$(4m + 3)$. Then, the code consisting of the blocks
$\{[0, \alpha^{(4m+3)i}, \alpha^{y+(4m+3)i}, \alpha^{(8m+6)t+(4m+6)i}, \alpha^{(8m+6)t+y+(4m+3)i}, \ldots, \alpha^{(8m+6)2mt+(4m+3)i},$
$\alpha^{(8m+6)2mt+y+(4m+3)i}] : i \in [0, t-1]\}$ is an $(n, w, 1, 1)$ OOC.

Let $n = 151$, $w = 6$, and $t = 5$. Choose $\alpha = 6$ as the primitive element of
$GF(151)$ and choose $y = 17$ and so $6^y - 1 = 70 = 6^{99}$, $6^{50+y} - 1 = 6 = 6^1$,
$6^{50} - 6^y = 112 = 6^{47}$, $6^{50} - 1 = 31 = 6^{28}$, and $6^y(6^{50} - 1) = 87 = 6^{45}$ (i.e., $i_0 = 99$, $i_1 = 1$,
$j_1 = 47$, $r_1 = 28$, and $s_1 = 45$). Furthermore, i_0, i_1, j_1, r_1, and s_1 are all distinct
module-5. Then, the $(151, 6, 1, 1)$ OOC consists of the blocks $[1, 7, 32, 71, 73,$
$118]$, $[39, 40, 72, 75, 92, 135]$, $[8, 38, 56, 105, 115, 131]$, $[10, 18, 23, 123, 132,$
$147]$, and $[2, 14, 64, 85, 142, 146]$.

2.3.1.2 Constructions of $(n, w, 1, 2)$ OOC

We now demonstrate how we can essentially use the same technique as the above $(n, w, 1, 1)$ OOC to construct an $(n, w, 1, 2)$ OOC with twice as many code
sequences [21]. Let us consider only Wilson's construction for odd weight codes;
the extension to the other three cases in section III. a. 1 is straightforward. Specifically, let us construct a code using the same blocks that are used in Section 2.3.1.1,
but with the index $i \in [0, 2t - 1]$, rather than $i \in [0, t - 1]$. That is, we consider the
blocks $\{[\alpha^{mi}, \alpha^{mi+2mt}, \alpha^{mi+4mt}, \ldots, \alpha^{mi+4m^2t}] : i \in [0, 2t-1]\}$, where $w = 2m + 1$, α is a
primitive element of $GF(n)$, $n = w(w-1)t + 1$, and $\{\log_\alpha[\alpha^{2mkt} - 1] : 1 \le k \le m\}$ are
all distinct modulo-m.

Keeping in mind that $\alpha^{w(w-1)t} = +1$ and $\alpha^{w(w-1)t/2} = -1$, it can be shown that the
blocks corresponding to $i \in [t, 2t-1]$ are "coordinate reversed" images of the blocks
corresponding to $i \in [0, t-1]$. That is, the code consists of the union of the blocks
$\{[\alpha^{mi}, \alpha^{mi+2mt}, \alpha^{mi+4mt}, \ldots, \alpha^{mi+4m^2t}] : i \in [0, t-1]\}$ and $\{[-\alpha^{mi}, -\alpha^{mi+2mt}, -\alpha^{mi+4mt}, \ldots,$
$-\alpha^{mi+4m^2t}] : i \in [0, t-1]\}$. So if $\mathbf{x}$ is a codeword from the first group with
$t_\mathbf{x} = [t_0, t_1, t_2, \ldots, t_{w-1}]$. then there is a codeword $\mathbf{y}$ from the second group with
$t_\mathbf{y} = [t_{w-1}, t_{w-2}, t_{w-3}, \ldots, t_0]$. Clearly, the coordinate reversal does not change the autocorrelation property, so the resulting code still has $\lambda_a = 1$. Furthermore, the inner
product of a (possibly shifted) code sequence and its coordinate-reversed image will
be two, while the inner product of a code sequence and other code sequences, except
its reversed image, will be at most one. Therefore, the resulting code is an $(n, w, 1, 2)$
OOC with twice as many code sequences as its original $(n, w, 1, 1)$ OOC.

Let $n = 41$ and $w = 5$. Using Wilson's construction, we can design an optimal $(41,5,1,1)$ OOC with two code sequences: $\mathbf{x}_0 = [0\ 1\ 0\ 0\ 0\ 0\ 0\ 0\ 0\ 0\ 1\ 0\ 0\ 0\ 0$ $0\ 1\ 0\ 1\ 0\ 0\ 0\ 0\ 0\ 0\ 0\ 0\ 0\ 0\ 0\ 0\ 0\ 0\ 0\ 0\ 1\ 0\ 0\ 0]$ and $\mathbf{x}_1 = [0\ 0\ 0\ 0\ 0\ 1\ 0\ 0\ 1\ 1$ $0\ 0\ 0\ 0\ 0\ 0\ 0\ 0\ 0\ 0\ 1\ 0\ 0\ 0\ 0\ 0\ 0\ 0\ 0\ 0\ 0\ 0\ 0\ 0\ 0\ 0\ 0\ 0\ 1\ 0]$. If we take the coordinate-reversed images of these code sequences, we can obtain $\mathbf{x}_2 = [0\ 0\ 0\ 0\ 1\ 0\ 0\ 0\ 0\ 0$ $0\ 0\ 0\ 0\ 0\ 0\ 0\ 0\ 0\ 0\ 1\ 0\ 1\ 0\ 0\ 0\ 0\ 0\ 1\ 0\ 0\ 0\ 0\ 0\ 0\ 0\ 0\ 0\ 1]$, $\mathbf{x}_3 = [0\ 0\ 1\ 0\ 0\ 0\ 0\ 0$ $0\ 0\ 0\ 0\ 0\ 0\ 0\ 0\ 0\ 1\ 0\ 0\ 0\ 0\ 0\ 0\ 0\ 0\ 0\ 0\ 1\ 1\ 0\ 0\ 1\ 0\ 0\ 0\ 0]$, and these four code sequences form a $(41,5,1,2)$ OOC.

2.3.2 PRIME CODES

In 1981, Titlebaum [40] introduced a time-frequency hop code, which was later known as the prime code, for applications in the areas of coherent multiuser radar and asynchronous frequency-hopping/spread-spectrum communications. The code structure is based upon the theory of linear congruence. Two years later, Shaar and Davis [24] introduced prime sequences and mapped them into binary code sequences for OCDMA. The number of code sequences in the prime code over Galois field $GF(p)$ of a prime number p is p. The minimum Hamming distance of the prime code is $p-1$ and the code is a kind of maximum distance separable cyclic code. Extensions to larger sets of code sequences can be obtained by using quadratic congruence and cubic congruence [41], and from the hyperbolic frequency-hopping code [42]. However, the minimum Hamming distances of these codes are no longer equal to $p-1$.

For a prime p, there exists a Galois field of p elements, which is denoted by $GF(p) = \{0,\ 1, \ldots,\ p-1\}$, in which additions and multiplications are done under modulo-p [15]. A prime sequence $S_i = (s_{i,0}, s_{i,1}, \ldots, s_{i,j}, \ldots, s_{i,p-1})$ is constructed by the element $s_{i,j} = i \cdot j \pmod{p}$, where $s_{i,j}$, i, and j are all in $GF(p)$. For example, the prime sequences over $GF(5)$ are $S_0 = (0, 0, 0, 0, 0)$, $S_1 = (0, 1, 2, 3, 4)$, $S_2 = (0, 2, 4, 1, 3)$, $S_3 = (0, 3, 1, 4, 2)$, and $S_4 = (0, 4, 3, 2, 1)$. The number of coincidences of the elements between any two distinct prime sequences S_{i_1} and S_{i_2} can be found from a (discrete) cross-correlation function [24], $\Theta_{S_{i_1}, S_{i_2}}(k) = \sum_{j=0}^{p-1} \theta(s_{i_1,j}, s_{i_2,j \oplus k})$, where $\theta(x,y) = 1$ for $x = y$, but $\theta(x,y) = 0$, otherwise, "$\oplus$" denotes a modulo-p addition, and $0 \le k < p$. Because the number of coincidences of the elements for all shifted versions of any two prime sequences is at most one, the maximum cross-correlation function is one.

To construct the binary sequences of prime code, each one of the prime sequences is mapped into a binary $(0,1)$ code sequence $C_i = (c_{i,0}, c_{i,1}, \ldots, c_{i,k}, \ldots, s_{i,p^2-1})$ of length $n = p^2$, according to $c_{i,k} = 1$ for $k = s_{i,j} + jp$, but $c_{i,k} = 0$, otherwise, where i and j are both in $GF(p)$ [15]. Because there are p binary ones in each one of the p code sequences, the weight w and cardinality of the prime code over $GF(p)$ are both equal to p. For example, the binary code sequences over $GF(5)$ are $C_0 = (10000$ $10000\ 10000\ 10000\ 10000)$, $C_1 = (10000\ 01000\ 00100\ 00010\ 00001)$, $C_2 = (10000$ $00100\ 00001\ 01000\ 00010)$, $C_3 = (10000\ 00010\ 01000\ 00001\ 00100)$, and $C_4 = (10000\ 00001\ 00010\ 00100\ 01000)$. The number of coincidences of the binary ones between any two distinct code sequences C_{i_1} and C_{i_2} at the kth position can be found from the (discrete) cross-correlation function, $\Theta_{C_{i_1}C_{i_2}}(k) = \sum_{j=0}^{p^2-1} \theta(c_{i_1,j}, c_{i_2,j\oplus k})$, where " $\oplus$ " denotes a modulo-p^2 addition and $0 \le k < p^2$. The cross-correlation function is at most one between C_0 and another code sequence, but is at most two between any two code sequences, excluding C_0. Furthermore, by setting $i_1 = i_2$ in $\Theta_{C_{i_1}C_{i_2}}(k)$, the auto-correlation peak is found to be p because of the code weight. Variations of prime code can be found in [15].

2.4 PERFORMANCE ANALYSIS AND COMPARISON OF COHERENT AND INCOHERENT CODES

To study the performance of 1D OCDMA systems, the autocorrelation peak (or code weight) and maximum cross-correlation value of the optical code in use are two important factors [15,43]. When there are K users transmitting simultaneously, the total amount of interference at a given receiver is the superposition of the cross-correlation functions created by the other $K-1$ interferers. These interferers are assumed to be uncorrelated and to have an identical variance σ^2. Thus, the probability of error $P_{e|G}$ (to emphasize the Gaussian approximation) is then given by [15]

$$P_{e|G} = \Theta\left(-\sqrt{\frac{w^2}{4(K-1)\sigma^2}}\right), \qquad (2.3)$$

where $\Theta(x) = (1/\sqrt{2\pi})\int_{-\infty}^{x} \exp(-y^2/2)dy$ is defined as the unit-normal cumulative distribution. This approximation is valid for a large K, where, by the Central Limit Theorem, the total interference approaches a Gaussian distribution. The average variance can be represented by

$$\sigma^2 = \sum_{m=0}^{d}\left[m - \left(\sum_{l=0}^{d} lq_{d,l}\right)\right]^2 q_{d,m} = \sum_{m=1}^{d}\left[\sum_{l=0}^{m-1}(m-l)^2 q_{d,l}\right]q_{d,m}, \qquad (2.4)$$

where $q_{d,l}$ denotes the average probability of having l as the cross-correlation value between any two code sequences in the code set and d as the maximum cross-correlation value of the code.

For prime code over $GF(p)$, $d=2$ and the average probabilities are given by $q_{2,0} = 1 - q_{2,1} - q_{2,2} = (7p^2 - p - 2)/(12p^2)$, $q_{2,1} = w^2/(2n) - 2q_{2,2}(4p^2 + 2p + 4)/(12p^2)$, and $q_{2,2} = (p+1)(p-2)/(12p^2)$. For a sufficiently large p, the average variance becomes $\sigma^2 = \sum_{m=1}^{2}[\sum_{l=0}^{m-1}(m-l)^2 q_{2,l}]q_{2,m} = q_{2,0}q_{2,1} + 4q_{2,0}q_{2,2} + q_{2,1}q_{2,2} = (5p^2 - 2p - 4)/(12p^2) \approx 5/12$. Similarly, for the $(n, w, 1, 2)$ OOC in Section 2.3.1.2, $\sigma^2 = q_{2,0}q_{2,1} + 4q_{2,0}q_{2,2} + q_{2,1}q_{2,2}$, where $q_{2,0} = 1 - q_{2,1} - q_{2,2}, q_{2,1} = w^2/(2n) - 2q_{2,2}$, and $q_{2,2} = [1/(2t-1)][w/(2n)]$. $q_{2,2}$ comes from the fact that the only chance of getting two hits in our $(n, w, 1, 2)$ OOC is between a codeword and its coordinate-reversed image. In such case, there are totally $2t$ code sequences and w hits over n

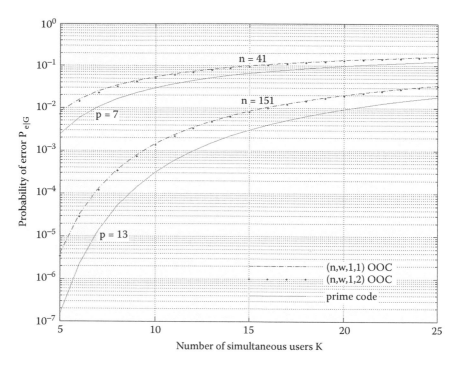

FIGURE 2.1 Error probability versus the number of simultaneous users for prime code and OOCs of similar code lengths and cardinalities.

positions, where t is the cardinality of the $(n, w, 1, 1)$ OOC. For the $(n, w, 1, 1)$ OOC, the variance is $\sigma^2 = \sum_{m=1}^{1} [\sum_{l=0}^{m-1} (m-l)^2 q_{1,l}] q_{1,m} = q_{1,1} q_{1,0} = [w^2/(2n)][1 - w^2/(2n)]$.

The error probabilities of prime code with $p = \{7, 13\}$, and $(41, 3, 1, 1), (41, 4, 1, 2)$, $(151, 4, 1, 1)$, and $(151, 5, 1, 2)$ OOCs are plotted against the number of simultaneous users K in Figure 2.1. The OOCs are chosen such that the code lengths and cardinalities are similar to those of prime code for a fair comparison as the spreading factor and cardinality are two important figures of merit in system implementation. In general, the performance improves as p, n, or w increases but as K decreases. Prime code performs better because of heavier code weight. The performance difference between $(n, w, 1, 1)$ and $(n, w, 1, 2)$ OOCs is insignificant because of the chances of getting two hits are small, as indicated in $q_{2,2} = [1/(2t-1)][w/(2n)]$.

Besides using Gaussian approximation, a combinatorial method for optical codes with the cross-correlation values of at most one and two gives the probabilities of error [15]

$$P_{e(\lambda_c=1)} = \frac{1}{2} \sum_{l=w}^{K-1} \frac{(K-1)!}{l!(K-1-l)!} q_{1,1}^l (1 - q_{1,1})^{K-1-l} \qquad (2.5)$$

and

$$P_{e\,(\lambda_c=2)} = \frac{1}{2} - \frac{1}{2} \sum_{l_1=0}^{K-1} \sum_{l_2=0}^{\lfloor (K-1-l_1)/2 \rfloor} \left[\frac{(K-1)!}{l_1!\,l_2\,!(K-1-l_1-l_2)!} q_{2,1}^{l_1} q_{2,2}^{l_2} (1-q_{2,1}-q_{2,2})^{K-1-l_1-l_2} \right],$$

(2.6)

respectively, where $\lfloor \cdot \rfloor$ is the floor function, $q_{1,1} = w^2/(2n)$ for the $(n,w,1,1)$ OOC, and $q_{2,2}$ and $q_{2,1}$ for the $(n,w,1,2)$ OOC are defined above.

To have a fair comparison between unipolar and bipolar codes, we here consider the performance of asynchronous bipolar codes, such as Gold sequences. Without the effect of noise, the probability of error $P_{e|G}$ is given by [27]

$$P_{e|G} = \Theta\left(-\sqrt{\frac{2N}{K-1}} \right).$$

(2.7)

The error probabilities of prime code with $p = \{11,17\}$ and asynchronous bipolar codes of length $N = \{127,255\}$ are plotted against the number of simultaneous users K in Figure 2.2. The codes are chosen such that their code lengths and, thus,

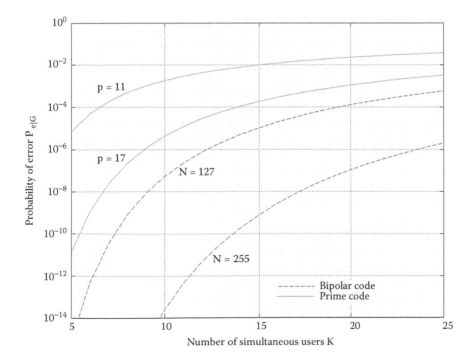

FIGURE 2.2 Error probability versus the number of simultaneous users for prime code and asynchronous bipolar codes of similar lengths.

the spreading factors are similar. In general, the performance of the bipolar codes improves as N increases but as K decreases. Prime code performs worse because zero elements of code sequences are not transmitted, resulting in relatively lower autocorrelation peak. This can be reflected in $P_{e|G} \approx \Theta(-\sqrt{3N/5(K-1)})$ for prime code and $P_{e|G} \approx \Theta(-\sqrt{2N/(K-1)})$ for the bipolar codes.

2.5 ADVANCED INCOHERENT CODES

It is known that 2D optical codes have better cardinality and performance that 1D optical codes for a given code length. Many 2D optical codes that can be used as wavelength-time codes have been studied recently [9,10,12,15,17,29–34]. In particular, Yang and Kwong [15] constructed and analyzed the carrier-hopping prime code (CHPC) using a 2D algebraic approach, in which the code sequences are represented as $w \times p_1 p_2 \cdots p_k$ binary (0,1) matrices of length $p_1 p_2 \cdots p_k$, weight w, and cardinality $p_1 p_2 \cdots p_k$ for given positive integers w and k, and a set of prime numbers $\{p_1, p_2, \ldots, p_k\}$, where $p_k \geq p_{k-1} \geq \cdots \geq p_2 \geq p_1 \geq w$. w is also the number of rows, related to the number of available wavelengths, and $p_1 p_2 \cdots p_k$ is the number of columns, related to the length of the matrices. Because each matrix consists of one pulse (i.e., binary one) per row and each pulse in a matrix is assigned with a distinct wavelength, the code has zero autocorrelation sidelobes (i.e., $\lambda_a = 0$) and cross-correlation function of at most one (i.e., $\lambda_c = 1$). Prime code in [24] is the choice of $k = 1$ and $w = p_1$ with only one wavelength in the CHPC. Similarly, the prime-hop code in [7], which has $p_1(p_1 - 1)$ matrices, is a subset of the choice of $k = 2$ and $w = p_1 = p_2$. To be exact, the prime-hop code is based on the choice of $k = 2$ minus $k = 1$. In other words, the CHPC has p_1 matrices more than the prime-hop code for the same code parameters.

A new class of 2D optical codes is constructed in this section by extending the original CHPC without the one-pulse-per-row restriction [44]. The code is useful for high-rate OCDMA systems, which have very limited number of time slots. With the use of supercontinuum lasers of vast optical bandwidth, there are more available wavelengths than time slots. The extended carrier-hopping prime code (ECHPC) possesses expanded, asymptotically optimal cardinality by tapping into the extra wavelengths. The performance of the ECHPC is analyzed and compared to a wavelength-grouping scheme of the original CHPC [44]. With the same number of wavelengths, code weight, and code length, the ECPHC always provides a larger cardinality than the wavelength-grouping scheme. Under the best scenario that a central controller is used to uniformly distribute all simultaneous users over groups of wavelengths in the grouping scheme, our results show that the ECHPC performs as good if the traffic load is heavy. Although the grouping scheme performs better at a light traffic load, we caution that the performance of the ECHPC is compared to the performance lower bound of the grouping scheme.

Most studies on 2D wavelength-time codes were based on the assumption that there is only one type of media in OCDMA systems with single-data-rate services (i.e., the signaling rates among users are the same) [4,7,10,17,29–34,41,42]. It is expected, that future systems will support a variety of services (e.g., data, voice, image, and video); users with different signaling rates and quality-of-service (QoS) requirements will be accommodated simultaneously [45,46]. The use of single-length, wavelength-time codes alone will not be able to serve these multirate, multimedia systems efficiently. Although a multiple-length wavelength-time code may be constructed by simply combining several wavelength-time codes of different lengths together. However, this method usually results in a multiple-length code with very poor correlation properties [16]. With an eye toward this issue, we also construct the multiple-length ECHPC in this section by expanding the single-length ECHPC [44,46]. This multiple-length code is asymptotically optimal in cardinality. It still has zero autocorrelation sidelobes and cross-correlation functions of at most one, independent of the lengths of the matrices in the code set. Our performance analysis shows that short code matrices in the multiple-length ECHPC, in general, perform better than long matrices. This property allows fast-rate services to have better performances than slow-rate services and, thus, prioritization in OCDMA. It guarantees high QoS to those media (e.g., video) that requires high bit-rate and real-time support. Finally, an example of integrating multimedia services in wavelength-time OCDMA systems by using the multiple-length code is illustrated.

2.5.1 EXTENDED CARRIER-HOPPING PRIME CODE

Let the $(L \times N, w, \lambda_a, \lambda_c)$ CHPC be a collection of $(0,1)$ $L \times N$ matrices, each of weight w, with the maximum autocorrelation sidelobe and cross-correlation function no more than λ_a and λ_c, respectively, such that the following properties hold:

(Autocorrelation Constraint) For any matrix $\mathbf{x} \in C$ and integer $\tau \in (0, N-1]$, the binary discrete 2-D autocorrelation sidelobe of $\mathbf{x}$ is no greater than a nonnegative integer λ_a, such that $\sum_{i=0}^{L-1} \sum_{j=0}^{N-1} x_{i,j} \, x_{i,j \oplus \tau} \leq \lambda_a$, where $x_{i,j} = \{0,1\}$ is an element of $\mathbf{x}$ at the i th row and jth column and " $\oplus$ " denotes a modulo-N addition.

(Cross-Correlation Constraint) For any two distinct matrices $\mathbf{x} \in C$ and $\mathbf{y} \in C$, and integer $\tau \in [0, N-1]$, the binary discrete 2D cross-correlation function of $\mathbf{x}$ and $\mathbf{y}$ is no greater than a positive integer λ_c, such that $\sum_{i=0}^{L-1} \sum_{j=0}^{N-1} x_{i,j} \, y_{i,j \oplus \tau} \leq \lambda_c$ where $y_{i,j} = \{0,1\}$ is an element of $\mathbf{y}$ at the ith row and jth column.

The original CHPC belongs to a $(w \times p_1 p_2 \cdots p_k, w, 0, 1)$ 2D optical code with $p_1 p_2 \cdots p_k$ matrices of length $N = p_1 p_2 \cdots p_k$ and weight w, where $L = w$ wavelengths are used, $p_k \geq p_{k-1} \geq \cdots \geq p_2 \geq p_1$ are prime numbers, $p_1 \geq w$, and k is the number of prime numbers in use [15]. If the number of available wavelengths L is more than the code weight w, say $L = wp'$ and p' is a prime number, a simple way to utilize the extra wavelengths is to separate them into p' groups (of w wavelengths each), and then to use the same $p_1 p_2 \cdots p_k$ matrices for each group, resulting in a total of $p' p_1 p_2 \cdots p_k$ matrices. (This method, here referred as a wavelength-grouping scheme, is similar to the use of wavelength-division multiplexing on top of OCDMA.) To achieve even larger cardinality, the ECHPC has each of the original $p_1 p_2 \cdots p_k$ code matrices taken as a seed from which $w^2 + p'$ groups of new matrices are generated in the following [32].

From [15], when $w = p_1$, given a positive integer k, code matrices, $x_{i_1, i_2, \ldots, i_k}$, with the ordered pairs*

$$\{[(0,0),(1, i_1 + i_2 p_1 + \cdots + i_k p_1 p_2 \cdots p_{k-1}),(2, 2 \otimes_{p_1} i_1 + (2 \otimes_{p_2} i_2) p_1 + \cdots$$
$$+ (2 \otimes_{p_k} i_k) p_1 p_2 \cdots p_{k-1}), \ldots, (p_1 - 1, (p_1 - 1) \otimes_{p_1} i_1 + ((p_1 - 1) \otimes_{p_2} i_2) p_1 + \cdots$$
$$+ ((p_1 - 1) \otimes_{p_k} i_k) p_1 p_2 \cdots p_{k-1})] : i_1 \in [0, p_1 - 1], i_2 \in [0, p_2 - 1], \ldots, i_k \in [0, p_k - 1]\}$$
$$(2.8)$$

form the $(p_1 \times p_1 p_2 \cdots p_k, p_1, 0, 1)$ CHPC, where " $\otimes_{p_j}$ " denotes a modulo-p_j multiplication for $j = \{1, 2, 3, \ldots, k\}$. Further, given prime numbers p' and w such that $p' \geq w$ and $wp' \leq p_1$, the w ordered pairs of new code matrices, $x_{i_1, i_2, \ldots, i_k, l_1, l_2}$ and $x_{i_1, i_2, \ldots, i_k, l_2}$, are obtained by choosing the ordered pairs from the original code matrix $x_{i_1, i_2, \ldots, i_k}$ with the first component of each ordered pair in the above equation given by

$$\{[l_1, (l_1 \oplus_w l_2) + w, (l_1 \oplus_w (2 \otimes_{p'} l_2)) + 2w, \ldots,$$
$$(l_1 \oplus_w ((w-1) \otimes_{p'} l_2)) + (w-1)w] : l_1 \in [0, w-1], l_2 \in [0, w-1]\}$$
$$(2.9)$$

and

$$\{[l_2 w, l_2 w + 1, \ldots, l_2 w + w - 1] : l_2 \in [0, p' - 1]\}, \qquad (2.10)$$

respectively, resulting in the $(wp' \times p_1 p_2 \cdots p_k, w, 0, 1)$ ECHPC with $(w^2 + p') p_1 p_2 \cdots p_k$ matrices of length $N = p_1 p_2 \cdots p_k$ and weight w, out of $L = wp'$ available wavelengths, where " $\otimes_{p'}$ " and " $\oplus_w$ " denote a modulo-p' multiplication and a modulo-w addition, respectively. The autocorrelation sidelobes of any matrix in the code set are zero and the cross-correlation function between any two distinct matrices in the code set is at most one. The new code always has a factor of $(w^2 + p')/p'$ more matrices than the wavelength-grouping scheme. Furthermore, there are $wp_1 p_2 \cdots p_k$ subsets of w matrices each and $p_1 p_2 \cdots p_k$ subsets of p' matrices each, respectively, from the two equations, such that all matrices within a subset have zero cross-correlation values. These "subset" partitions give improved performance, while the cardinality is substantially expanded.

Using $w = p' = 3$ and $p_1 = p_2 = 11$ as an example, the original CHPC has 121 matrices, x_{i_1, i_2} (for $i_1 \in [0, 10]$ and $i_2 \in [0, 10]$), represented by the ordered pairs $[(0,0), (1, i_1 + 11 i_2), (2, 2 \otimes_{11} i_1 + (2 \otimes_{11} i_2)11), \ldots, (10, 10 \otimes_{11} i_1 + (10 \otimes_{11} i_2)11)]$. Separating the $wp' = 9$ wavelengths into three groups, this gives a total of 363 code matrices of weight 3 and length 121. However, following the new construction, each

---

* For ease of representation, every matrix can equivalently be written as a set of v ordered pairs (i.e., one ordered pair for every binary one), where an ordered pair (f_v, t_h) records the vertical (v) and horizontal (h) displacements of a binary one from the bottom-leftmost corner of a matrix. In other words, f_v represents the transmitting carrier and t_h shows the (time or chip) position of a binary one in the matrix.

original matrix can generate 12 new groups of matrices, resulting in 1,452 matrices of weight 3 and length 121. Each of these new matrices, denoted as $\mathbf{x}_{i_1,i_2,l_1,l_2}$ and $\mathbf{x}_{i_1,i_2,l_2}$ (for $i_1 \in [0,10]$, $i_2 \in [0,10]$, $l_1 \in [0,2]$, and $l_2 \in [0,2]$), is represented by three ordered pairs chosen from the ordered pairs of the original matrix $\mathbf{x}_{i_1,i_2}$ with the first component of each ordered pair given by $[l_1,(l_1 \oplus_3 l_2)+3, (l_1 \oplus_3 (2 \otimes_3 l_2))+6]$ and $[3l_2, 3l_2+1, 3l_2+2]$. For example, $\mathbf{x}_{2,5,1,2}$ has the ordered pairs $[(1,57), (3,50), (8,82)]$, $\mathbf{x}_{2,5,2,2}$ has the ordered pairs $[(2,114), (4,107), (6,89)]$, and $\mathbf{x}_{2,5,1,1}$ has the ordered pairs $[(1,57), (5,43), (6,89)]$.

The cardinality upper bound of the $(L \times N, w, \lambda_a, \lambda_c)$ ECHPC can be derived by modifying the OOC Johnson bound in [18,20], such that

$$\Phi(L \times N, w, \lambda, \lambda) \leq \frac{L(LN-1)(LN-2)\cdots(LN-\lambda)}{w(w-1)(w-2)\cdots(w-\lambda)}, \quad (2.11)$$

where the maximum autocorrelation sidelobe and cross-correlation values are assumed to be identical (i.e., $\lambda = \lambda_a = \lambda_c$). Thus, the upper bound for the $(wp' \times p_1 p_2 \cdots p_k, w, 0, 1)$ ECHPC with $L = wp'$, $N = p_1 p_2 \cdots p_k$, $\lambda_a = 0$, and $\lambda_c = 1$ can be modified to be $\Phi(wp' \times p_1 p_2 \cdots p_k, w, 0, 1) \leq \Phi(wp' \times p_1 p_2 \cdots p_k, w, 1, 1) \leq wp'$ $(wp' p_1 p_2 \cdots p_k - 1)/[w(w-1)] = p'^2 p_1 p_2 \cdots p_k + (p'^2 p_1 p_2 \cdots p_k - 1)/(w-1)$. If $p' = w$, the upper bound becomes $\Phi(w^2 \times p_1 p_2 \cdots p_k, w, 0, 1)(w^2 p_1 p_2 \cdots p_k - 1)/(w-1)$. Compared to the actual cardinality (i.e., $(w^2 + p')p_1 p_2 \cdots p_k$), the upper bound is larger by a factor of about $1/w^2$. The factor is nearly equal to zero for a large w. The ECHPC is therefore asymptotically optimal.

To analyze the performance of the $(wp' \times p_1 p_2 \cdots p_k, w, 0, 1)$ ECHPC, the probability of hitting one of the pulses (i.e., a binary one) in an address matrix with a pulse in a received matrix is given by

$$q = \frac{w^2 F}{2LN} = \frac{wF}{2p' p_1 p_2 \cdots p_k}, \quad (2.12)$$

where 1/2 denotes equiprobable 0-1 data-bit transmission and F represents the ratio of the number of matrices contributing one hit (in the cross-correlation function) to the total number of interfering matrices F is used to account for the unique cross-correlation property of the ECHPC since matrices contain no common wavelengths and have zero cross-correlation value if they all come from the same subset. Thus,

$$F = \sum_{\text{all subsets}} P(\text{a matrix causes one hit} \mid \text{the matrix is from a subset})$$

$$\times P \text{ (the subset is chosen from the code set } \Phi) \quad (2.13)$$

$$= \frac{\Phi - 1 - (w-1)p_1 p_2 \cdots p_k}{\Phi - 1} \cdot \frac{w^2}{w^2 + p'} + \frac{\Phi - 1 - (p'-1)p_1 p_2 \cdots p_k}{\Phi - 1} \cdot \frac{p'}{w^2 + p'}$$

$$= 1 - \frac{(w^3 - w^2 + p'^2 - p')p_1 p_2 \cdots p_k}{(\Phi - 1)(w^2 + p')}, \quad (2.14)$$

where $\Phi = (w^2 + p')p_1 p_2 \cdots p_k$ is the code cardinality and $\Phi - 1$ represents the total number of possible interfering matrices. The error probability of the ECHPC, which has cross-correlation value of at most one, is given by [15,43]

$$P_e = \frac{1}{2} \sum_{i=w}^{K-1} \binom{K-1}{i} q^i (1-q)^{K-1-i}, \tag{2.15}$$

where K is the number of simultaneous users.

For the grouping scheme of utilizing the wp' wavelengths by separating them into p' groups of w wavelengths each, matrices from different groups will not interfere with each other because they have different sets of wavelengths, even though each group uses the same matrices from the $(w \times p_1 p_2 \cdots p_k, w, 0, 1)$ original CHPC. If all these p' groups of matrices are uniformly distributed to the K simultaneous users by a central controller, there will be at most $\lceil K/p' \rceil$ users transmitting at the same time in each group, where $\lceil \cdot \rceil$ is the ceiling function. The error probability is then given by [15]

$$P_e = \frac{1}{2} \sum_{i=w}^{\lceil K/p' \rceil - 1} \binom{\lceil K/p' \rceil - 1}{i} q^i (1-q)^{\lceil K/p' \rceil - 1 - i}, \tag{2.16}$$

where $q = w^2/(2LN) = w/(2p_1 p_2 \cdots p_k)$ and a central controller is used to uniformly distributed simultaneous users over the p' groups.

The error probabilities, P_e, of the $(wp' \times p_1 p_2 \cdots p_k, w, 0, 1)$ ECHPC and the grouping scheme with p' groups of the $(w \times p_1 p_2 \cdots p_k, w, 0, 1)$ original CHPC versus the number of simultaneous users K for $k = 1$ are plotted in Figure 2.3. In general, the error probabilities worsen as K increases, but improve as the number of wavelengths, code weight, or code length increases. The steplike solid curves represent the performance lower bound of the grouping scheme, which can only be achieved theoretically with the use of a central controller so that all simultaneous users are uniformly distributed over p' groups. The curves stop at the points (i.e., error free) where the number of simultaneous users in each group is less than the code weight (i.e., $\lceil K/p' \rceil \leq w$) as the cross-correlation functions of the original CPHC are at most one. Note that the performance of the ECHPC, (i.e., dashed curves) is here compared to the performance lower bound of the grouping scheme. The ECHPC will have a better performance than the grouping scheme without the central controller. The solid curves are increasingly better than the dashed curves at small K when p' increases because the grouping scheme allows the simultaneous users to spread into more groups. On the other hand, the solid curves converge with the dashed curves when the traffic load is heavy (i.e., $\lceil K/p' \rceil \gg wK/p'$), showing comparable performance, even only comparing to the lower bound of the grouping scheme.

2.5.2 MULTIPLE-LENGTH EXTENDED CARRIER-HOPPING PRIME CODE

An $(m \times N, w, A, B, D)$ multiple-length ECHPC, C, is a collection of $(0,1)$ code matrices with a set of code lengths $N = \{n_0, n_1, \ldots, n_i, \ldots, n_{k-1}\}$, a number of available

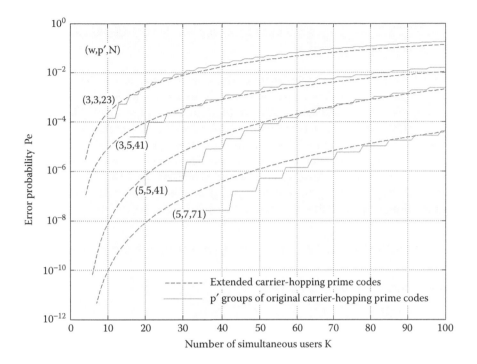

FIGURE 2.3 Error probability versus the number of simultaneous users for an OCDMA system with the $(wp' \times p_1, w, 0, 1)$ ECHPC and the grouping scheme with p' groups of the $(w \times p_1, w, 0, 1)$ original CHPC. (A central controller is assumed in the grouping scheme.) (Yang, G.-C., Kwong, W.C. 2004. *IEEE Trans. Commun.*, 53(5):876–881. © 2005 IEEE. Reprinted with permission.)

wavelengths m, a code weight w, a set of autocorrelation constraints $A = \{\lambda_a^{(0)}, \lambda_a^{(1)}, \dots, \lambda_a^{(i)}, \dots, \lambda_a^{(k-1)}\}$, a set of cross-correlation constraints $B = \{\lambda_c^{(0)}, \lambda_c^{(1)}, \dots, \lambda_c^{(i)}, \dots, \lambda_c^{(k-1)}\}$, and a set of matrix-cardinality distributions $D = \{d_0, d_1, \dots, d_i, \dots, d_{k-1}\}$, where k is the number of different code lengths in C, such that

(Length Distribution) Every code matrix in C has a length n_i contained in the length set N. There are $d_i |C|$ code matrices of length n_i, where d_i indicates the fraction of matrices of length n_i in C and $|C|$ represents the code cardinality.

(Autocorrelation Constraint) For any code matrix $\mathbf{x} \in C$ of length n_i, its autocorrelation sidelobes are bound by $\lambda_a^{(i)}$ such that $\sum_{s=0}^{m-1} \sum_{t=0}^{n_i-1} x_{s,t} \, x_{s,t \oplus_i \tau} \leq \lambda_a^{(i)}$ for any integer delay $\tau \in (0, n_i)$, where $x_{s,t} = \{0, 1\}$ is an element in $\mathbf{x}$ and "$\oplus_i$" denotes a modulo-n_i addition.

(Intra-cross-correlation Constraint) For any two distinct matrices $\mathbf{x} \in C$ and $\mathbf{y} \in C$ of the same length n_i, their cross-correlation function is bound by $\lambda_c^{(i)}$ such that $\sum_{s=0}^{m-1} \sum_{t=0}^{n_i-1} x_{s,t} \, y_{s,t \oplus_i \tau} \leq \lambda_c^{(i)}$ for any $\tau \in [0, n_i)$, where $y_{s,t} = \{0, 1\}$ is an element of $\mathbf{y}$.

(Inter-cross-correlation Constraint) For any two distinct matrices $\mathbf{x} \in C$ of length n_i and $\mathbf{y} \in C$ of length n_j, their cross-correlation function is bound by λ such that $\sum_{s=0}^{m-1}\sum_{t=0}^{n_j-1} x_{s,t}\, y_{s,t\oplus_j\tau} \leq \lambda$ for any $\tau \in [0, n_j)$ and $\sum_{s=0}^{m-1}\sum_{t=0}^{n_j-1} x_{s,t\oplus_j\tau}\, y_{s,t} \leq \lambda$ for any $\tau \in [0, n_i)$, where "$\oplus_j$" denotes a modulo-n_j addition and $n_i \neq n_j$.

Each matrix in the multiple-length CHPC consists of an $m \times n_i$ 2D $(0,1)$ pattern with weight w, comprising w ones in m rows (related to the number of available wavelengths) and n_i columns (related to the number of time slots). The multiple-length CHPC with zero autocorrelation sidelobes and cross-correlation functions of at most one are generated by the introduction of a set of positive integers $\{t_1, t_2, ..., t_{k-1}\}$ to control the number of matrices in each length, out of k different lengths in the code set. Also, given a set of prime numbers $\{p_1, p_2, ..., p_k\}$, such that $p_1 \geq t_1 \geq t_2 \geq \cdots \geq t_{k-1} \geq 0$ and $p_k \geq p_{k-1} \geq p_{k-2} \geq \cdots \geq p_1$, code matrices, $\mathbf{x}_{i_1}$, with the ordered pairs [15]

$$\{[(0, 0), (1, i_1), (2, 2\otimes_{p_1} i_1), ..., (p_1 - 1, (p_1 - 1)\otimes_{p_1} i_1)] : i_1 \in [t_1, p_1 - 1]\} \quad (2.17)$$

matrices, $\mathbf{x}_{i_1, i_2}$, with the ordered pairs

$$\{[(0, 0), (1, i_1 + i_2 p_1), (2, 2\otimes_{p_1} i_1 + (2\otimes_{p_2} i_2)p_1), ...,$$

$$(p_1 - 1, (p_1 - 1)\otimes_{p_1} i_1 + ((p_1 - 1)\otimes_{p_2} i_2)p_1)] : i_1 \in [t_2, t_1 - 1], i_2 \in [0, p_2 - 1]\} \quad (2.18)$$

$\vdots$

and matrices, $\mathbf{x}_{i_1, i_2, ..., i_k}$, with the ordered pairs

$$\{[(0, 0), (1, i_1 + i_2 p_1 + \cdots + i_k p_1 p_2 \cdots p_{k-1}), (2, 2\otimes_{p_1} i_1 + (2\otimes_{p_2} i_2)p_1 + \cdots$$

$$+(2\otimes_{p_k} i_k)p_1 p_2 \cdots p_{k-1}), ..., (p_1 - 1, (p_1 - 1)\otimes_{p_1} i_1 + ((p_1 - 1)\otimes_{p_2} i_2)p_1 + \cdots$$

$$+((p_1 - 1)\otimes_{p_k} i_k)p_1 p_2 \cdots p_{k-1})] : i_1 \in [0, t_{k-1} - 1], i_2 \in [0, p_2 - 1], ..., i_k \in [0, p_k - 1]\}$$

$$\quad (2.19)$$

form the multiple-length CHPC of weight $w = p_1$ with $p_1 - t_1$ matrices of length p_1, $p_2(t_1 - t_2)$ matrices of length $p_1 p_2$, $p_2 p_3(t_2 - t_3)$ matrices of length $p_1 p_2 p_3$, ..., $p_2 p_3 \cdots p_{k-1}(t_{k-2} - t_{k-1})$ matrices of length $p_1 p_2 p_3 \cdots p_{k-1}$, and $t_{k-1} p_2 p_3 \cdots p_k$ matrices of length $p_1 p_2 p_3 \cdots p_k$, respectively, where "$\otimes_{p_j}$" denotes a modulo-p_j multiplication for $j = \{1, 2, 3, ..., k\}$.

If the number of available wavelengths m is more than the code weight w, say $m = wp'$ and p' is a prime number, we can construct the multiple-length ECHPC [46]. In general, each of the original multiple-length CHPC matrices, denoted as $\mathbf{x}_{i_1, i_2, ..., i_i}$ (with $i_1 \in [t_i, t_{i-1}]$ (or $i_1 \in [0, t_{k-1} - 1]$), $i_2 \in [0, p_2 - 1]$, ..., and $i_i \in [0, p_i - 1]$, for $i \in [1, k]$), can be taken as seeds from which $w^2 + p'$ groups of new matrices are generated.

Given prime numbers p' and w such that $p' \geq w$ and $wp' \leq p_1$, the w ordered pairs of code matrices, $\mathbf{x}_{i_1, i_2, \ldots, i_k, l_1, l_2}$ and $\mathbf{x}_{i_1, i_2, \ldots, i_k, l_2}$, are obtained by choosing the ordered pairs from the code matrix $\mathbf{x}_{i_1, i_2, \ldots, i_k}$ with the first component of each ordered pair in the above equations given by

$$\{[l_1, (l_1 \oplus_w l_2) + w, (l_1 \oplus_w (2 \otimes_{p'} l_2)) + 2w, \ldots, (l_1 \oplus_w ((w-1) \otimes_{p'} l_2))$$
$$+ (w-1)w] : l_1 \in [0, w-1], l_2 \in [0, w-1]\} \tag{2.20}$$

and

$$\{[l_2 w, l_2 w + 1, \ldots, l_2 w + w - 1] : l_2 \in [0, p' - 1]\}, \tag{2.21}$$

respectively, resulting in the multiple-length ECHPC of weight w with $(w^2 + p')(p_1 - t_1)$ matrices of length p_1, $(w^2 + p')(t_1 - t_2)p_2$ matrices of length $p_1 p_2$, $(w^2 + p')(t_2 - t_3)p_2 p_3$ matrices of length $p_1 p_2 p_3$, $\ldots, (w^2 + p')(t_{k-2} - t_{k-1})$ $p_2 p_3 \cdots p_{k-1}$ matrices of length $p_1 p_2 p_3 \cdots p_{k-1}$, and $(w^2 + p')t_{k-1}p_2 p_3 \cdots p_k$ matrices of length $p_1 p_2 p_3 \cdots p_k$, out of $m = wp'$ available wavelengths, where "$\otimes_{p'}$" denote a modulo-p' multiplication and "$\oplus_w$" denote a modulo-w addition. The autocorrelation sidelobe of any matrix in the code set is zero. The intra- and inter-cross-correlation functions of any two distinct matrices in the code set are both at most one. There are $w(p_1 - t_1)$, $w(t_1 - t_2)p_2$, $\ldots$, and $wt_{k-1}p_2 p_3 \cdots p_k$ subsets of w matrices each, and $p_1 - t_1$, $(t_1 - t_2)p_2$, $\ldots$, and $t_{k-1}p_2 p_3 \cdots p_k$ subsets of p' matrices each, respectively, from both equations, such that all matrices within a subset have zero cross-correlation values.

Using $k = 2$, $w = p' = 3$, $p_2 = p_1 = 11$, and $t_1 = 6$ as an example, the double-length CHPC gives $p_1 - t_1 = 5$ matrices of length 11, $\mathbf{x}_{i_1}$ (for $i_1 \in [6, 10]$, represented by the ordered pairs $[(0,0), (1, i_1), (2, 2 \otimes_{11} i_1), \ldots, (10, 10 \otimes_{11} i_1)]$, and $t_1 p_2 = 66$ matrices of length 121, $\mathbf{x}_{i_1, i_2}$ (for $i_1 \in [0, 5]$ and $i_2 \in [0, 10]$), represented by the ordered pairs $[(0,0), (1, i_1 + 11i_2), (2, 2 \otimes_{11} i_1 + (2 \otimes_{11} i_2)11), \ldots, (10, 10 \otimes_{11} i_1 + (10 \otimes_{11} i_2)11)]$. Following the construction, each original matrix can generate $w^2 + p' = 12$ new groups of matrices. Thus, the double-length ECHPC has $(w^2 + p')(p_1 - t_1) = 60$ matrices of length 11, $\mathbf{x}_{i_1, l_1, l_2}$ and $\mathbf{x}_{i_1, l_2}$ (for $i_1 \in [6, 10]$. $l_1 \in [0, 2]$, and $l_2 \in [0, 2]$), and $(w^2 + p')t_1 p_2 = 792$ matrices of length 121, $\mathbf{x}_{i_1, i_2, l_1, l_2}$ and $\mathbf{x}_{i_1, i_2, l_2}$ (for $i_1 \in [0, 5]$, $i_2 \in [0, 10]$, $l_1 \in [0, 2]$, and $l_2 \in [0, 2]$), represented by 3 ordered pairs chosen from the ordered pairs of the original matrix $\mathbf{x}_{i_1}$ and $\mathbf{x}_{i_1, i_2}$ with the first component of each ordered pair given by $[l_1, (l_1 \oplus_3 l_2) + 3, (l_1 \oplus_3 (2 \otimes_3 l_2)) + 6]$ and $[3l_2, 3l_2 + 1, 3l_2 + 2]$, respectively. For example, $\mathbf{x}_{i_1, l_1, l_2} = \mathbf{x}_{6,1,2}$ is a 9×11 matrix with the ordered pairs $[(1,6), (3,7), (8,4)]$, and $\mathbf{x}_{i_1, i_2, l_1, l_2} = \mathbf{x}_{2,6,2,2}$ is a 9×121 matrix with the ordered pairs $[(2,15), (4,30), (6,34)]$, both of weight $w = 3$.

The cardinality upper bound of the multiple-length ECHPC is given by modifying [15, Theorem 2.10], such that

$$\Phi \leq \frac{p'\left(wp' \prod_{j=1}^{k} p_j - 1\right)}{\sum_{i=1}^{k} d_i \left(\prod_{j=i+1}^{k} p_j w - 1\right)}, \tag{2.22}$$

where p_i is the prime number in use and d_i is the ratio of the number of matrices of length n_i to the total number of matrices in the code set for $i = \{1, 2, 3, ..., k\}$. The total number of matrices in the multiple-length ECHPC is $\Phi_{multiple} = (w^2 + p')[(p_1 - t_1) + p_2(t_1 - t_2) + \cdots + p_2 p_3 \cdots p_k t_{k-1}]$. Then, we have $d_1 = (w^2 + p')(p_1 - t_1)/\Phi_{multiple}$, $d_2 = (w^2 + p')p_2(t_1 - t_2)/\Phi_{multiple}$, $\ldots$, and $d_k = (w^2 + p')p_2 p_3 \cdots p_k t_{k-1}/\Phi_{multiple}$. After some manipulations, $\Phi_{multiple}/\Phi \approx (w^2 + p')(p_1 - t_1)(p'^2 p_1)^{-1}[1 - (wp_2 p_3 \cdots p_k)^{-1}] + (w^2 + p')(t_1 - t_2)(p'^2 p_1)^{-1}[1 - (wp_3 p_4 \cdots p_k)^{-1}] + \cdots + (w^2 + p')(t_{k-2} - t_{k-1})(p'^2 p_1)^{-1}[1 - (wp_k)^{-1}] + (w^2 + p')t_{k-1}(p'^2 p_1)^{-1}[1 - w^{-1}]$, which is nearly equal to one for a large w and p' close to w, resulting in asymptotic optimality.

The performance of every user in an OCDMA system employing the single-length ECPHC is assumed identical. However, with the multiple-length ECHPC in use, the performance of a user also depends on the length of its address matrix and the lengths of the arriving matrices. While the analysis based on MAI generated by matrices of the same length can be treated as traditional cross-correlation functions [15,43], special cares are required for MAI generated by matrices of different lengths. In particular, it is complex to evaluate the interference at a user with an address matrix longer than the arriving matrix. Here, we consider the performance of multiple-length ECHPC for the case of $k = 2$ (i.e., double-media services). Let the code lengths be $n_s = p_1$ and $n_l = rn_s = p_2 p_1$ (i.e., $r = p_2$), and the cardinalities are $\Phi_s = (w^2 + p')(p_1 - t_1)$ and $\Phi_l = (w^2 + p')t_1 p_2$ for the short and long matrices, respectively. Each matrix has w pulses (i.e., wavelengths), out of $m = wp'$ possible wavelengths. If the length of the address matrix is r times longer than the arriving matrix, the inter-cross-correlation process at this user will involve $r + 1$ consecutive cross-correlation functions. Since the code weight is w, the resulting periodic inter-cross-correlation function can be as large as w. This results in strong interference, even though the aperiodic cross-correlation function between the long matrix and the short matrix is at most one. To alleviate this potential problem, the multiple-length ECHPC is specially designed so that the number of hits between a long matrix and the multiple copies of a short matrix is at most one.

By modifying the hit-probability equation of the original ECHPC, the probability of obtaining one hit between two same-length matrices is given by

$$q_l = \frac{w^2 F_l}{2mn_l} = \frac{w}{2p'n_l} \cdot \left[1 - \frac{(w^3 - w^2 + p'^2 - p')t_1 p_2}{(\Phi_l - 1)(w^2 + p')} \right] \tag{2.23}$$

for the long matrices and

$$q_s = \frac{w^2 F_s}{2mn_s} = \frac{w}{2p'n_s} \cdot \left[1 - \frac{(w^3 - w^2 + p'^2 - p')(p_1 - t_1)}{(\Phi_s - 1)(w^2 + p')} \right] \tag{2.24}$$

for the short matrices, where 1/2 denotes equiprobable 0-1 data-bit transmissions. F_l and F_s represent the ratio of the number of matrices contributing one hit (in the cross-correlation function) to the total number of interfering matrices of length n_l and n_s, respectively. The terms $(w^3 - w^2 + p'^2 - p')t_1 p_2 p_3 \cdots p_h/(w^2 + p')$ and $(w^3 - w^2 + p'^2 - p')(p_1 - t_1)p_2 p_3 \cdots p_g/(w^2 + p')$ represent the numbers of matrices that do not have any wavelength overlap (i.e., zero cross-correlation value) with a given address matrix. Since the number of hits between a long matrix and $r + 1$ copies of a short matrix is no greater than one, the probability of obtaining one hit is given by

$$q_{l,d} = \frac{w^2 F_s}{2 m n_s} = \frac{w}{2 p' n_s} \cdot \left[1 - \frac{(w^3 - w^2 + p'^2 - p')(p_1 - t_1)}{(\Phi_s - 1)(w^2 + p')} \right], \qquad (2.25)$$

where "d" stands for the cross-correlation function of two different-length matrices. The probability of obtaining one hit between a short matrix and a long matrix is given by

$$q_{s,d} = \frac{w^2 F_l}{2 m n_l} = \frac{w}{2 p' n_l} \cdot \left[1 - \frac{(w^3 - w^2 + p'^2 - p')t_1 p_2}{(\Phi_l - 1)(w^2 + p')} \right]. \qquad (2.26)$$

The probabilities of error of users with address matrices of length n_l and n_s can be obtained by modifying the error-probability equation of the original ECHPC as

$$P_{e,l} = \frac{1}{2} \sum_{l_l + l_s = w}^{K_l + K_s - 1} \left[\binom{K_l - 1}{l_l} (q_l)^{l_l} (1 - q_l)^{K_l - 1 - l_l} \binom{K_s}{l_s} (q_{l,d})^{l_s} (1 - q_{l,d})^{K_s - l_s} \right] \qquad (2.27)$$

and

$$P_{e,s} = \frac{1}{2} \sum_{l_l + l_s = w}^{K_l + K_s - 1} \left[\binom{K_l}{l_l} (q_{s,d})^{l_l} (1 - q_{s,d})^{K_l - l_l} \binom{K_s - 1}{l_s} (q_s)^{l_s} (1 - q_s)^{K_s - 1 - l_s} \right], \qquad (2.28)$$

respectively, where K_l and K_s represent the numbers of simultaneous users using long and short matrices.

An $(\{23,529\},4,\{1,1\},\{1,1,1\},\{252/5565,5313/5565\})$ double-length ECHPC with $k = 2$, $r = p_1 = p_2 = 23$, $w = 4$, $p' = 5$, and $t_1 = 11$ is used as a numerical example. Shown in Figure 2.4 are the error probabilities versus the number of simultaneous users K_s transmitting short matrices. While the solid curves correspond to the performance of users with short address matrices (i.e., $P_{e,s}$), the dash-dotted curves show the performance of users with long address matrices (i.e., $P_{e,l}$). As a whole, the performance worsens as the total number of simultaneous users increases. Users with short address matrices always perform better than those with long address matrices, which can be explained by stronger interference created by multiple copies

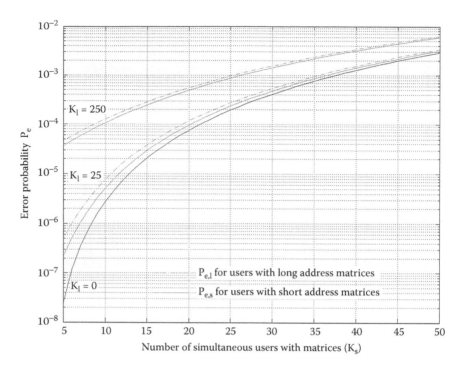

FIGURE 2.4 Error probability versus the number of simultaneous users K_l transmitting long matrices (of length $nl = 529$) for the double-media OCDMA systems with the ({23,529},4,{1,1},{1,1,1},{252/5565,5313/5565}) double-length EGCPC. (K_s represents the number of simultaneous users transmitting short matrices.)

of short matrices in the inter-cross-correlation process. However, the improvement is rather small because a very short code length $n_s = 23$ is used, which has poor performance in the first place. The solid curve with $K_l = 0$ shows the performance of a single-length ECHPC with the short matrices only. The figure shows that short matrices are dominating the performance and both short and long address users see strong interference from short matrices. Assume that there are $K_l + K_s = 30$ simultaneous users in total. For example, $K_s = 5$ and $K_l = 25$, we can seen from the figure that $P_{e,s} = 2 \times 10^{-7}$ and $P_{e,l} = 5 \times 10^{-7}$. That is, the short matrices perform better than the long matrices. This contradicts the results in conventional single-length OCDMA codes that long matrices were known to perform better than short matrices.

2.5.3 Application to Multirate, Multimedia OCDMA Systems

To illustrate an application of the multiple-length ECHPC in OCDMA, we assume three types of services (e.g., digitized voice, data, and video) in Figure 2.5. Real-time video transmission, which has a continuous traffic pattern, is usually given the highest priority and bit rate. It is assumed that video bit rate $1/T_v$ is an integral multiple of that of the data service $1/T_d$ (i.e., $1/T_v = p_2/T_d$) which, in turn, is a multiple of that of the voice service $1/T_s$ (i.e., $1/T_v = p_3/T_s$), where p_2 and p_3 are

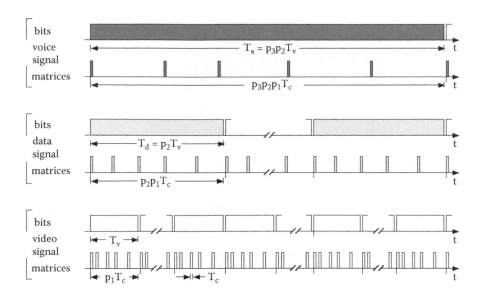

FIGURE 2.5 Relationship of bit durations and code lengths, among digitized voice, data, and video services. (The pulses are assumed in different wavelengths.)

the expansion factors defined in the construction. For the service with the highest bit rate (i.e., video, in this case), we assume to use matrices of length p_1, where $T_v = p_1 T_c$ and T_c is the width of time slots. As the same width T_c is used for all services, the medium-rate service (i.e., data) requires matrices of p_2 times longer than that of the video service for supporting the rate $1/T_d = 1/(p_2 T_v) = 1/(p_2 p_1 T_c)$. Similarly, the lowest-rate service (i.e., voice) requires matrices of times longer than that of the video service for supporting the rate $1/T_s = 1/(p_3 p_2 T_v) = 1/(p_3 p_2 p_1 T_c)$. By using the same time-slot width for every services, we can use laser sources with the same pulse-width and simplify the timing requirement with the use of a single clock.

To pick code matrices for these three services, by assuming that $T_c \approx 4.55$ nsec, $w = 3$, $p' = 3$, $p_1 = p_2 = p_3 = 13$, $t_1 = 3$, and $t_2 = 1$, we can construct a triple-length ECHPC with $(w^2 + p')t_2 p_2 p_3 = 2,028$ matrices of length $p_1 p_2 p_3 = 2,197$ for the voice services at a bit rate of about 100 Mbit/sec, $(w^2 + p')(t_1 - t_2)p_2 = 312$ matrices of length $p_1 p_2 = 169$ for the data services at a bit rate of about 1.25 Gbit/sec, and finally $(w^2 + p')(p_1 - t_1) = 120$ matrices of length $p_1 = 13$ for the voice services at a bit rate of about 16.5 Gbit/sec. Since our analysis shows that the shortest matrices have the best performance, the QoS for the critical real-time video transmission is guaranteed. This unique priority feature is not supported in conventional single-length codes.

2.6 SUMMARY

In this chapter, we studied various optical codes for OCDMA. The constructions of three families of 1D bipolar codes, maximal-length sequences, Walsh codes, and Gold sequences, and two families of 1D unipolar codes, OOCs and prime code, were reviewed. Their performances were analyzed and compared. Afterwards, recently developed 2D ECHPC was constructed and analyzed. This wavelength-time code is particularly suitable for high bit-rate OCDMA in which the number of time slots is very restricted. For future multirate, multimedia OCDMA applications, we showed the construction of 2D multiple-length ECHPC. Finally, an example illustrating an application of the multiple-length code in multirate, multimedia OCDMA systems was given.

ACKNOWLEDGMENT

The work on the Advanced Incoherent Codes section is supported in part by the U.S. Defense Advanced Research Projects Agency (DARPA) under grant MDA972-03-1-0006, the National Science Council of Republic of China under grant NSC 93-2213-E-005-002, the Presidential Research Award of Hofstra University, and the Faculty Development and Research Grants of Hofstra University.

REFERENCES

[1]   Prucnal, P. R., Santoro, M. A., Fan, T. R. (1986). Spread spectrum fiber-optic local area network using optical processing. *J. Lightwave Technol.* 4(5):547–554.

[2]   Salehi, J. A. (1989). Code division multiple-access techniques in optical fiber networks: Part 1: Fundamental principles. *IEEE Trans. Commun.* 37(8):824–833.

[3]   Kwong, W. C., Perrier, P. A., Prucnal, P. R. (1991). Performance comparison of asynchronous and synchronous code-division multiple-access techniques for fiber-optic local area networks. *IEEE Trans. Commun.* 39(11):1625–1634.

[4]   Park, E., Mendez, A. J., Garmire, E. M. (1992). Temporal/spatial optical CDMA networks: Design, demonstration, and comparison with temporal networks. *IEEE Photon. Technol. Lett.* 4(10):1160–1162.

[5]   Marhic, M. E. (1993). Coherent optical CDMA networks. *J. Lightwave Technol.* 11(5/6):895–864.

[6]   Zaccarin, D., Kavehrad, M. (1993). An optical CDMA system based on spectral encoding of LED. *IEEE Photon. Technol. Lett.* 4(4):479–482.

[7]   Tancevski, L., Andonovic, I. (1994). Wavelength hopping/time spreading code division multiple access systems. *Electron. Lett.* 30(17):1388–1390.

[8]   Griffin, R. A., Sampson, D. D., Jackson, D. A. (1995). Coherence coding for photonic code-division multiple access networks. *J. Lightwave Technol.* 13(2):204–205.

[9]   Yang, G. -C., Kwong, W. C. (1997). Performance comparison of multiwavelength CDMA and WDMA+CDMA for fiber-optic networks. *IEEE Trans. Commun.* 45(11): 1426–1434.

[10] Kwong, W. C., Yang, G. -C. (1998). Image transmission in multicore-fiber code-division multiple-access networks. *IEEE Commun. Lett.* 2(9):285–288.

[11] Sardesai, H. P., Chang, C. -C., Weiner, A. M. (1998). A femtosecond code-division multiple-access communication system test bed. *J. Lightwave Technol.* 16(11): 1953–1964.

[12] Mendez, A. J., Gagliardi, R. M., Feng, H. X. C., Heritage, J. P., Morookian, J. -M. (2000). Strategies for realizing optical CDMA for dense, high-speed, long span, optical network applications. *J. Lightwave Technol.* 18(12):1685–1696.

[13] Yu, K., Shin, J., Park, N. (2000). Wavelength-time spreading optical CDMA system using wavelength multiplexers and mirrored fiber delay lines. *IEEE Photon. Technol. Lett.* 12(9):1278–1280.

[14] Chen, L. R. (2001). Flexible fiber Bragg grating encoder/decoder for hybrid wavelength-time optical CDMA. *IEEE Photon. Technol. Lett.* 13(11):1233–1235.

[15] Yang, G. -C., Kwong, W. C. (2002). *Prime Codes with Applications to CDMA Optical and Wireless Networks*. Norwood: Artech House.

[16] Inaty, E., Shalaby, H. M. H., Fortier, P., Rusch, L. A. (2002). Multirate optical fast frequency hopping CDMA system using power control. *J. Lightwave Technol.* 20(2):166–177.

[17] Mendez, A. J., Gagliardi, R. M., Hernandez, V. J., Bennett, C. V., Lennon, W. J. (2003). Design and performance analysis of wavelength/time (W/T) matrix codes for optical CDMA. *J. Lightwave Technol.* 21(11):2524–2533.

[18] Chung, F. R. K., Salehi, J. A., Wei, V. K. (1989). Optical orthogonal codes: Design, analysis, and applications. *IEEE Trans. Info. Theory* 35(3):595–604.

[19] Chung, H., Kumar, P. V. (1990). Optical orthogonal codes: New bounds and an optimal construction. *IEEE Trans. Info. Theory* 36(4):866–873.

[20] Yang, G. -C., Fuja, T. (1995). Optical orthogonal codes with unequal auto- and cross-correlation constraints. *IEEE Trans. Info. Theory* 41(1):96–106.

[21] Yang, G. -C. (1995). Some new families of optical orthogonal codes for code-division multiple-access fibre-optic networks. *IEEE Proc. Commun.* 142(6):363–368.

[22] Yang, G. -C. (1996). Variable-weight optical orthogonal codes for CDMA networks with multiple performance requirements. *IEEE Trans. Commun.* 44(1):47–55.

[23] Fuji-Hara, R., Miao, Y. (2000). Optical orthogonal codes: Their bounds and new optimal constructions. *IEEE Trans. Info. Theory* 46(7):2396–2406.

[24] Shaar, A., Davies, P. (1983). Prime sequences: Quasi-optimal sequences for or channel code division multiplexing. *Electron. Lett.* 19(21):888–890.

[25] Huang, W., Andonovic, I., Tur, M. (1998). Decision direct PLL used for coherent optical pulse CDMA in the presence of multiuser interference laser phase noise and shot noise. *J. Lightwave Technol.* 16(10):1786–1794.

[26] Komo, J. J., Yuan, C. -C. (1989). Evaluation of code division multiple access systems. In: *Proceedings IEEE Energy and Information Technologies in the Southeast* 2:849–854.

[27] Lam, A. W., Tantaratana, S. (1994). *Theory and Application of Spread-Spectrum Systems — A Self Study Course*, IEEE.

[28] Dinan, E. H., Jabbari, B. (1998). Spreading codes for direct sequence CDMA and wideband CDMA cellular networks. *IEEE Commun. Mag.* 36(9): 48–54.

[29] Yang, G. -C., Kwong, W. C. (1996). Two-dimensional spatial signature patterns. *IEEE Trans. Commun.* 44(2):184–191.

[30] Tancevski, L., Andonovic, I. (1996). Hybrid wavelength hopping/time spreading schemes for use in massive optical LANs with increased security. *J. Lightwave Technol.* 14(12):2636–2647.

[31] Kwong, W. C., Yang, G. -C., Baby, V., Bres, C. -S., Prucnal, P. R. (2005). Multiple-wavelength optical orthogonal codes under prime-sequence permutations for optical CDMA. *IEEE Trans. Commun.*, 53(1): 117–123.

[32] Yang, G. -C., Kwong, W. C. (2004).) A new class of carrier-hopping codes for code-division multiple-access optical and wireless systems. *IEEE Commun. Lett.* 8(1):51–53.

[33] Yim, R. M. H., Chen, L. R., Bajcsy, J. (2002). Design and performance of 2-D codes for wavelength-time optical CDMA. *IEEE Photon. Technol. Lett.* 14(5):714–716.

[34] Kwong, W. C., Yang, G. -C. (2004). Extended carrier-hopping prime codes for wavelength-time optical code-division multiple-access. *IEEE Trans. Commun.* 52(7): 1084–1091.

[35] EIA/TIA-95 Rev A. (1995). Mobile station-base station compatibility standard for dual-mode wideband spread spectrum cellular system.

[36] Dixon, R. C. (1994). *Spread Spectrum Systems with Commercial Applications.* New York: John Wiley & Sons.

[37] Bose, R. C. (1939). On the construction of balanced incomplete block design. *Ann. Eugenics* 9:353–399.

[38] Wilson, R. M. (1972). Cyclotomy and difference families in elementary abelian groups. *J. Number Theory* 4:17–47.

[39] Hanani, H. (1961). The existence and construction of balanced incomplete block designs, *Ann. Math. Statist.* 32:361–386.

[40] Titlebaum, E. L. (1981). Time-frequency hop signals: Part I: Coding based upon the theory of linear congruences. *IEEE Trans. Aerospace Electron. Syst.* 17(4):490–494.

[41] Maric, S. V., Titlebaum, E. L. (1990). Frequency hop multiple access codes based upon the theory of cubic congruences. *IEEE Trans. Aerospace Electron. Syst.* 26(6): 1035–1039.

[42] Maric, S. V., Titlebaum, E. L. (1992). A class of frequency hop codes with nearly ideal characteristics for use in multiple-access spread-spectrum communications and radar and sonar systems. *IEEE Trans. Commun.* 40(9):1442–1447.

[43] Azizoglu, M. Y., Salehi, J. A., Li, Y. (1992). Optical CDMA via temporal codes. *IEEE Trans. Commun.* 40(7):1162–1170.

[44] Yang, G. -C., Kwong, W. C. (2005). Performance analysis of extended carrier-hopping prime codes for optical CDMA. *IEEE Trans. Commun.,* 53(5): 876–881.

[45] Kwong, W. C., Yang, G. -C. (2002). Design of multilength optical orthogonal codes for optical code-division multiple-access multimedia networks. *IEEE Trans. Commun.* 50(8):1258–1265.

[46] Kwong, W. C., Yang, G. -C. (2004). Multiple-length extended carrier-hopping prime codes for optical CDMA systems supporting multirate, multimedia services. *J. Lightwave Technol,* in press.

3 Information Capacity of Nonlinear Fiber-Optical Systems: Fundamental Limits and OCDMA Performance

Evgenii Narimanov

CONTENTS

3.1 INTRODUCTION

The performance of any communication system is fundamentally limited by the available bandwidth, signal to noise ratio of the received signal, and the codes used to relate the original information to the transmitted signal. An attempt to operate above the corresponding limit inevitably leads to increasingly frequent errors and the corresponding loss of the information.

This limit can be expressed by using the concept of "information capacity" originally introduced within the framework of information theory [1]. The information capacity R is defined as the maximum possible bit rate for error-free transmission in the presence of noise, and depends on the parameters of the "communication channel" (e.g., optical fiber) and on the particular encoding algorithm.

While the use of more advanced codes may improve the system performance, the bandwidth and the signal-to-noise ratio in the communication channel put a fundamental limit on information capacity. This limit (often referred to as the "channel capacity,"[*]) is defined as the maximum value of information capacity calculated over all possible information encoding algorithms:

$$C = \max[R] \tag{3.1}$$

The knowledge of this limit (which only depends on the communication channel) allows to evaluate the performance of the system, and find the guidance to its improvement.

In particular, for a *linear* communication channel with additive noise, and a total signal power constraint at the input, the fundamental limit to the information capacity is given by the celebrated Shannon formula [1]

$$C = W \log\left(1 + \frac{P_0}{P_N}\right) \tag{3.2}$$

where W is the channel bandwidth, P_0 is the average signal power, and P_N is the average noise power.

In comparison, the standard on-off-keying (OOK) procedure where the information in encoded via a "train" of nonoverlapping pulses—with the presence of the pulse in a given time corresponding to "1" and its absence to "0," leads to the transmission rate of one bit per time slot. To avoid the interference of pulses from the adjacent slots, the size of the time slot is generally chosen to be no less than the pulse width τ, leading to the transmission rate of up to $1/\tau$, bits/sec. Since the corresponding bandwidth is on the order of $1/\tau$, OOK algorithm performance is by a factor of $\sim \log_2(\text{SNR})$ less than the fundamental limit (where SNR is the signal-to-noise ratio).

The latter conclusion immediately points to an approach that would improve the code performance. Instead of just two power "levels" (pulse/no pulse), one could use a multilevel code where different values of pulse power would correspond to different transmitted "numbers." For the detector to reliably distinguish between different "messages," the difference between these power levels should not be less than the noise power, leading to the upper limit of $\sim P_0/P_N$ different power levels. As the amount of the transmitted information scales as the $\log_2$ of the number of

---

[*] We shall avoid using this term to prevent confusion with the "information capacity" that we introduced earlier.

possible "messages," the use of multilevel codes instead of OOK leads to an information capacity improvement by a factor of $\log_2 SNR$, thus reaching the fundamental limit of Equation 3.2.

However, the representation of the fundamental limit to the information capacity in the standard form of Equation 3.2 is unsuitable for applications to fiber-optical systems, as it was obtained based on the assumption of *linearity* of the communication channel, while modern fiber optics systems operate in a substantially nonlinear regime. Since the optical transmission lines must satisfy very strict requirements for bit-error-rate (10^{-9} to 10^{-15}), the pulse amplitude should be large enough so that it can be effectively detectable. The increase of the number of multiple-access channels (using e.g., wavelength-, time- or code-division) [2] in modern fiber-optical communication systems also leads to a substantial increase of the electric field intensity in the fiber. As a consequence, the Kerr nonlinearity of the fiber refractive index $n = n_0 + \gamma P$ (where P is the pulse intensity) and resulting four-wave mixing processes strongly affect the information capacity. However, so far nonlinear-optical effects were only considered for the *lower bound* estimate to the maximum information capacity [7]—which was later shown to differ not only quantitatively but also *qualitatively* from the actual information capacity [10].

As both the actual information capacity, and its fundamental limit are necessary for OCDMA performance consideration and system improvement, the objective of the present chapter is twofold:

- Demonstrate the fundamental limit to the information capacity of a nonlinear fiber-optical system
- Show the actual information capacity for existing coherent and incoherent CDMA formats

The chapter is organized as follows. In Section 3.2, we review the relevant properties of the fiber-optical transmission line as a nonlinear communication channel. Section 3.3 is devoted to the calculation of the fundamental limit to the information capacity of such a channel. Section 3.4 deals with the actual information capacity corresponding to existing coherent and incoherent formats, while Section 3.5 introduces a novel approach to OCDMA codes with improved performance. Finally, Section 3.6 summarizes the chapter.

3.2 FIBER OPTICAL COMMUNICATION SYSTEM AS AN INFORMATION CHANNEL

In the present section, we review the properties of the signal propagation in fiber-optical systems, and the relevant concepts from information theory that describe information transmission in such communication channels.

3.2.1 Light Propagation in Nonlinear Fiber-Optical Systems

We consider a typical fiber-optical communication system, which consists of a sequence of N fibers each followed by an amplifier (see Figure 3.1). The amplifiers

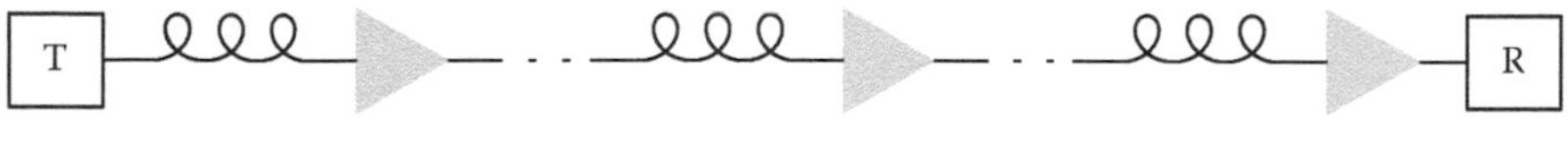

FIGURE 3.1 The schematic representation of a fiber-optical communication channel.

have to be introduced in order to compensate for the power loss in the fiber. An inevitable consequence of such design, however, is the generation of the noise in the system, coming from the spontaneous emission in the optical amplifiers. For simplicity, we will assume that all the fibers and the amplifiers of the link are identical.

The time evolution of the electric field in the fiber $E(z, t)$, where z is the distance along the fiber, can be accurately described in the "envelope approximation"[2], when

$$E(z, t) = A(z, t)\exp(i(\beta_0 z - \omega_0 t) + \text{c.c.}$$ (3.3)

where the function A represents the slowly (compared to the light frequency) varying amplitude of the electric field in the fiber. The evolution of $A(z, t)$ is described by the equation [2]

$$\frac{\partial A}{\partial z} + \beta_1 \frac{\partial A}{\partial t} + \frac{i}{2}\beta_2 \frac{\partial^2 A}{\partial t^2} + \frac{\alpha}{2} A = i\gamma \mid A \mid^2 A$$ (3.4)

Here the coefficients β describe the frequency dependence of the wavenumber

$$\beta(\omega) = \beta_0 + \beta_1\omega + \frac{\beta_2}{2}\omega^2 + O[\omega^3]$$ (3.5)

where ω is measured from the center of the band ω_0.

Equation 3.4 neglects effects such as stimulated Raman scattering and stimulated Brillouin scattering [2], compared to the Kerr nonlinearity of the refraction index of the fiber, represented by the term $\gamma \mid A \mid^2 A$. Note however, that these higher-order nonlinear effects can also be treated within the framework described in the present section.

The optical amplifiers incorporated into the communication system (see Figure 3.1) compensate for the power losses in the fiber, but due to spontaneous emission each of them will inevitably introduce noise

$$n(t) = \exp(-i\omega_0 t)\int d\omega \, n_\omega \exp(-i\omega t)$$ (3.6)

into the channel, with the average power [3]

$$I_1 = aG\hbar\omega_0 W$$ (3.7)

where G is the amplifier gain, $\hbar$ is the Plank's constant, W is the bandwidth, and a is a numerical constant (which is generally close to 2).

Generally, even in a single optical amplifier, the noise distribution at any given frequency $\omega_0 + \omega$ within the channel bandwidth $n(\omega)$ is close to a Gaussian:

$$p_n[n(\omega)] \sim \exp\left[-\frac{|n(\omega)|^2}{P_N^\omega} \right] \qquad (3.8)$$

This is even more so in a system with many independent amplifiers, due to the Central Limit Theorem.

If the envelope function just before the amplifier is $A(t) \equiv A_\omega^0 \exp(i\omega t)$, then immediately after the amplifier

$$A_\omega = \exp\left(\frac{\alpha}{2} d \right) A_\omega^0 + n_\omega \qquad (3.9)$$

where d is the span of a single fiber.

Equations 3.4 and Equation 3.9 define the evolution of the electric field envelope over one "fiber-amplifier" link of the communication system. The total "input-output" relation for the fiber-optical communication channel can be obtained by solving the corresponding equations for all N iterations of the single fiber-amplifier unit.

3.2.2 INFORMATION-THEORETICAL DESCRIPTION

According to the basic result of Shannon's information theory [1], the maximum amount of the information that can be transmitted through the communication system per unit time (i.e., the information capacity), is given by the maximum value of the mutual information per second, calculated for the input distribution $p(x(t))$ corresponding to the modulation format used in the encoder:

$$R = H[y(t)] - \langle H[y(\omega) \mid x(\omega)] \rangle_{p_x} \qquad (3.10)$$

where $x(t)$ and $y(t)$ are, respectively, the input and output signals of the communication channel. For a simple single-user, single fiber span, system $x(t) \equiv A(0,t)$ and $y(t) = A(L,t)$, where L is the length of the fiber.

Here, the information entropy $H[y(\omega)]$ is the measure of the information received at the output of the communication channel. It is not, however, equal to the information capacity, as due to the presence of the noise in the channel for any output signal there is some uncertainty of what was originally sent. The conditional entropy $H[y(t) \mid x(\omega)]$ at the output *for a given* input $x(\omega)$ represents this uncertainty.

The information entropies $H[y(t)]$ and $H[y(t) \mid x(t)]$ are defined in terms of the corresponding distributions $p(y)$ and $p(y \mid x)$ via the standard relation

$$H \equiv -\int Dy(t)\, p[y(t)] \log(p[y(t)]) \qquad (3.11)$$

where $p \equiv p_y(y)$ for the entropy $H[y(t)]$, and $p \equiv p(y|x)$ for the entropy $H[y(t)|x(t)]$, and the functional integral is defined in the standard way

$$\int D\xi(t) \equiv \lim_{M \to \infty} c_M \left[\prod_{m=1}^{M} \int d\xi(t_m) \right]$$ (3.12)

where is a normalization constant.

Equivalently, the information entropy can be also defined in the frequency domain

$$H \equiv -\int Dy(\omega) p[y(\omega)] \log(p[y(\omega)])$$ (3.13)

where

$$\int D\xi(\omega) \equiv \lim_{M \to \infty} c_M \left[\prod_{m=1}^{M} \int d\xi(\omega_m) \right]$$ (3.14)

If the communication system uses the standard OOK approach, the output signal can be expressed as

$$y(t) = su(t) + n(t)$$ (3.15)

where $s \in \{0,1\}$ is the transmitted (binary) information, $u(t)$ is a known functional form corresponding to the (possibly distorted due to fiber dispersion and nonlinearity) "pulse" and $n(t)$ is the noise (due to the combined effect of spontaneous emission in the amplifiers, detector dark current, etc., and the crosstalk between different information channels).

While for a nonlinear fiber-optical system the received "pulse" profile $u(t)$ may be strongly distorted due to nonlinear and dispersive effects, it can be calculated (or determined experimentally) in advance and on its own does not lead to system performance deterioration.

In contrast, it is the statistical properties of the effective "noise" $n(t)$ that determine the information transmission (and loss). The optical nonlinearity, mixing the signals of different communication channels (either time, wavelength, or code-divided), leads to the effective increase of the noise power, thus suppressing the information capacity.

Assuming detection at the peak power, Equation 3.15 can be reduced to the discrete channel

$$y = su_{\text{peak}} + n$$ (3.16)

If the distribution function $p(n)$ of the noise n is Gaussian

$$p(n) = \frac{1}{\sqrt{2\pi P_N}} \exp\left(-\frac{n^2}{2P_N}\right) \tag{3.17}$$

and the noise is *uncorrelated* with the signal

$$\langle n(t)u(t)\rangle = 0 \tag{3.18}$$

the calculation of the information capacity is straightforward and yields

$$R = \frac{1}{T} - \frac{1}{T} F_1\left(\frac{P_{peak}}{2P_N}\right) \tag{3.19}$$

where $1/T$ is the bit rate, $P_{peak} \equiv u^2_{peak}$ is the peak signal power, and

$$F_1(x) = \int_{-\infty}^{+\infty} d\xi \frac{\exp(-\xi^2)}{\sqrt{\pi}} \log_2\left[1 + \exp\left(-x + \sqrt{x}\xi\right)\right] \tag{3.20}$$

The universal function F_1, shown in Figure 3.2, is positive definite and is generally smaller than unity. In the limit $x \gg 1$ corresponding to large signal-to-noise ratio, it is exponentially small and, as follows from Equation 3.19, the information capacity

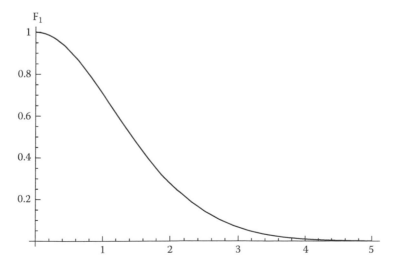

FIGURE 3.2 The universal function F_1 of Equation 3.20.

$R \to 1/T$. In the opposite limit $x \ll 1$ of vanishing signal power, F_1 asymptotically approached unity, resulting is the complete loss of information capacity: $R \to 0$.

As we shall see in the next sections, the Gaussian approximation for the noise distribution is generally well justified. While the noise due to spontaneous emission in the amplifiers is indeed Gaussian (see Equation 3.8 and its discussion), the noise due to the channel crosstalk approaches Gaussian in the limit of many channels due to the Central Limit Theorem.

Less justified is the assumption of uncorrelated noise. In fact, the nonlinearity will inevitably mix the transmitted pulse with the noise that originates from the spontaneous emission, and with the signals from the other channels, thus effectively increasing the noise power for $s = 1$ as compared to $s = 0$. This effect, leading to correlations between the signal and the noise, may or may not be comparable to the other contributions to the noise, depending to the numbers of multiplexed channels, spontaneous emission, etc. When the correlations are important, Equation 3.21 should be replaced by

$$R = \frac{1}{T} - \frac{1}{2T}\left[F_2\left(\eta, \frac{P_{peak}}{2P_N}\right) + F_2\left(\frac{1}{\eta}, \frac{P_{peak}}{2P_N}\right)\right] \tag{3.21}$$

where $\eta \equiv P_N(s = 1)/P_N(s = 0)$ is the ratio of the average noise powers for the presence ($s = 1$) and absence ($s = 0$) of the OOK pulse, and

$$F_2(\eta, x) = \int_{-\infty}^{+\infty} d\xi \frac{\exp(-\xi^2)}{\sqrt{\pi}} \log_2\left[1 + \sqrt{\eta}\exp\left((1-\eta)\xi^2\right.\right.$$
$$\left.\left. + \sqrt{2\eta(1+\eta)x}\xi - (1+\eta)x\right)\right] \tag{3.22}$$

Note that in the limit of vanishing signal-noise correlations ($\eta \to 1$) we obtain $F_2(\eta \to 1, x) \to F_1(x)$.

As the transmitter sends the information at the rate of 1 bit per the time interval T, the difference between $1/T$ and the actual information capacity R of the communication channel, represents the error rate of the communication system. The corresponding error probability is then equal to

$$P_{err} = \frac{1}{2}(1 - TR) \tag{3.23}$$

$$= \begin{cases} \dfrac{1}{2}F_1\left(\dfrac{P_{peak}}{2P_N}\right), & \text{uncorrelated noise} \\[4mm] \dfrac{1}{4}\left(F_2\left(\eta, \dfrac{P_{peak}}{2P_N}\right) + F_2\left(\dfrac{1}{\eta}, \dfrac{P_{peak}}{2P_N}\right)\right), & \text{correlated noise} \end{cases} \tag{3.24}$$

where the factor of 1/2 in Equation 3.23 accounts for the fact that no transmitted information corresponds to 50% error probability.

While the Equation 3.24 yields closed-form expression for the information capacity, they may seem relatively complicated for practical use (especially when a quick estimate is desired). Furthermore, in practice one is generally interested in knowing the information capacity of a (more or less) working system, when the signal power dominates that of the noise: $P_{peak} > P_N$. In this limit, Equation 3.24 for e.g., uncorrelated noise can be reduced to

$$P_{err} \simeq \frac{3}{2} \text{Erfc}\left(\frac{1}{2}\sqrt{\frac{P_{peak}}{2P_N}} \right) \tag{3.25}$$

$$\simeq 3\sqrt{\frac{2}{\pi}\frac{P_N}{P_{peak}}} \exp\left(-\frac{P_{peak}}{8P_N} \right) \tag{3.26}$$

In Figure 3.3 we compare Equation 3.24 with its "low noise approximations," Equation 3.25 and Equation 3.26. As clearly seen from this comparison, both Equation 3.25 and Equation 3.26 show excellent agreement with the exact result for the signal-to-noise ratio above 10 dB.

The fundamental limit to the information capacity of a given communication channel is defined as the maximum over all input distributions (and thus over all possible information coding algorithms)

$$C = \max_{p_x} R \equiv \max_{p_x}\{H[y] - \langle H[y\,|\,x]\rangle_{p_x}\} \tag{3.27}$$

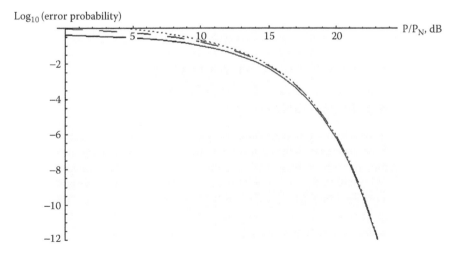

FIGURE 3.3 The comparison of the exact expression for the error probability, Equation 3.24, for uncorrelated noise (solid line), with the low noise approximations of Equation 3.25 (dashed line) and Equation 3.26 (dotted line).

For any communication link, the signal power is limited by the system hardware. Therefore, the maximum of the mutual information in (2) should be found under the constraint of the fixed total power P_0 at the input:

$$P_0 = \int Dx(\omega) \, |x(\omega)|^2 \, p_x[x(\omega)] \qquad (3.28)$$

If the propagation in the communication channel is described by a linear equation, then the input-output relation for the system is given by

$$y(\omega) = K(\omega)x(\omega) + n(\omega) \qquad (3.29)$$

where $n(\omega)$ is the noise in the channel. In this approximation, the problem of finding the maximum of the mutual information can be solved exactly, with the corresponding "optimal" input distribution p_x being Gaussian [1]. If the amplifiers compensate exactly for the power losses in the fibers, and no nonlinear effects are present, the fundamental limit on the information capacity is given by the Shannon formula, Equation 3.2.

As follows from Equation 3.2, the better bit rates (using e.g., multilevel codes) can be obtained for the higher signal-to-noise ratio P_0/P_N. With this in mind, fiber-optics communication systems are designed to operate with the pulses of high power. As a result, the fiber-optics links operate in a regime in which, due to the Kerr nonlinearity, the refraction index of the fiber strongly depends on the local electric field intensity. Therefore, the modern high-capacity, fiber-optical communication system becomes, in fact, essentially a nonlinear communication channel, and cannot be adequately described within the framework of the standard *linear* theory. The next section describes the corrections to the fundamental limit on the information capacity, originating from the nonlinearity of the fiber.

3.3 THE FUNDAMENTAL LIMIT TO THE INFORMATION CAPACITY OF A NONLINEAR FIBER-OPTICAL SYSTEM

Equation 3.4, which described the pulse propagation in optical fiber, is, in fact, the well-studied [4] nonlinear Schroedinger equation, with the time and distance variables interchanged. While some partial solutions of this equation are known, corresponding to solitons [4,5], in general (e.g., for a nonzero loss coefficient α) this equation is nonintegrable. More importantly, in order to find the fundamental limit C to the information capacity of a given channel, one has to calculate the maximum of the information capacity over *all* possible distributions of the input signal. In our case, this implies solving a set of N essentially nonlinear equations [4] for *arbitrary* initial conditions (where N is the number of fiber spans in the system). Even knowing some partial solutions, doing such calculation exactly for an essentially nonlinear system is not possible in a closed form.

However, the problem has a natural perturbation parameter, namely γ. In fact, the fiber Equation 3.4 is already an approximation, derived in the limit, when the change in the effective refraction index due to pulse propagation, described by the nonlinear term $i\gamma \mid A \mid^2 A$, is *small* compared to the "unperturbed" value of the index of refraction n_0. It is therefore possible to develop a perturbative technique where the solution of the nonlinear evolution equation is represented as a power series in γ, and derive the corrections to the maximum information capacity due to fiber nonlinearity.

3.3.1 THE PERTURBATIVE CALCULATION OF THE INFORMATION CAPACITY

The technique that we use in the present section involves a perturbative computation of the relevant mutual information and subsequent optimization [10]. The first step in this calculation is to find the "input-output" relation for the communication channel.

If one is able to calculate the "output" signal $y(\omega)$ in terms of the "input" $x(\omega)$ and the noise contributions of each of the amplifiers $n_\omega^{\{\alpha\}}, \alpha = 1,\ldots,N,$

$$y(\omega) = \Phi\left(x(\omega); n_\omega^{\{1\}},\ldots,n_\omega^{\{N\}}\right) \tag{3.30}$$

then the conditional distribution $p(y \mid x)$ can be simply calculated as follows:

$$p(y \mid x) = \left\{\prod_{\alpha=1}^{N-1}\int Dn_\omega^{\{\alpha\}} p_n\left[n_\omega^{\{\alpha\}}\right]\right\} p_n\left[y(\omega) - \Phi\left(x(\omega); n_\omega^{\{1\}},\ldots,n_\omega^{\{N\}}\right)\right] \tag{3.31}$$

where p_n is the distribution function of the noise, produced by a single amplifier. The output distribution $p_y(y)$ can then be directly related to the input distribution $p_x(x)$ via the standard relation

$$p_y(y) = \int Dx(\omega)p[y(\omega) \mid x(\omega)] \, p_x[x(\omega)] \tag{3.32}$$

Using Equation 3.31 and Equation 3.32, one is able to express the mutual information in terms of a single distribution p_x. The calculation of the channel capacity then reduces to a standard problem of finding the maximum of a (nontrivial) functional.

Solving Equation 3.4 separately for each power of γ, and using Equation 3.9, for the input-output relation of a single fiber-amplifier unit $\Phi_\omega^{(n)}$, defined as

$$A_\omega^{(n)} = \Phi_\omega^{(n)}\left(A_\omega^{(n-1)}\right), \tag{3.33}$$

we obtain:

$$\Phi_\omega^{(n)}\left(A_\omega^{(n-1)}\right) = n_\omega + \left[A_\omega^{(n-1)} + \sum_{\ell=1}^{\infty} \gamma^\ell F_\omega^{(\ell)}\left(A_\omega^{(n-1)}\right)\right] \exp(-i\kappa_\omega d) \qquad (3.34)$$

where d is the length of a single fiber, and

$$\kappa_\omega = \beta_1\omega - \frac{1}{2}\beta_2\omega^2 \qquad (3.35)$$

The procedure for the calculation of the functions $F_\omega^{(\ell)}$, described in detail in the Appendix A of Ref. [10], can be carried to an arbitrary order ℓ.

The further calculation then involves the following steps:

- Iterating Equation 3.34 N times, to obtain the "input-output" relation for the whole communication system $\Phi_\omega[x(\omega); n_\omega^{(1)},\ldots,n_\omega^{(N)}]$
- Substituting the result into Equation 3.31 and Equation 3.32 to obtain the conditional distribution $p(x \mid y)$ and the output distribution $p_y(y)$ in terms of the input distribution $p_x(x)$ as expansions in powers of γ
- Calculating the entropies $H[y(\omega)]$ and $H[x(\omega) \mid y(\omega)]$, and the information capacity R
- Finding the maximum of R over all possible input distributions $p_x(x)$

Following these steps, the calculation of the maximum of the information capacity becomes a straightforward (albeit quite tedious) procedure [10]. We obtain:

$$C = W \log\left(1 + \frac{P_0}{P_N}\right) - \Delta C_1 - \Delta C_2 + \mathcal{O}(\gamma^4) \qquad (3.36)$$

where in the limit of large signal-to-noise ratio $P_0 \gg P_N$

$$\Delta C_1 = N^2 W \left(\frac{\gamma P_0}{\alpha}\right)^2 Q_1\left(\alpha d, \frac{\beta_2^2 W^4}{\alpha^2}\right) \qquad (3.37)$$

$$\Delta C_2 = \frac{4}{3}(N^2 - 1)W\left(\frac{\gamma P_0}{\alpha}\right)^2 Q_2\left(\alpha d, \frac{\beta_2^2 W^4}{\alpha^2}\right) \qquad (3.38)$$

and the functions Q_1 and Q_2 are defined as follows:

$$Q_1(u,z) = \int_{-1/2}^{1/2} dx_1 \int_{-1/2}^{1/2} dx_2 \int_{1/2}^{\bar{x}} dx\, f(u,z;x_1,x_2,x) \qquad (3.39)$$

$$Q_2(u,z) = \int_{-1/2}^{1/2} dx_1 \int_{-1/2}^{1/2} dx_2 \int_{-1/2}^{1/2} dx\, f(u,z;x_1,x_2,x) \qquad (3.40)$$

where $\bar{x} \equiv \max [1/2, 1/2 + x_1 + x_2]$, and

$$f(u, z; x_1, x_2, x) \equiv \frac{|1 - \exp(-u - iv^2)|^2}{1 + v^2} \tag{3.41}$$

where

$$v \equiv z(x - x_1)(x - x_2) \tag{3.42}$$

Equation 3.36 yields the result for the fiber optics channel capacity in the second order in the nonlinearity γ. In the next subsection we will discuss the physical origins of the corrections ΔC_1 and ΔC_2.

3.3.2 DISCUSSION

In the spirit of the Shannon formula, the decrease of the capacity of a communication channel with a fixed bandwidth can be attributed to (1) the effective suppression of signal power, and (2) the enhancement of the noise. The corrections to the channel capacity, derived in the previous section, can be interpreted as resulting precisely from these two effects.

The four-wave scattering [2] induced by the fiber nonlinearity inevitably leads to the processes that generate photons with the frequencies outside the channel bandwidth. Such photons are not recorded by the "receiver," and are therefore lost for the purpose of the information transmission. This corresponds to an effective power loss from the bandwidth, and should therefore lead to a decrease of the channel capacity. Since for small nonlinearity this power loss $\Delta P \sim \gamma^2$, the dimension analysis implies

$$\Delta P = \frac{\gamma^2 P_0^2}{\alpha^2} F\left(\frac{\beta W^2}{\alpha}\right) \tag{3.43}$$

Such scattering processes are suppressed, when the scattering leads to a substantial change of the total momentum $\delta\kappa[\omega_1, \omega_2 \to \omega_3, \omega]$, so that the corresponding scattering rate

$$S[\omega_1, \omega_2 \to \omega_3, \omega] \sim \frac{\delta(\omega_1 + \omega_2 - \omega_3 - \omega)}{1 + (\delta\kappa / \kappa_0)^2} \tag{3.44}$$

In the spirit of the uncertainty relation, $\kappa_0 \sim 1/L_{\text{eff}}$, where L_{eff} corresponds to the length of the concentration of the power of the signal in the fiber. For a small absorption coefficient $\alpha \ll 1/d$ the distance L_{eff} is of the order of the fiber length d, while in the opposite limit $\alpha \gg 1$ the effective length $L_{\text{eff}} \approx 1/\alpha$.

Using Equation 3.35, and the energy conservation $\omega_3 = \omega_1 + \omega_2 - \omega$, the momentum change $\delta\kappa[\omega_1, \omega_2 \to \omega_3, \omega]$ can be expressed as

$$\delta\kappa = \beta_2(\omega - \omega_1)(\omega - \omega_1) \tag{3.45}$$

Substituting Equation 3.45 into Equation 3.46, for the channel capacity loss due to the bandwidth power "leakage", in the limit $P_0 \gg P_N$, and $\alpha d \gg 1$, we obtain

$$\Delta C_P \sim W\frac{\Delta P}{P} \sim W\frac{\gamma^2 P_0^2}{\alpha^2}\int_W d\omega_1 \int_W d\omega_2 \int_W d\omega_3 \int_{\omega \notin W} d\omega\, S[\omega_1, \omega_2 \to \omega_3, \omega]$$

$$= W\frac{\gamma^2 P_0^2}{\alpha^2}\int_W d\omega_1 \int_W d\omega_2 \int_{\omega \in W} \frac{d\omega}{1 + (\beta_2/\alpha)^2(\omega - \omega_1)^2(\omega - \omega_2)^2} \tag{3.46}$$

which in the appropriate limit is consistent with ΔC_1.

In Figure 3.4 we plot the dependence of ΔC_1 on the dimensionless parameter $\beta_2 W^2/\alpha$. Since momentum change $\delta\kappa$ is proportional to β_2, the increase of the dispersion leads to a strong suppression of the power leakage from the bandwidth window, and of the corresponding correction to the channel capacity.

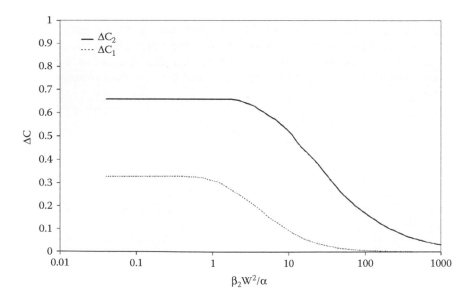

FIGURE 3.4 The corrections to the channel capacity, ΔC_1 and ΔC_2, in units of $WN^2\gamma^2 P_0^2\alpha^{-2}$ shown as functions of $|\beta_2|W^2/\alpha$ in the limit $\alpha d \gg 1$, $N \gg 1$. The correction ΔC_2 is represented by the solid line, while ΔC_1 corresponds to the dashed line. Note that ΔC_1, which describes the effect of the power leakage from the bandwidth, is more strongly affected by the dispersion.

In a communication system with many "fiber-amplifier" units, the fiber nonlinearity leads not only to the mixing of the signals at different frequencies, but also to the mixing of the signal with the noise [6]. Qualitatively, this would correspond to an effective enhancement of the noise power in the system, and therefore to a loss of the channel capacity. This effect is not present, when the system has only one "fiber-amplifier" link, which explains the appearance of the $(N-1)$ factor in ΔC_2.

The effective noise enhancement is caused by the scattering processes, which involve a "signal photon" and a photon, produced due to spontaneous emission in one of the amplifiers. The total power of this extra noise can be expressed as

$$\frac{\Delta P_N}{P_0} \sim \gamma^2 \frac{P_0 P_N}{\alpha^2} \int_W d\omega_1 \int_W d\omega_2 \int_W d\omega_3 \int_W d\omega \, S[\omega_1, \omega_2 \to \omega_3, \omega] \qquad (3.47)$$

The corresponding correction to the capacity

$$\Delta C_N \sim W \frac{\Delta P_N}{P_N} \sim W \frac{\gamma^2 P_0^2}{\alpha^2} \int_W d\omega_1 \int_W d\omega_2 \int_W d\omega$$

$$\times \frac{1}{1+(\beta_2/\alpha)^2(\omega-\omega_1)^2(\omega-\omega_2)^2} \qquad (3.48)$$

where we assumed $\alpha d \gg 1$. In this limit, Equation 3.48 is up to a constant factor identical to ΔC_2.

The dependence of ΔC_2 on $\beta_2 W^2/\alpha$ is shown in Figure 3.4. Note that ΔC_2 also decreases with the increase of the dispersion, but more slowly than ΔC_1. Since the scattering processes that contribute to ΔC_1, need to "move" one of the frequencies out of the bandwidth window, they generally involve a substantial change of the total momentum, and are therefore more strongly affected by the dispersion.

The two physical effects described above determine the fundamental limit to the bit rate for a fiber optics communication system. As follows from our analysis (see Figure 3.4), the relative contributions of ΔC_1 and ΔC_2, often referred to as the "four-wave mixing," can be suppressed by choosing a fiber with a large dispersion, or when using a larger bandwidth.

The physical processes described above can substantially influence the performance of current fiber-optical communication systems. In Figure 3.5, for a link with $N = 80$ fiber spans each with the effective length $L_{eff} \equiv \alpha^{-1} = 100$ km, effective bandwidth $W = 50GHz$ and average noise power $P_N \equiv \int d\omega P_\omega^N = 0.1$ mW, we compare the fundamental limit to the information calculated in the present section, with the standard Shannon result, and with the recently derived lower bound [7] based on Gaussian approximation to the joint input-output correlation function (see Appendix B of Ref. [10]). As seen in Figure 3.5, the information capacity significantly deteriorates at the higher signal powers $P_0 > 10$ mW.

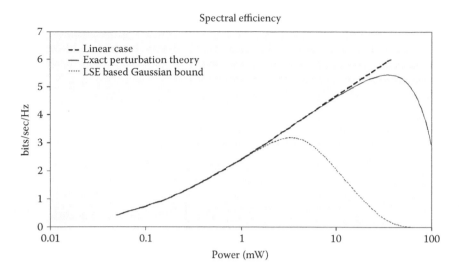

FIGURE 3.5 The comparison of the channel capacity calculated using the Shannon formula (dashed line), Gaussian lower bound of Ref. [7] (dotted line), and the exact result of the present work (solid line). Note that although the Gaussian estimate captures the qualitative features of the exact result, it severely underestimates its value. The parameters of the fiber system used for the comparison are given in the text.

Note that the cross-phase modulation [2], which severely limits the performance of advanced wavelength-division multiplexing systems [8] (WDM), does not affect the fundamental limit to the information capacity. The reason for this seemingly contradictory behavior is that in a WDM system the "receiver," tuned to a particular WDM channel, has no information on the signals at the other channels. Therefore, even in the absence of the "genuine" noise, the nonlinear interaction between different channels, leading to a change in the signal in any given channel, will be an effective noise source, thus limiting the communication rate. This limit however is not fundamental, and can be overcome—for a system with "lumped" amplification by using a detector that probes all information channels, or—in case of distributed (such as Raman) amplification—by phase conjugation in the midpoint of the fiber link.

The fundamental limits on the channel capacity established in the present section have significant implications for the development of practical communication systems. First, this calculation rigorously proves that (Kerr) nonlinearity always reduces the channel capacity. One of the implications of this general statement is that the performance of any soliton-based fiber optics communication system [2] will be inferior to that of its linear counterpart. Second, this calculation demonstrates how the performance impairment due to nonlinear effects can be reduced (e.g., by increasing dispersion strength when dispersion-compensation is available). Finally, it identifies the actual physical processes responsible for the performance degradation due to nonlinearity in a single user system. This will help the designers to distinguish the processes that fundamentally degrade the system performance from those that may be circumvented. In particular, as we shall see in the next section, it is the strong sensitivity of the coherent CDMA to cross-phase modulation that makes it inferior to incoherent CDMA for long-range fiber-optical networks.

3.4 THE INFORMATION CAPACITY OF FIBER-OPTICAL CDMA SYSTEMS

In the present section, we consider the effect of fiber nonlinearity on the information capacity of fiber-optical CDMA systems, both coherent and incoherent. First, we will consider the limit when the dispersion broadening *within a single fiber span* is small compared to the bit interval T. This approximation is justified if either (1) the dispersion coefficient β_2 is small (corresponding to dispersion-shifted fibers), or (2) dispersion compensation is employed within each fiber-amplifier unit of the system. While the first case is rather unusual in modern fiber-optical communication systems, the dispersion compensation (using e.g., dispersion-compensating fiber), often referred to as the "dispersion management," is now quite common. Finally, we will discuss the effect of higher dispersion, and conclude the section with the comparison of the coherent and incoherent CDMA.

3.4.1 COHERENT CDMA

In the coherent approach to optical CDMA, the information is first encoded in a pulse train using standard OOK. Then different frequency subbands of the signal are given different (but unique to a particular user) phase shifts, leading to substantial spreading of the pulses and the resulting decrease of the peak power. The detector with the complementary filter compensates the phase shifts *for the given channel*, while the signals of other channels remain broadened and only produce a noisy background. To increase the security of the multiuser communication system, the phase shifts can be chosen at random.

If the function $\phi_\mu(\omega)$ represents the frequency-dependent phase shift in the channel μ, in the limit of small dispersion for the "output" signal (after the complementary filter) $y_\mu(t)$ we obtain:

$$y_\mu(t) = \frac{1}{(2\pi)^2} \int d\omega' \int d\omega'' \int dt_1 \int dt_2 \, x(t_1) \exp(i\phi_{\omega'} - \phi_{\omega''})$$

$$\times \exp(i\omega'(t_2 - t) + i\omega(t' - t'') + i\theta(t_2)) \tag{3.49}$$

where $x(t)$ is the original OOK input signal (*before* the phase shifts are introduced at the encoder), and $\theta(t)$ is the nonlinear phase shift:

$$\theta(t) = \gamma N d_{\text{eff}} \, |A(t)|^2 \tag{3.50}$$

Here, the effective length of a *single* fiber span

$$d_{\text{eff}} = \frac{1 - \exp(-\alpha d)}{\alpha} \tag{3.51}$$

and $E(t) = A(t)\exp(i\omega_0 t)$ is the total optical field. The function $A(t)$ can be decomposed into the sum of the signal at the *given channel*

$$A_\mu(t) = \int \frac{1}{(2\pi)} \int d\omega \int dt' x(t') \exp(i\phi_\omega - i\omega t') \tag{3.52}$$

and the total contribution of the other channels and additive noise (due to spontaneous emission etc.):

$$n_{\text{eff}}(t) = \sum_{\nu \neq \mu} A_\nu(t) + n(t) \tag{3.53}$$

When many uncorrelated CDMA channels are active, the Central Limit Theorem implies that the statistical properties of n_{eff} can be adequately described by Gaussian distribution with delta-correlations (as the correlation time scales as $1/M$, where M is the number of independent channels). Therefore,

$$\langle \exp(i\theta(t_2) - i\theta(t_1)) \rangle = \frac{1}{1 + \gamma^2 \langle |n_{\text{eff}}|^2 \rangle d_{\text{eff}}^2}$$

$$\times \exp\left[-\frac{2\gamma^2 \langle |n_{\text{eff}}|^2 \rangle d_{\text{eff}}^2}{1 + \gamma^2 \langle |n_{\text{eff}}|^2 \rangle d_{\text{eff}}^2} \frac{1}{M - 1} \right] \tag{3.54}$$

As follows from Equation 3.49 and Equation 3.54, the output signal power

$$\langle |y(t)|^2 \rangle = |x(t)|^2 \exp\left[-\frac{2\gamma^2 \langle |n_{\text{eff}}|^2 \rangle d_{\text{eff}}^2}{1 + \gamma^2 \langle |n_{\text{eff}}|^2 \rangle d_{\text{eff}}^2} \frac{1}{M - 1} \right] \tag{3.55}$$

At the same time, the effective noise power can be evaluated as

$$\langle |n_{\text{eff}}(t)|^2 \rangle = P_N + P_{\text{peak}}(M - 1)\frac{\tau}{2T} \tag{3.56}$$

where $1/T$ is the bit rate in a *single* channel, and τ is the OOK pulse width (*before the phase shifts at the CDMA encoder*). Here, the first term corresponds to the "additive" noise due to spontaneous emission in the optical amplifiers, etc., while the second contribution describes the effect of the crosstalk between different CDMA channels. As the application of the phase shifts at the encoder spreads the signal over the time interval T, the corresponding power drops by a factor of $\sim T/\tau$. Depending on the definition of the "pulse width" τ (e.g., at the half-maximum, or at the level of $1/e$, etc.) and on the pulse shape (Gaussian, $1/\cosh$, etc.), the exact relation for the average power in a given channel may also involve a coefficient of order unity. To avoid this unnecessary complication, we define τ so that this coefficient is exactly equal to one. Finally, the factor $(M-1)$ represents the fact that out of all of M active CDMA channels, only $(M-1)$ contributes to the effective noise, and the factor $1/2$ accounts for 50% probability of the bit " 0 " corresponding to no signal in the given channel and thus no contribution to the effective noise.

For the effective signal-to-noise ratio in coherent CDMA we therefore obtain:

$$
\mathrm{SNR} = \frac{P_{\mathrm{peak}}}{P_N + P_{\mathrm{peak}}(M-1)\frac{\tau}{2T}}
$$

$$
\times \frac{\exp\left[-\frac{\gamma^2 d_{\mathrm{eff}}^2 P_{\mathrm{peak}}^2 N^2 (M-1)\tau^2/(2T^2)}{1+\gamma^2 d_{\mathrm{eff}}^2 P_{\mathrm{peak}}^2 N^2 (M-1)^2\tau^2/(4T^2)}\right]}{1+\gamma^2 d_{\mathrm{eff}}^2 N^2 P_{\mathrm{peak}}^2 (M-1)^2 \tau^2/(4T^2)} \tag{3.57}
$$

where P_{peak} is the peak power of the "unmodulated" OOK signal.

Note that the argument in the exponential in Equation 3.57 is always less than $2/(M-1)$ and in the practical limit $M \gg 1$ is small. The signal-to-noise ratio can be therefore approximated by

$$
\mathrm{SNR} = \frac{P_{\mathrm{peak}}}{P_N + P_{\mathrm{peak}}(M-1)\frac{\tau}{2T}}
$$

$$
\times \frac{1}{1+\gamma^2 d_{\mathrm{eff}}^2 N^2 P_{\mathrm{peak}}^2 (M-1)^2 \tau^2/(4T^2)} \tag{3.58}
$$

As the correlations between the noise and the signal, which arise due to the contribution of a *single* channel to the overall phase $\theta(t)$, for $M \gg 1$ can be neglected, for the information capacity we obtain:

$$
R = \frac{1}{T}\left\{1 - \log_2\left[1 + \exp\left(-\frac{\mathrm{SNR}_0}{1+\gamma^2 d_{\mathrm{eff}}^2 N^2 P_{\mathrm{peak}}^2 (M-1)^2 \tau^2/(4T^2)}\right)\right]\right\} \tag{3.59}
$$

where SNR_0 is the effective signal-to-noise ratio in the absence of nonlinear effects:

$$
\mathrm{SNR}_0 = \frac{P_{\mathrm{peak}}}{P_N + P_{\mathrm{peak}}(M-1)\frac{\tau}{2T}} \tag{3.60}
$$

When no error-correction is used, the logarithmic term in Equation 3.59 has the meaning of the lower bound on the average bit-error-rate probability (BER):

$$
\mathrm{BER} = \log_2 \left[1 + \exp\left(-\frac{\mathrm{SNR}_0}{1 + \gamma^2 d_{\mathrm{eff}}^2 N^2 P_{\mathrm{peak}}^2 (M-1)^2 \tau^2 / (4T^2)} \right) \right]
$$

$$
\simeq \frac{1}{\ln 2} \exp\left(-\frac{\mathrm{SNR}_0}{1 + \gamma^2 d_{\mathrm{eff}}^2 N^2 P_{\mathrm{peak}}^2 (M-1)^2 \tau^2 / (4T^2)} \right) \qquad (3.61)
$$

As follows from Equation 3.60 and Equation 3.61, when the number of simultaneous users

$$
M > M_c \equiv \frac{2T/\tau}{\gamma L P_{\mathrm{peak}}} \qquad (3.62)
$$

the nonlinear effects lead to quick deterioration of the signal-to-noise ratio and the resulting exponential increase in the average error probability. Equation 3.62 sets a limit for the number of simultaneous users M in terms of the bit rate $1/T$, used bandwidth ($\sim 1/\tau$), signal power P_{peak} and the communication distance $L = Nd_{\mathrm{eff}}$ before signal regeneration; e.g., for the typical values of $d_{\mathrm{eff}} = 100$ km, $\gamma = 0.002$ W^{-1} m^{-1}, $P_{\mathrm{peak}} = 20$ mW, $T/\tau = 10$, we find $M_c = 5$.

3.4.2 INCOHERENT CDMA

In the incoherent approach to the CDMA, the original OOK-modulated signal is divided into several parts and each part is delayed by the amount determined by the code used. At the end of the transmission line, a complementary set of time delays corresponding to the given CDMA channel is assigned, and the signal is detected with *intensity*-sensitive receiver.

The way the OOK signal is "divided" depends on the particular realization of the incoherent CDMA. In the earlier versions of this approach [11], the signal was literally separated into E_t equal parts by a simple $1 : E_t$ splitter. The relative simplicity of this approach however, is overshadowed by severe restrictions on the number of supported users (e.g., $K \sim \sqrt{T/\tau}$ for the prime codes used in Ref. [11]) and poor signal-to-noise performance (as the complementary detector only recovers $1/E_t$ fraction of the original signal power). A more recent approach, related to the "carrier-hopping prime codes" (see Chapter 2 for a detailed description), demonstrated a much better performance. In this method, the "division" is performed in the frequency domain—which allows independent control of the time delays and thus a complete power recovery at the detector. As demonstrated in Chapter 2, the carrier-hopping approach also allows to support a much larder number of simultaneous users and higher bit rates than earlier "single-wavelength" implementations of the incoherent OCDMA.

The carrier-hopping approach is remarkably similar to the standard methods of the coherent CDMA. In fact, the only fundamental difference is the incoherent detection, which is sensitive to the sum of the intensities, rather than the fields, of the "parts" of the original signal. This difference, however, will have dramatic effect on the system performance in the presence of the fiber nonlinearity.

In this section, we consider the carrier-hopping incoherent CDMA system with N_λ wavelength. That is, the original OOK signal is passed through a filter (e.g., prism- or grating-based) that separates N_λ components different by their central wavelength, followed by individual delay lines for each wavelength. The resulting signal is then "assembled" (using e.g., a simple $N_\lambda : 1$ coupler) and injected into the fiber—see Figure 3.7 (panel 3). The detector includes the "complementary" filter (which is physically similar to the encoder except the length of the delay line for the wavelength λ is $L - x_\lambda$ instead of x_λ, followed by a photodiode.

As the detector is only sensitive to the *intensity* of the individual "components," the difference in their phases accumulated over the propagation distance, which was so critical (and detrimental!) for the coherent CDMA, is now irrelevant. Already for this reason incoherent CDMA is more robust to the nonlinear effects.

However, the fiber nonlinearity, due to the effects of the four-wave mixing, leads to the spectral broadening in each of the wavelength channels. The resulting "spilling" of the power from one channel to another (which included the contributions from all CDMA channels that contribute to the optical field at a given time), yet again leads to the "crosstalk" between different CDMA channels and resulting performance degradation.

In the limit of small dispersion, the single channel spectral width $\Delta\omega$ defined as the RMS value $\sqrt{\langle (\omega - \omega_0)^2 \rangle}$ in a given wavelength channel (where ω_0 is its central frequency), is determined by

$$\Delta\omega^2 = \int dt_1 \int dt_2 u^*(t_1) u(t_2) \exp(i\gamma d_{\text{eff}} N) \qquad (3.63)$$

where $u(t)$ is the (electric field) profile of the individual pulse, and the $\theta(t)$ is non-linear phase shift defined by Equation 3.50.

The total field in the fiber $A(t)$ that determines $\theta(t)$, again can be decomposed into the contribution of the given channel A_μ and the total contribution of the other channels and additive noise: $n_{\text{eff}}(t) = \sum' A_\nu(t) + n(t)$. For carrier-hopping incoherent CDMA systems,

$$A_\mu(t) = s_\mu \sum_\lambda u(t - t_{\mu\lambda}) \exp(-i\omega_\lambda (t - t_{\mu\lambda})) \qquad (3.64)$$

where the "bit" $s \in \{0,1\}$, the carrier frequency ω_λ is measured with respect to ω_0, and the time delays $t_{\mu\lambda}$ are defined by the particular code.

Assuming that the system is asynchronous, the average of $\Delta\omega^2$ over the relative time delays of *different users* and the bit information transmitted by *other users*

$(\nu \neq \mu)$, can be calculated exactly. For the Gaussian pulses $u \propto \exp(-t^2/2\tau^2)$, it can be easily shown by a simple and straightforward calculation[*] that

$$\Delta\omega = \Delta\omega_0 \left[1 + \gamma^2 d_{\text{eff}}^2 N^2 \frac{P_{\text{peak}}^2}{N_\lambda^2} \left(\frac{4}{3\sqrt{3}} + \frac{\tau}{T} N_\lambda (M-1) F_\lambda \right) \right]^{1/2} \qquad (3.65)$$

where

$$F_\lambda = \frac{4\sqrt{2}+3}{2\sqrt{2}} \sqrt{\pi} + \frac{\pi}{12} \frac{\tau}{T} (M-1) \left(1 + \frac{1}{N_\lambda} \right) \left(1 + \frac{2}{N_\lambda} \right) (\Omega\tau)^2 \qquad (3.66)$$

and Ω is the frequency spacing between different carriers.

The first term in the round brackets in Equation 3.65 describes the spectral broadening in the case of a single "user" in a single wavelength channel, it was calculated in Ref. [12]. The last contribution in Equation 3.65, defined by the function F_λ, represents the effect of other users and other wavelength channels. For closely spaced wavelength channels $\Omega\tau \sim 1$ and, since in practical systems the number of simultaneous users M is generally smaller than the number of "time slots" T/τ, the first term in F_λ (see Equation 3.66) gives the dominant contribution, so that $F_\lambda \simeq 5.4$.

Assuming that the frequency filter function can be accurately approximated by Gaussian,[**] for the effective signal-to-noise ratio at the output we obtain:

$$\text{SNR} = \frac{N_\lambda P_{\text{peak}}}{\frac{\tau}{2T} N_\lambda M G_\lambda \left(\frac{\Omega^2}{2(\Delta\omega_0^2 + \Delta\omega^2)} \right) P_{\text{peak}} + \sqrt{\frac{\Delta\omega^2 + \Delta\omega_0^2}{2\Delta\omega\Delta\omega_0}} P_N} \qquad (3.67)$$

where the function

$$G_\lambda(Q) = \sum_{m=-N_\lambda}^{N_\lambda} \left(1 - \frac{|m|}{N_\lambda} \right) \exp(-Q^2 m^2) \qquad (3.68)$$

describes the crosstalk between different channels. For small interchannel coupling $(Q \gg 1)$ which is of practical use (as it describes the system that has not yet lots most of its power to noise), one can use the "near-neighbor approximation"

$$G_\lambda(Q) \simeq 1 + 2 \left(1 - \frac{1}{N_\lambda} \right) \exp(-Q^2) \qquad (3.69)$$

while in the opposite limit $(Q \ll 1)$ $G \rightarrow N_\lambda(1 - N_\lambda^2 Q^2/12)$.

---

[*] Which took me just several days to complete and easily fits into not more than ~20 pages of closely spaced notes.

[**] The use of other filter profiles, e.g., square functions, yields similar results.

Note that the same "power spill" effect leads to the relative enhancement of the additive noise (described by the square root in front of the additive noise power in Equation 3.67): while some of the signal power falls "between" the wavelengths and is therefore lost after the filtering, the uniform spectrum of the additive noise is not affected.

If none of the time delays $t_{\mu\lambda}$ for a given user are close to zero or identical within the single pulse width τ (which is satisfied for most codes), no signal pulse overlaps in time with its "family members" that belong to the same "symbol" of the same user. In the limit of many simultaneous users ($M \gg 1$) *or* many wavelength channels ($N_\lambda \gg 1$) the contribution of this single pulse to the overall phase in a *different frequency channel* can be neglected, which makes the crosstalk noise uncorrelated with the signal. In the (practical) limit when the effective noise is dominated by the multiuser interference ($P_{peak} N_\lambda M \gg P_N T / \tau$), we obtain

$$R = \frac{1}{T}\left\{1 - \log_2\left[1 + e^{-2T/M\tau G_\lambda\left(\frac{(\Omega/2\Delta\omega_0)^2}{1+\gamma^2 d_{eff}^2 N^2 P_{peak}^2\left(2/3\sqrt{3}+N_\lambda(M-1)F_\lambda\tau/T\right)}\right)}\right]\right\} \quad (3.70)$$

The corresponding bit-error-rate is then given by

$$\mathrm{BER} \simeq \frac{1}{\ln 2}\exp\left(-\frac{2T}{M\tau G_\lambda\left(\frac{(\Omega/2\Delta\omega_0)^2}{1+\gamma^2 d_{eff}^2 N^2 P_{peak}^2\left(2/3\sqrt{3}+N_\lambda(M-1)F_\lambda\tau/T\right)}\right)}\right) \quad (3.71)$$

As follows from Equation 3.71, the nonlinear effects become important when

$$M > M_c \equiv \frac{TN_\lambda}{\tau}\left(\frac{\Omega\tau}{\gamma L P_0}\right)^2 \quad (3.72)$$

where the total signal power at the detector $P_0 = N_\lambda P_{peak}$. At this point the nonlinear effects lead to quick deterioration of the signal-to-noise ratio and the resulting exponential increase in the average error probability. The Equation 3.62 sets a limit for the number of simultaneous users M in terms of the bit rate $1/T$, used bandwidth ($\sim 1/\tau$), the number of wavelengths channels used N_λ, total signal power P_0, the communication distance $L = N d_{eff}$ before signal regeneration, and the frequency spacing $\Omega\tau$. In particular, for $L = 100$ km, $\gamma = 0.002 W^{-1}m^{-1}$, $P_0 = 20$ mW, $T/\tau = 10$, $N_\lambda = 10$, and $\Omega\tau = 2$, we find $M_c = 25$. This should be compared with the corresponding limit for coherent CDMA, which—for precisely the same parameters—was found to be $M < 5$.

3.4.3 THE EFFECT OF DISPERSION

As already discussed in Section 3.3.2, the fiber dispersion leads to the suppression of four-wave mixing processes. At zero dispersion, the (photon) energy conservation

$$\hbar\omega_1 + \hbar\omega_2 \rightarrow \hbar\omega_3 + \hbar\omega_4$$

immediately implies the phase-matching requirement

$$\Delta k \equiv (k_3 + k_4) - (k_1 + k_2) = \frac{n}{c}(\omega_3 + \omega_4 - \omega_1 - \omega_2) = 0$$

The presence of dispersion, leading to different values of the refraction index n at different frequencies, brakes the phase-matching requirement and thus causes the suppression of four-wave mixing processes. The monotonic decrease of the nonlinear corrections Q_1 and Q_2 to the fundamental limit of the information capacity C with dispersion in Figure 3.4 is precisely due to this effect.

Therefore, the pulse spectrum broadening caused by the four-wave scattering processes will be suppressed by fiber dispersion. As a result, the effect of fiber nonlinearity on the *incoherent* CDMA systems will weaken with dispersion, and Equation 3.71 derived in Section 3.4.2, could be considered as the upper bound to the corresponding bit-error-rate.

In contrast, the nonlinear processes that dominate the performance degradation of the *coherent* CDMA, are those of the *cross-phase modulation*, where the nonlinear scattering mostly leads to the change of the *phase* of the signals and the original frequencies ω_1 and ω_2, without a substantial change to the photon frequency (energy). As a result, for $\omega_3 \approx \omega_1$ and $\omega_4 \approx \omega_2$, the phase-matching requirement is not strongly violated even in the presence of dispersion. Therefore we don't expect that the fiber dispersion (provided it is properly compensated at the receiver) will strongly affect the performance of the coherent CDMA systems.

There is, however, yet another nonlinear effect present for high dispersion, when the pulses broaden to the point when the pulse width becomes of the order of the bit interval. Nonlinear scattering between the photons of the same frequencies but originating from different pulses leads to the creation of what is known as the "ghosts" (or the so-called shadows) [13]. After the dispersion compensation, these ghost/shadow pulses are observed in the time slots where "zeros" (i.e., no signals) were transmitted, and thus lead to the increase of the bit-error-rate.

3.4.4 PERFORMANCE COMPARISON AND DISCUSSION

While fiber nonlinearity leads to the performance degradation for both coherent and incoherent versions of the optical CDMA, quantitatively the effect is quite different. The advantage of the coherent approach is that during the propagation the signals are broadened, thus decreasing the average power and the power fluctuations compared to the original peak values. This explains the factor τ^2/T^2 in the denominator of Equation 3.61, and accounts for the robust performance of coherent fiber-optical CDMA systems for small number of users.

While a similar effect is also present in the incoherent CDMA—where the peak power is decreased by the factor equal to the number of wavelength/frequency channels N_λ, the number of time slots within the single bit interval T/τ is generally larger than N_λ and thus the effective "nonlinear" peak power is larger in the incoherent approach.

However, while the performance deterioration in coherent CDMA systems is primarily due to the cross-phase modulation (which brakes the phase relations between different "slices" of the signal and thus prevents its recovery), their incoherent counterparts are mostly affected by four-wave mixing (needed to broaden the frequency channels to the point where they overlap and "crosstalk"). If the incoherent frequency channels are strongly separated $(\Omega\tau \gg 1)$, the nonlinearity will have little or no effect of the system performance—see Equation 3.71 and Equation 3.68. Also, as we already discussed in Section 3.4.3), the fiber dispersion will stronger suppress the nonlinear effect in incoherent CDMA systems.

As a result, a properly designed incoherent CDMA system will show a better bit-error-rate performance than its coherent counterpart. This is illustrated in Figure 3.6, where we compare variation of the BER with the power-distance product $\gamma P_{peak} L_{eff}$, for 10-user coherent (blue curve) and incoherent (red line) CDMA systems, each with 41 time slots per bit period.

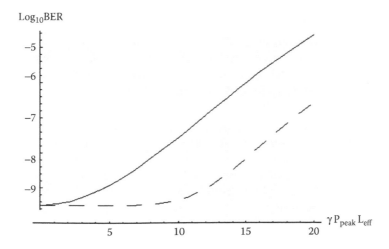

FIGURE 3.6 Bit-error-rate BER as a function of the peak power—distance product $\gamma P_{peak} L_{eff}$, for coherent (solid line) and incoherent (dashed line) CDMA systems. The system parameters are as follows: the number of simultaneous users $N_u = 10$, the number of "time slots" per bit period $T/\tau = 41$, the number of wavelength channels (for incoherent CDMA) $N_\lambda = 8$. The noise due to spontaneous emission etc. is assumed to be much smaller than the cross-talk between different users.

This comparison, however, could be misleading, as for the same transmission rate and error performance, the coherent CDMA system would use smaller aggregate bandwidth than its incoherent counterpart. In particular, the incoherent system in Figure 3.6 uses eight different wavelength channels, each with the bandwidth equal to that of the entire coherent system. One should therefore have in mind that a properly designed coherent system will have a better spectral efficiency (bit rate divided by bandwidth)—thus, the coherent CDMA is superior for applications that require high spectral efficiency and only moderate powers and propagation distances.

3.5 ADVANCED CODING TECHNIQUES FOR PERFORMANCE IMPROVEMENT

While incoherent fiber-optical CDMA systems with well-separated frequency channels based on carrier-hopping codes, show acceptable bit-error-rate performance, their spectral efficiency leaves much to be desired. For example, the spectral efficiency for incoherent system used for the Figure 3.6, is only 3.2%.

To improve the performance of the incoherent CDMA systems, new coding techniques are desired. From the general principles, the desired new codes, in addition to higher spectral efficiency, should:

- Preserve the quasi-orthogonality of the codes assigned to different users to keep the error rate within the desired bounds,
- Keep the average and peak power levels to the minimum, to avoid the nonlinear impairments that strongly affect the system performance for many simultaneous users
- Limit the complexity of the encoder and receiver to the minimum, with most of the required operations preferably performed in the optical domain
- Add extra functionality to the system

The second in this list of requirements strongly indicates that such "optimal" codes should involve some kind of the pulse position modulation (PPM)[14] approach, which is known to be optimal if low *signal power per bit* is desired. A direct application of the PPM however will require some kind of synchronization and thus substantially complicate the system.

Instead, we propose to use the M-ary approach, where each user is assigned a set of M codes, each corresponding to a particular "digit." In the M-ary system with $M = 8$ each user would transmit one *byte* (instead of one bit) per period. This M-ary approach first proposed for OCDMA in Ref. [17], would increase the information content of a every "message" a factor by $\log_2 M$, with the corresponding improvement of the spectral efficiency

$$R = \frac{K \log_2 M}{TW} \tag{3.73}$$

where K is the number of simultaneous users, T is the signaling period ("symbol interval"), and W is the system bandwidth.

Note that in every signaling period each user may transmit one and only one codeword—so the total received optical power *per bit*

$$P_b = P_b^{OOK} / \log_2 M \qquad (3.74)$$

where P_b^{OOK} is the power per bit in the corresponding on-off-keying (OOK) CDMA system.

The use of the M-ary codes can also improve the security of CDMA communications. If the message "zero," instead of being represented by an empty time slot, is assigned to a particular codeword, the detector needs no advance knowledge on the time interval between the the the arriving "symbols." As a result, the signaling period can be randomly varied—with resulting dramatical reduction in the probability of interception.

The incorporation of M-ary approach into the design of OCDMA system is only possible if the total number of (quasi-orthogonal) codes (referred to as the "code cardinality" C) is equal or greater than $M * K$ where K is the number of users. For a CDMA system performing at high bit rates (which implies short bit intervals and small number of time slots per bit interval), the existing techniques provide only a limited number of independent quasi-orthogonal codes. For example, the cardinality for the (single-wavelength) direct-sequence encoding scales as $\sqrt{T/\tau}$, while for the standard carrier-hopping prime code $C \sim T/\tau$, where T/τ is number of the time slots in the symbol interval [15]. Therefore, in order to take full advantage of the M-ary approach, we need to introduce a new family of two-dimensional codes with high cardinality.

Our approach to the codeword design, illustrated in Figure 3.7 (panel 4), uses the ideas from both frequency-hopping codes [16] and direct sequence encoding [15] *at each wavelength*. First, the signal is uniformly (i.e., equally at all frequency channels) divided into E_t parts which are receive different time delays (see the bottom panel of Figure 3.7). This stage of the encoder is essentially equivalent to the direct sequence encoder (Figure 3.7, panel 2) with the only difference being the use multiwavelength source. At this point, the signal is represented by a sequence of E_t broadband pulses. Then, yet another set of time delays introduced independently in N_λ different wavelength channels. This second state is essentially an implementation of the frequency-hopping CDMA code (Figure 3.7, panel 3), and increases the number of pulses by the factor of N_λ. As a result, the code word of the proposed (E_t, N_λ) code consists of $N_\lambda * E_t$ pulses.

Note that the choices of the time delays for the first and second stages of the (E_t, N_λ) transmitter are completely independent. One can choose the times delays corresponding to quasi-orthogonal prime codes [15] based on two different prime numbers and different "time slot" intervals. As a result, the cardinality of the family of (E_t, N_λ) codes

$$C_{en} = C_T C_\lambda \propto \frac{T}{\tau} \sqrt{\frac{T}{\tau}} \qquad (3.75)$$

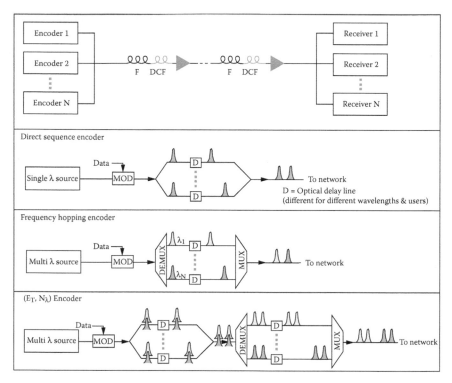

FIGURE 3.7 (From top to bottom) Panel 1: the schematics of an asynchronous fiber-optical CDMA system; F and DCF correspondingly represent regular and dispersion-compensating fibers. Panel 2: the (single wavelength channel) direct sequence encoder. Panel 3: the transmitter for carrier-hopping CDMA codes. Panel 4: (E_t, N_λ) transmitter.

where $C_T(\sim \sqrt{T/\tau})$ is the cardinality of the code family used for the "direct sequence stage" and $C_\lambda(\sim \sqrt{T\backslash\tau})$ is the code cardinality of the code family used at the "frequency-hopping stage." The increased cardinality of the proposed (E_t, N_λ) code family will then allow a direct implementation of the M-ary approach.

Similarly to the existing coding techniques for CDMA systems, the bit-error-rate (BER) performance of the proposed (E_t, N_λ) codes (which we will simply refer to as the "en-codes") suffers from the multiple-access interference (MAI). In the limit where the noise due to spontaneous emission in optical amplifiers, detector dark current, etc. can be neglected compared to the effect of MAI, for the average BER we obtain:

$$\text{BER} = \frac{1}{2}\left[\sum_{m=N_\lambda}^{U} \frac{n!}{m!(n-m)!}\left(\frac{E_t N_\lambda \tau}{2T}\right)^m \left(1 - \frac{E_t N_\lambda \tau}{2T}\right)^{n-m}\right]^{E_t} \quad (3.76)$$

where U is the number of simultaneous users and τ is the single pulse width. In Figure 3.8, we compare the BER performance for (E_t, N_λ) code with $N_\lambda = 8$ different

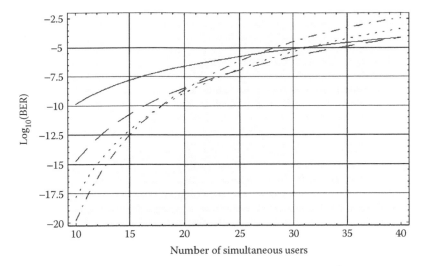

FIGURE 3.8 Bit-error-rate for (E_t, N_λ) codes vs. number of simultaneous users; for $E_t = 1$ (solid line), $E_t = 2$ (dashed line), $E_t = 3$ (dotted line), $E_t = 4$ (dashed-dotted line). In all cases, the number of wavelength channels $N_\lambda = 8$ and the number of time slots per symbol $T/\tau = 101$.

wavelength channels and different values of E_t, as a function of the number of simultaneous users. As follows from Figure 3.8, *BER* is non-monotonic in E_t —in other words, for a given number of simultaneous users there is an optimal choice of E_t that depends on the number of wavelength channels and the ratio of the bit/symbol interval to the single pulse width. Note, however, that the choice of $E_t = 1$, corresponding to the standard wavelength-hopping prime code generally leads to inferior performance if small BER is required.

Finally, the greater set of possible code words in the en-code family can further improve the system security if a large number of codes is assigned to some (or all) individual users. In this case, dynamical redistribution of M-ary "symbols" between different available "code words" will further reduce the probability of the message interception.

3.6 SUMMARY AND CONCLUSIONS

In this chapter, we considered the limits on the information capacity of fiber-optical CDMA systems. We have demonstrated that the fiber nonlinearity leads to the deterioration of the information capacity, and developed the theory of this effect.

For fiber-optical CDMA, we derived the analytical expressions for the information capacity and the bit-error-rate for both coherent and incoherent systems taking into account the effect of fibre nonlinearity as well as additive noise and channel crosstalk. We demonstrated that a properly designed incoherent system is more robust to the effect of nonlinearity in terms of bit-error-rate performance.

Finally, we presented novel code techniques designed for performance improvement of the incoherent CDMA systems.

REFERENCES

[1]   C. E. Shannon. A mathematical theory of communication. *The Bell System Technical Journal* 27:379–423, 623–656 (1948).

[2]   G. P. Agrawal. *Nonlinear fibre Optics*. Academic Press, San Diego, 1995.

[3]   E. Desurvire, *Erbium Doped Amplifiers*, Wiley, New York, 1994.

[4]   R. K. Dodd, J. C. Eilbeck, J. D. Gibbon, and H. C. Morris, *Solitons and Nonlinear Wave Equations*, Academic Press, New York, 1984.

[5]   G. L. Lamb, Jr. *Elements of Soliton Theory*. Wiley, New York, 1980.

[6]   J. P. Gordon and L. F. Mollenauer, Phase noise in photonic communication systems using linear amplifiers. Optics Letters, 23:1351–1353. (Dec. 1990).

[7]   P. P. Mitra and J. B. Stark. Nonlinear limits to the information capacity of optical fibre communications. *Nature* 411:1027–1030 (2001).

[8]   G. P. Agrawal. *Fibre-Optic Communication Systems*. John Wiley and Sons, New York, 1997.

[9]   E. Telatar, unpublished.

[10] E. E. Narimanov and P. Mitra. The channel capacity of a fibre optics communication system: perturbation theory. *Journal of Lightwave Technology* 20:530–537 (2002).

[11] P. R. Prucnal, M. Santoro, and F. Tan. Spread spectrum fibre-optic local area network using optical processing. *Journal of Lightwave Technology* 4:547–554 (1986).

[12] S. C. Pinault and M. J. Potasek, Frequency broadening by self-phase modulation in optical fibers *Journal of the Optical Society of America* B 2:1318–1319 (1985).

[13] P. V. Mamyshev and N. A. Mamysheva. Pulse-overlapped dispersion-managed data transmission and intrachannel four-wave mixing. *Optics Letters* 24:1454–1456 (1999).

[14] R. M. Gagliardi and S. Karp. *Optical Communications*, Chapter 8. John Wiley & Sons, New York, 1976.

[15] G. C. Yang and W. C. Kwong, *Prime Codes with Applications to CDMA Optical and Wireless Networks*. Artech House, Boston, 2002.

[16] E. S. Shivaleela, K. N. Sivarajan, and A. Selvarajan. One-coincidence sequences with specified distance. *Journal of Lightwave Technology* 16:501–50 (1998).

[17] E. E. Narimanov, submitted to *Optics Letters*.

4 Optical Code-Division Multiple-Access Enabled by Fiber Bragg Grating Technology

Lawrence R. Chen

CONTENTS

4.1 INTRODUCTION

Code-division multiple-access (CDMA) is a "tell-and-go" access strategy in which data from multiple users are transmitted concurrently over the network and differentiated with code (or user) specific detection [1–3]. CDMA does not require the overhead associated with synchronization or network protocols and is well-suited for network applications involving bursty, asynchronous traffic and where aggregate network capacity varies. Furthermore, fiber-optic CDMA (FOCDMA) is well-suited in situations where optical diffuse transmission is not possible (e.g., interconnections over long distances), as well as in applications where high-bandwidth interconnections with low bit-error-rate (BER) are required and mobility is not an issue.

Broadly speaking, optical CDMA (OCDMA) can be divided into the following categories depending on the coding approach: (1) direct-sequence or temporal encoding (also known as spread-spectrum encoding) [4,5], (2) spectral amplitude encoding [6–9], (3) spectral phase encoding (also known as spread-time encoding) [10–12], (4) two-dimensional (2D) spatial encoding (also known as spread-space encoding) [13–15], and (5) hybrid coding approaches involving combinations of any of the four previous ones [16–21]. These coding approaches can be further divided into those that are coherent and those that are incoherent. Coherent approaches use (ultra)short broadband optical pulses and typically seek to manipulate the phase of the pulses in the encoding and decoding processes. Incoherent approaches use incoherent sources such as broadband ASE noise and decoding is based on power summation rather than manipulation of field quantities. Finally, for completeness, we note that OCDMA approaches can also be synchronous, i.e., a system clock is assumed throughout the network to synchronize transmission from the users, though these have not been studied as extensively as asynchronous ones and will not be considered further.

In early direct sequence OCDMA schemes, delay line networks were used to encode an ultrashort optical pulse into a low intensity pulse train, i.e., replicas of the ultrashort pulse were suitably positioned into various time slots or time chips (the bit window is divided into time chips). For matched filtering at the decoder, a second delay line network having a complementary delay response was used so that the pulses in the train are despread into the same time slot. Unmatched filtering would cause the pulses in the low intensity pulse train to be redistributed thereby forming a background interference signal (decoding is essentially an intensity correlation process). Although relatively simple in concept, the incoherent nature of the approach prohibited truly orthogonal transmission and required the development of optical orthogonal codes in order to minimize multiuser interference. These codes have very long lengths and small weight (sparse codes) and support very few users resulting in poor spectral efficiency. Moreover, implementation of the codes requires pulses with a duration much smaller than the bit window, which makes processing more difficult.

Spectral amplitude encoding schemes involve dispersing the frequency content of a broadband signal followed by spatial filtering with an amplitude mask. The various frequency encoded patterns that can be generated correspond to the different user codes. When combined with balanced or differential detection, the codes can be designed to eliminate multiuser interference and achieve truly orthogonal transmission. Bipolar signaling can also be used to further improve system performance. Spectral phase encoding is similar to spectral amplitude encoding in that the spectral content of a broadband signal (typically from an ultrashort pulse) is first dispersed. Encoding involves manipulating the phases of the spectral content of the ultrashort pulse, and in particular, applying different phase shifts to the different frequency components, thereby transforming the pulse into a low pseudo-noise burst. Decoding involves a conjugate phase process matched to that used at the encoder: for matched filtering, the phase encoding will be removed and the high-peak ultrashort pulse will be reconstructed; unmatched filtering will cause the signals to be further scrambled and will form a background interference at the receiver. Generally, properly reconstructed pulses are distinguished from the low level pseudo-noise using nonlinear optical detection. Unless properly accounted for, spectral phase encoding approaches suffer severe transmission impairments, especially from chromatic fiber dispersion, due to the large bandwidth of the signals involved.

Spatial OCDMA was first developed to achieve multiplexing functionality in optical parallel interconnections based on free space optics or fiber image guides. Spatial optical orthogonal signature patterns, which are collections of (0,1) 2D matrices with good correlation characteristics, are used to encode the information (e.g., images). Hybrid approaches seek to exploit the simultaneous use of two or more domains, for example, space and time or wavelength and time, to provide increased flexibility in the design of suitable codes with good correlation properties, increase the cardinality of code families, and improve system performance.

Many OCDMA codes and schemes have been proposed, developed, and analyzed theoretically. There have also been numerous demonstrations of encoding and decoding of optical signals. However, existing proposals often involve bulk optic components that may be difficult to package or ensure stability for proper operation, expensive components, and very complex implementations. Furthermore, there have been relatively few multiuser system demonstrations.

In recent years, FBGs have emerged as an enabling technology for many lightwave communications applications [22,23]. Their wavelength selective properties make them ideal for a number of filtering applications including channel add/drop and wavelength routing. They have been used to define the lasing wavelength in single and multiwavelength fiber lasers, in external cavity lasers, and as pump stabilizers. Their dispersive properties make them ideal for dispersion compensation. More complex FBG designs and FBG-based devices have been developed for advanced signal processing functions, including pulse shaping. This short list is far from complete and in fact, the applications and possibilities in designs of new grating structures are seemingly endless and continually attract considerable research interest. It is thus of no surprise that there have been several proposals of using FBGs to implement various forms of FOCDMA [24–45].

This chapter discusses the potential role of FBGs in enabling FOCDMA and is organized as follows. In section 2, we provide a brief overview of FBG technology. In section 3, we describe the use of FBGs for implementing several different types of FOCDMA with an emphasis on the grating properties, requirements, and modeling. Implementations considered include spectral amplitude encoding with balanced detection, spectral phase encoding, and hybrid 2D wavelength-time FOCDMA. In all cases, sample results, simulations, and experiments where available, are presented; limitations and comparisons are discussed. Finally, a summary is given in section 4.

4.2 FIBER BRAGG GRATING TECHNOLOGY

4.2.1 OVERVIEW

A fiber grating consists of a periodic modulation (perturbation) of the refractive index in the longitudinal direction of the core of an optical fiber. Fiber gratings can be classified in two types: those that couple counterpropagating waves (and thus typically operate in reflection) and are referred to as short-period or fiber Bragg gratings (FBGs), and those that couple copropagating waves (and, thus, typically operate in transmission) and are referred to as long-period gratings. If we consider an FBG in single mode fiber, then light guided in the fundamental LP_{01} mode will be reflected and scattered by the refractive index perturbation. If the wavelength of the incident light satisfies the Bragg condition $\lambda = 2n_{eff}\Lambda$ where n_{eff} is the effective refractive index of the propagating mode and Λ is the grating period, then the reflected/scattered waves will add coherently, otherwise it will become progressively out of phase with each reflection and eventually cancel out. In other words, if the Bragg condition is satisfied, there is a coherent reflection of the input light, otherwise it is mostly transmitted. Figure 4.1 illustrates the principle of operation of an FBG.

Readers interested in learning more about the photosensitive phenomenon, which is the mechanism responsible for allowing the formation of gratings in fiber, as well as a detailed description of different grating fabrication techniques are referred to the excellent discussion and overview presented in [23].

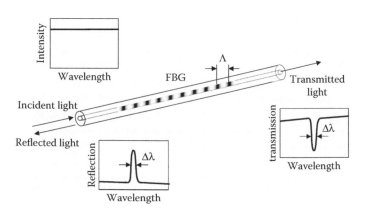

FIGURE 4.1 Schematic representation and principle of operation of an FBG.

4.2.2 THEORY OF FBGS

An FBG is a periodic perturbation of the refractive index in the core of an optical fiber. For simplicity, we assume that this perturbation results in an effective refractive index of the propagating fundamental mode extending over the distance $0 \leq z \leq L$ as follows [46]:

$$\delta n_{eff}(z) = \overline{\delta n}_{eff}(z)\left\{1 + \upsilon(z)\cos\left(\frac{2\pi z}{\Lambda_0} + \phi(z)\right)\right\} \tag{4.1}$$

where $\overline{\delta n}_{eff}(z)$ is average effective index (DC index); $\upsilon(z)$ is the fringe visibility or apodization function; Λ_0 is the nominal, or resonant, grating period, $\phi(z)$ described the spatially-varying phase (chirp) of the grating; and L is the grating length. The above expression assumes that the grating distribution is uniform in the transverse direction.

Assume that an electromagnetic wave of frequency ω and described by

$$E(z,t) = \hat{x}E(z)e^{-j\omega t} + c.c. \tag{4.2}$$

propagates in the periodic medium with a spatial dependence

$$E(z) = A(z)e^{j\beta z} + B(z)e^{-j\beta z} \tag{4.3}$$

where $A(z)$ and $B(z)$ are the complex envelopes of the forward and backward propagating waves, respectively, and β is the propagation constant. The grating causes a coupling between the forward and backward propagating waves and the interaction can be described in terms of the following coupled-mode equations [46]:

$$\frac{dR}{dz} = j\hat{\sigma}R(z) + j\kappa S(z) \tag{4.4a}$$

$$\frac{dS}{dz} = -j\hat{\sigma}S - j\kappa^* R(z) \tag{4.4b}$$

where $R(z) = A(z)\exp(j\delta z - \phi/2)$, $S(z) = B(z)\exp(-j\delta z + \phi/2)$, δ is the detuning parameter defined by

$$\delta = \beta - \frac{\pi}{\Lambda_0} = 2\pi n_{eff}\left(\frac{1}{\lambda} - \frac{1}{2 n_{eff}\Lambda_0}\right) \tag{4.5}$$

and

$$\hat{\sigma} = 2\pi n_{eff}\left(\frac{1}{\lambda} - \frac{1}{2 n_{eff}\Lambda_0}\right) + \frac{2\pi}{\lambda}\overline{\delta n}_{eff}(z) - \frac{1}{2}\frac{d\phi}{dz}. \tag{4.6}$$

Equation 4.6 represents the general DC self-coupling coefficient and κ is the AC coupling coefficient which, for a single mode FBG, is given by

$$\kappa(z) = \kappa^*(z) = \frac{\pi}{\lambda} \upsilon \overline{\delta n}_{eff}(z) \tag{4.7}$$

For a uniform FBG, $\overline{\delta n}_{eff}$ and all other grating parameters are independent of z and the coupled-mode solutions can be solved in closed form when appropriate boundary conditions are imposed. The boundary conditions are based on the assumption that there is a forward propagating wave from $z = -\infty$ [$R(0) = 1$] and no backward propagating wave for $z \geq L$ [$S(L) = 0$]. Equation 4.4, along with these boundary conditions, constitutes a boundary value problem. The propagation of waves through the uniform grating can be described by a 2×2 fundamental matrix, $[F]^{uni}$:

$$\begin{pmatrix} R(L) \\ S(L) \end{pmatrix} = [F]^{uni} \begin{pmatrix} R(0) \\ S(0) \end{pmatrix} \tag{4.8}$$

where

$$F^{uni} = \begin{pmatrix} \cosh(\gamma L) - j\dfrac{\hat{\sigma}}{\gamma}\sinh(\gamma L) & -j\dfrac{\kappa}{\gamma}\sinh(\gamma L) \\[2ex] j\dfrac{\kappa}{\gamma}\sinh(\gamma L) & \cosh(\gamma L) + j\dfrac{\hat{\sigma}}{\gamma}\sinh(\gamma L) \end{pmatrix} \tag{4.9}$$

and $\gamma = \sqrt{\kappa^2 - \hat{\sigma}^2}$. The reflection ($r$) and transmission ($t$) coefficients can then be calculated from the following:

$$r(\lambda) = \frac{S(0)}{R(0)} \tag{4.10}$$

$$t(\lambda) = \frac{R(L)}{R(0)} \tag{4.11}$$

Finally, the reflectivity $R(\lambda)$ and transmissivity $T(\lambda)$ are given by $R(\lambda) = |r|^2$ and $T(\lambda) = |t|^2$. Note that if we assume a lossless grating and that the grating only couples the forward and backward propagating fundamental mode, then $R(\lambda) + T(\lambda) = 1$. Using Equations 4.8–4.10, we immediately get the following analytic expression for the reflectivity:

$$R(\lambda) = \frac{\kappa^2 \sinh^2(\gamma L)}{\kappa^2 \cosh^2(\gamma L) - \hat{\sigma}^2} \tag{4.12}$$

At the Bragg wavelength, $\hat{\sigma} = 0$ and the grating reflectivity attains its peak value:

$$R_{max}(\lambda = \lambda_B) = \tanh^2(\kappa L) \tag{4.13}$$

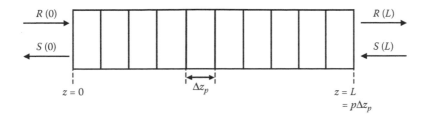

FIGURE 4.2 Decomposition of a FBG in terms of multiple fundamental matrices.

For a nonuniform grating, all parameters ($\overline{\delta n}_{e\!f\!f}$, υ, and ϕ) can be position dependent. In this case, we can determine the spectral response of the grating numerically by first dividing the grating into p discrete segments of length Δz_p, each of which can be treated as a uniform segment, see Figure 4.2. The F matrix defined in Equation 4.9 relates the input and output fields of the individual grating segments (the length Δz_p rather than L is used) and once these are known, the total grating structure is then a cascade of the p discrete segments. The input and output fields are related by an expression similar to Equation 4.8:

$$\begin{pmatrix} R(L) \\ S(L) \end{pmatrix} = [F]_p^{uni}[F]_{p-1}^{uni}\dots[F]_2^{uni}[F]_1^{uni}\begin{pmatrix} R(0) \\ S(0) \end{pmatrix} = [F]^{non-uni}\begin{pmatrix} R(0) \\ S(0) \end{pmatrix}. \qquad (4.14)$$

The reflectivity and transmissivity can be found as above.

Another useful property of FBGs are their group delays and dispersions. The group delay and dispersion of a light wave reflected from a grating can be determined from the phase of the reflection coefficient r. If $\theta_r = \text{phase}[r] = \arctan[\text{Im}(r)/\text{Re}(r)]$, then we can expand it as a Taylor series about the resonant frequency ω_0. Since the first derivative $\frac{d\theta_r}{d\omega}$ is directly proportional to ω, then from linear systems theory, it can be identified as a time delay. Thus, the relative time delay for wavelengths reflected by a grating is

$$\tau_r[\text{ps}] = \frac{d\theta_r}{d\omega} = -\frac{\lambda}{2\pi c^2}\frac{d\theta_r}{d\lambda}. \qquad (4.15)$$

The corresponding dispersion, which is the rate of change of delay with wavelength, is

$$D_r[\text{ps/nm}] = \frac{d\tau_r}{d\lambda} = \frac{2\pi c}{\lambda^2}\frac{d^2\theta_r}{d\omega^2}. \qquad (4.16)$$

Similar expressions can be found for the group delay and dispersion in transmission using the phase of the transmission coefficient θ_t.

The transfer matrix method requires certain conditions to be satisfied in order to accurately simulate the grating response. For gratings whose parameters are position dependent (such as apodized or chirped gratings), it is not true that increased accuracy can be obtained by dividing the grating into more sections. In fact, the minimum section length is usually many grating periods, with the maximum number of segments limited by ensuring that the slowly varying envelope approximation is satisfied.

The reflection and transmission responses (amplitude and phase) are dependent on the grating profile described by Equation 4.1. Common grating profiles include the following:

- Uniform gratings in which $\overline{\delta n}_{eff}\,(z)$, $\upsilon(z)$, and $\phi(z)$ are constant,
- Chirped gratings in which $\phi(z)$ is nonzero and the grating period varies with position,
- Apodized gratings in which $\overline{\delta n}_{eff}\,(z)$ varies,
- Apodized gratings with no DC background index in which $\overline{\delta n}_{eff}$ is constant and $\upsilon(z)$ varies,
- Phase-shifted gratings in which discrete phase shifts are introduced in the grating structure, and
- Superstructure gratings in which $\overline{\delta n}_{eff}$ is constant and typically $\upsilon(z)$ varies periodically.

The gratings profiles are illustrated in Figure 4.3. For an excellent overview of the spectral characteristics of these grating profiles, the reader is referred to [46]. In this chapter, we will also describe other grating structures that are used to implement various types of FOCDMA.

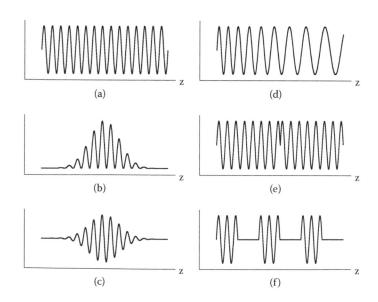

FIGURE 4.3 Refractive index profiles of typical FBGs: (a) uniform, (b) apodized with DC index change, (c) apodized with no DC index change, (d) chirped, (e) phase-shifted, and (f) superstructure.

4.3 FBGS FOR FOCDMA

4.3.1 SPECTRAL AMPLITUDE ENCODING FOCDMA

4.3.1.1 Overview

Spectral amplitude encoding (SAE) FOCDMA was first investigated by Zaccarin and Kavehrad [6,7]. A typical SAE FOCDMA system is shown in Figure 4.4. In SAE, the frequency content of an incoherent broadband optical source are spatially dispersed and filtered using a bulk 4-f optical system. A pair of lenses and uniform diffraction gratings are arranged to (1) spatially disperse the spectral content of the incoherent broadband input, (2) filter the spatially dispersed spectral content in amplitude using suitable amplitude masks, and (3) recombine the filtered spectral content. The receiver comprises two spectral filters and two photodetectors connected in a differential or balanced configuration. One of the filters (direct filter) has the same spectral amplitude response as that used in the transmitter to encode the signal. The second filter (complementary filter) has the complementary response. The codes are designed such that, in conjunction with differential detection, multiuser interference is cancelled. For example, let $X = (x_0, x_1, \ldots, x_{N-1})$ and $Y = (y_0, y_1, \ldots, y_{N-1})$ be two sequences with a cross-correlation defined by $\theta_{XY}(k) = \sum_{i=0}^{N-1} x_i y_{i+k}$. Furthermore, let $\bar{X}$ denote the complement of sequence X whose elements are obtained from X by $\bar{x}_i = 1 - x_i$. If the sequences satisfy the property $\theta_{XY} = \theta_{\bar{X}Y}$, then a receiver that computes $\theta_{XY} - \theta_{\bar{X}Y}$ will reject the interference from the user having sequence Y. In other words, for unmatched detection, half of the transmitted power will match the direct filter and the other half will match the complementary filter so that the output current of the differential detector is zero. Note that the use of differential detection allows for the implementation of bipolar codes in which "1" bits are coded with the amplitude response $X(\omega)$ whereas "0" bits are coded with the response $\bar{X}(\omega)$.

In terms of unipolar implementations, various codes have been considered including, among others, M-sequences, Hadamard codes, unipolar representations of bipolar codes [a bipolar sequence $Z = (z_0, z_1, \ldots, z_{N-1})$ can be represented by its unipolar version $Z^u = (z_0^u, z_1^u, \ldots, z_{2N-1}^u)$ where $y_{2i}^u = \begin{cases} 1, y_i=1 \\ 0, y_i=-1 \end{cases}$ and $y_{2i+1}^u = \begin{cases} 0, y_i=1 \\ 1, y_i=-1 \end{cases}$], and modified quadratic congruence (MQC) codes.

An alternative to the bulk 4-f optical system implementation to perform spectral amplitude filtering is to use FBGs as the wavelength selective filters to define the spectral slices of a given code. Several FBG structures can be used. Among those that have been studied or demonstrated for SAE FOCDMA are a serial or linear array of FBGs [31–33] and superimposed FBGs (SI-FBGs) [43].

A linear array of FBGs is a concatenation of FBGs, each centered at a frequency (wavelength) defined by the code. In terms of spectral content, the reflection of a broadband pulse (signal) representing a "1" bit from the linear array will correspond directly to the desired code (the grating reflection response corresponds to the direct filter); on the other hand the transmission will correspond to the code complement (the grating transmission response corresponds to the complementary filter). The wavelengths of each FBG can be individually tuned (for example using localized thermal means or applied strain), which allows the encoder/decoder to be dynamically reconfigured. Note, however, that due to the physical separation of the gratings in the

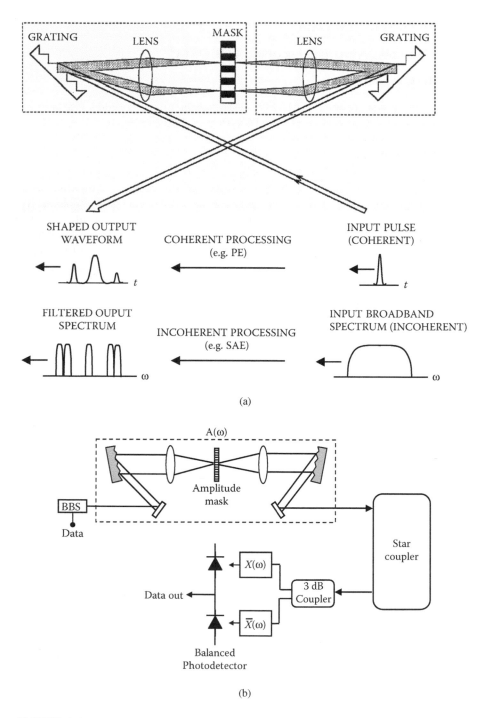

FIGURE 4.4 (a) 4-*f* pulse shaper used to spatially process the spectral content of a broadband optical input signal. (b) Block diagram of SAE FOCDMA system.

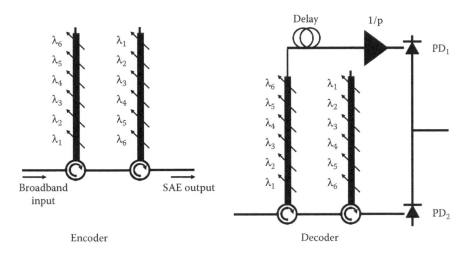

FIGURE 4.5 Encoder and decoder based on a linear array of FBGs for implementing SAE FOCDMA.

encoder, the wavelengths will also be separated, or dispersed, in time (the same problem would exist in the decoder). This has a similar effect to chromatic fiber dispersion and, in fact, would compound the dispersion problem. One possible means to overcome this problem is to incorporate a second identical linear array comprising the same FBGs but in the reverse order to compensate the time delays incurred. Such an encoder/decoder structure is shown in Figure 4.5. Modeling a linear array of FBGs is relatively straightforward using the fundamental matrix formalism. The physical separation between the gratings simply adds an additional wavelength-dependent phase response, which gives rise to the difference in group delay.

With SI-FBGs, the gratings are written at the same physical location in the fiber [47]. We can tailor the reflection response of the SI-FBGs to match the direct or complementary filter by controlling the periods of the superimposed gratings. Note that with SI-FBGs, we avoid the problem of temporally separating the wavelength slices. However, in contrast to the serial array, the gratings in the superimposed structure cannot be individually tuned. Rather the entire grating structure can be thermally tuned or strained so that the complete reflection response shifts uniformly to longer or shorter wavelengths.

4.3.1.2 Some General Considerations

We now discuss some general considerations in the design of filters based on linear arrays of FBGs or SI-FBGs for implementing SAE FOCDMA.

The weight of the code determines the number of FBGs required in the encoder/decoder structure.

If we consider fixed (static) encoders and decoders, then the wavelength range that can be used to define codes is constrained to the available bandwidth from the broadband source.

If we allow tunable (dynamic or reconfigurable) encoders and decoders, then the wavelength range that can be used to define the codes is limited either by the available bandwidth from the broadband source as above, or by the wavelength tuning range of the FBGs. While tuning bandwidths in excess of 90 nm have been reported [48,49], typical tuning ranges of commercially available devices are on the order of a few nanometers.

In the bulk optics implementation, the code length depends on the spatial separation of the spectral content of the broadband input signal, i.e., the number of spatially resolvable frequency components. In the FBG-based approach, the code length depends on the spectral efficiency that can be achieved, or in other words, the number of wavelength slices that the FBGs can define within the available source bandwidth and/or FBG tuning range. Ideally, gratings with a boxlike amplitude response are required in order to maximize spectral efficiency and minimize cross-talk between adjacent wavelength slices. Uniform FBGs typically have large sidelobes, which can only be reduced using apodization. Simple apodization profiles, such as Gaussian with $\upsilon(\bar{z}) = \exp(-a\bar{z}^2)$, Blackman with $\upsilon(\bar{z}) = \frac{1+1.19\cos(2\pi\bar{z})+0.19\cos(4\pi^2\bar{z}^2)}{2.38}$, and Gaussian-sinc with $\upsilon(\bar{z}) = \exp(-a\bar{z}^2)\frac{\sin(2\pi b\bar{z})}{2\pi b\bar{z}}$, where $\bar{z} = \frac{z-L/2}{L}$ and L is the grating length, are effective at reducing the sidelobes. Figure 4.6 compares the reflectivity of a uniform FBG and apodized FBGs with the above profiles, assuming in all cases a grating length $L = 1.0$ cm, a peak index modulation $\delta n_{eff} = 5 \times 10^{-4}$, and where applicable, $a = 10$ and $b = 2$. Even more boxlike responses can be obtained using specialized profiles optimized using inverse design algorithms [50,51].

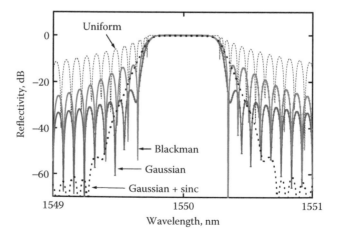

FIGURE 4.6 (Color Figure 4.6 follows page 240.) Comparison of the spectral response of a uniform FBG to an FBG with Gaussian, Blackman, or Gaussian-sinc apodization.

The change in peak reflection wavelength $\Delta\lambda_{temp}$ of an FBG as a function of a change in temperature ΔT can be given by

$$\frac{\Delta\lambda_{temp}}{\lambda} = (\xi + \alpha)\Delta T \tag{4.17}$$

where ξ is the thermo-optic coefficient and α is the thermal expansion coefficient with typical values of $8.31 \times 10^{-6}/°C$ and $0.55 \times 10^{-6}/°C$, respectively, in germanosilicate glass [52]. The corresponding wavelength shift is ≈ 14 pm/°C. Since this value is very small, strain is usually applied to obtain larger wavelength shifts. A strain can be applied to an FBG by stretching the fiber (this can be accomplished with mechanical means or by bonding the grating on a piezoelectric rod and applying a voltage). In this case, the change in peak reflection wavelength $\Delta\lambda_{strain}$ for a fiber stretching ΔL can be given by

$$\frac{\Delta\lambda_{strain}}{\lambda} = (1 - P_\alpha)\frac{\Delta L}{L} \tag{4.18}$$

where P_α is the effective strain optic coefficient with a value of ≈ 0.22 for typical fiber [52].

Based on the above considerations, the choice of grating structure (linear array or superimposed) for implementing SAE FOCDMA depends on the type of codes that are used. For example, let us first consider MQC codes. MQC codes are defined by an odd prime number p. The code weight is $w = p + 1$ which corresponds to the number of FBGs required in the grating structure. The available source bandwidth is divided into w subbands, each containing a single "1", or wavelength slice, and $p - 1$ "0"'s. Thus, the code length is $p^2 + p$. The upper bound on the number of codes that can be generated is p^2. Since the codes are not cyclic and for each different code, the wavelength slice per subband can vary, then to implement a reconfigurable encoder/decoder, we should use a linear array of FBGs since each grating can be independently tuned. For a given source spectral bandwidth, the tuning range of the FBG sets the code length and hence the number of users that can be supported. In other words, the tuning range would effectively define the number p and the total bandwidth that can be spanned is then $w \times p$. As an example, if we assume that we can stretch a fiber in the amount $\frac{\Delta L}{L} = 0.005$, then for an operating wavelength of 1550 nm, the peak wavelength shift (tuning range) is $\Delta\lambda = 6$ nm. For a source with a bandwidth of 50 nm, we have $p = 7$ which results in an upper bound of 49 users. A simulation that shows the BER performance of MQC codes that can be implemented with the encoder/decoder structures shown in Figure 4.5 and that takes into account factors such as thermal and shot noise is shown in Figure 4.7. These codes show improved performance compared to Hadamard codes with the same effective power and bandwidth in terms of the number of simultaneous users that can operate at a given BER.

On the other hand, the cyclic nature of certain codes, such as M-sequences, make SI-FBGs more attractive for the simple reason that one grating structure can be used to implement more than one code. Recall that for SI-FBGs, applying a longitudinal strain will cause all of the reflected wavelengths to shift uniformly. It is precisely this uniform wavelength shift that will correspond to a new code. Thus when constructing an SAE FO-CDMA system, multiple users can use the same grating

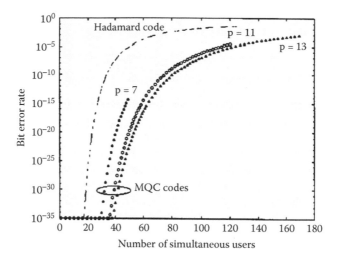

FIGURE 4.7 Performance comparison of MQC codes. BER vs. number of simultaneous users. (Reprinted from Wei, Z. et al., *IEEE-OSA J. Lightwave Technol.*, 19(9): 1274–1281. With permission from IEEE © 2004.)

structure and the different user codes are obtained by strain tuning. For example, consider the family of *M*-sequences with $N = 15$ given by the starting code $(1,1,1,1,0,1,0,1,1,0,0,1,0,0,0)$. The different codes and their wavelength patterns are illustrated in Table 4.1. It is clear that the codes can be divided into groups, where each

---

TABLE 4.1
Family of *M*-Sequences with *N* = 15 and Starting Code
(1,1,1,1,0,1,0,1,1,0,0,1,0,0,0)

	λ_1	λ_2	λ_3	λ_4	λ_5	λ_6	λ_7	λ_8	λ_9	λ_{10}	λ_{11}	λ_{12}	λ_{13}	λ_{14}	λ_{15}
Code 1	1	1	1	1	0	1	0	1	1	0	0	1	0	0	0
Code 2	0	1	1	1	1	0	1	0	1	1	0	0	1	0	0
Code 3	0	0	1	1	1	1	0	1	0	1	1	0	0	1	0
Code 4	0	0	0	1	1	1	1	0	1	0	1	1	0	0	1
Code 5	1	0	0	0	1	1	1	1	0	1	0	1	1	0	0
Code 6	0	1	0	0	0	1	1	1	1	0	1	0	1	1	0
Code 7	0	0	1	0	0	0	1	1	1	1	0	1	0	1	1
Code 8	1	0	0	1	0	0	0	1	1	1	1	0	1	0	1
Code 9	1	1	0	0	1	0	0	0	1	1	1	1	0	1	0
Code 10	0	1	1	0	0	1	0	0	0	1	1	1	1	0	1
Code 11	1	0	1	1	0	0	1	0	0	0	1	1	1	1	0
Code 12	0	1	0	1	1	0	0	1	0	0	0	1	1	1	1
Code 13	1	0	1	0	1	1	0	0	1	0	0	0	1	1	1
Code 14	1	1	0	1	0	1	1	0	0	1	0	0	0	1	1
Code 15	1	1	1	0	1	0	1	1	0	0	1	0	0	0	1

group requires only one encoder structure (grating) to implement the corresponding codes. For this case, there are eight independent wavelength patterns so that eight different grating encoder structures are required to implement all 15 codes. Reducing the number of different encoder structures will have advantages in terms of simplifying overall system implementation and fabrication of the encoders/decoders. Note also that SI-FBGs can provide limited reconfigurability in that an encoder structure can be tuned to all the codes within its corresponding group.

Although more complex approaches to model SI-FBGs exist [53], they can be modeled using the fundamental matrix formalism as follows. The refractive index modulation representing N superimposed gratings can be written as

$$\delta n_{eff}(z) = \sum_{i=1}^{N} \overline{\delta n}_{eff,i}(z) \left\{ 1 + \upsilon_i(z) \cos\left(\frac{2\pi z}{\Lambda_{0,i}} + \phi_i(z) \right) \right\}. \tag{4.19}$$

We can rewrite the above equation as

$$\delta \tilde{n}_{eff}(z) = \delta n'_{eff}(z) \left\{ 1 + \upsilon'(z) \cos\left(\frac{2\pi z}{\Lambda_{min}} + \phi'(z) \right) \right\} \tag{4.20}$$

where $\Lambda_{min} = \min\{\Lambda_{0,1}, \cdots, \Lambda_{0,N}\}$. The values of $\delta n'_{eff}(z), \upsilon'(z)$, and $\phi'(z)$ can only be determined numerically. Once they are obtained, then the refractive index modulation of the SI-FBG is in the form of Equation 4.1 so that the grating response can be computed using the fundamental matrix formalism discussed in Section 4.2.2. To determine these quantities, we first determine the solutions of the equation $\frac{d(\delta n_{eff}(z))}{dz} = 0$ to obtain the local extrema within each grating subdivision Δz_j. We denote A_j^+ and A_j^- as the local maxima and extrema, respectively, and define an upper envelope function $f^+(z_j) = A_j^+$ as well as a lower envelope function $f^-(z_j) = A_j^-$. We then find that

$$\delta n'_{eff}(z_j) = \frac{f^+(z_j) + f^-(z_j)}{2} \tag{4.21}$$

and

$$\upsilon'(z_j) = \frac{f^+(z_j) - f^-(z_j)}{2\delta n'_{eff}(z_j)}. \tag{4.22}$$

Finally, the phase term $\phi'(z)$ accounts for variations in the grating phase given by $\frac{2\pi z}{\Lambda_{min}}$ in Equation 4.20 and the real grating phase in Equation 4.19 and is given by

$$\phi'(z) = \arccos\left(\frac{1}{\upsilon(z)} \left[\frac{\delta n_{eff}(z)}{\delta n'_{eff}(z)} - 1 \right] \right) - \frac{2\pi z}{\Lambda_{min}}. \tag{4.23}$$

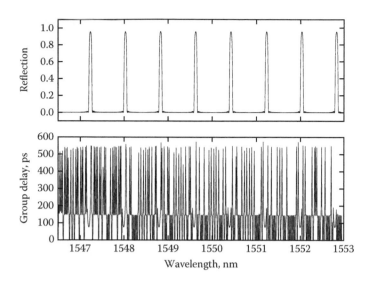

FIGURE 4.8 Simulated reflectivity and group delay of 8 superimposed gratings.

Figure 4.8 shows the simulated reflectivity and group delay of eight SI-FBGs spectrally separated by 0.8 nm (100 GHz), where each grating is 3 cm long with a Gaussian apodization ($a = 4.5$) and a peak index modulation of 8×10^{-4}. The wavelength separation is 0.8 nm (100 GHz). The group delay response shows clearly that all of the wavelengths are reflected at the same position, i.e., they are colocated in the fiber.

4.3.1.3 Experimental Results

Recently, Magné and colleagues demonstrated a four-user SAE FOCDMA system based on M-sequences with $N = 15$ implemented using SI-FBGs [43]. The codes were defined over the wavelength range from 1542 nm to 1548 nm, which corresponded to the flat region of the superfluorescent fiber source that was used as the broadband input. The gratings had a sinc main-lobe apodization; each had a peak reflectivity of 85 ± 5%, a 3 dB bandwidth of 12.5 ± 1.5 GHz (0.1 nm), a peak frequency accuracy of ±0.25 GHz, and were nominally separated by 50 GHz (0.4 nm). Figure 4.9 shows the experimental setup including the reflection response of the SI-FBG encoders. The broadband source was NRZ modulated with a $2^{23} - 1$ pseudo-random bit sequence (PRBS) at 155 Mb/sec before being encoded by the four different encoders. A differential detector with an 830 MHz bandwidth was used in the receiver. Figure 4.10 shows the eye diagrams and BER measurements for different transmission scenarios. The noise on the "1" levels is typical of beat noise and is clearly a limitation for system performance (note the BER floor, even for only two transmitting users). The impact of beat noise on system performance has been studied and is an important consideration in the development of SAE FOCDMA systems [54]. A detailed discussion, however, is beyond the scope of this chapter.

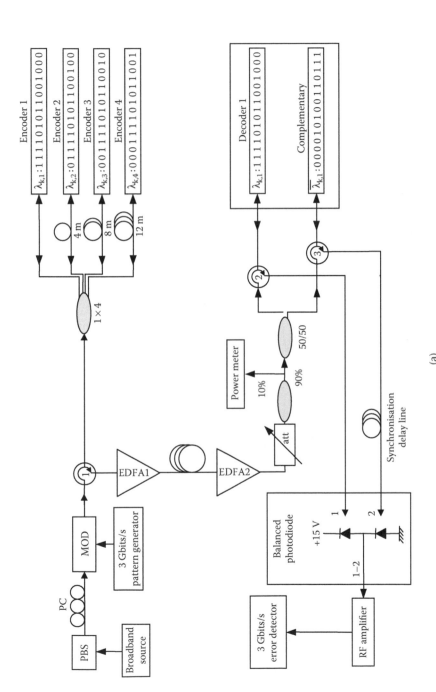

(a)

FIGURE 4.9 (a) Experimental setup of 4-user SAE FOCDMA demonstration system employing *M*-sequences and balanced detection. (b) Reflection responses of the SI-FBG encoders and decoders.

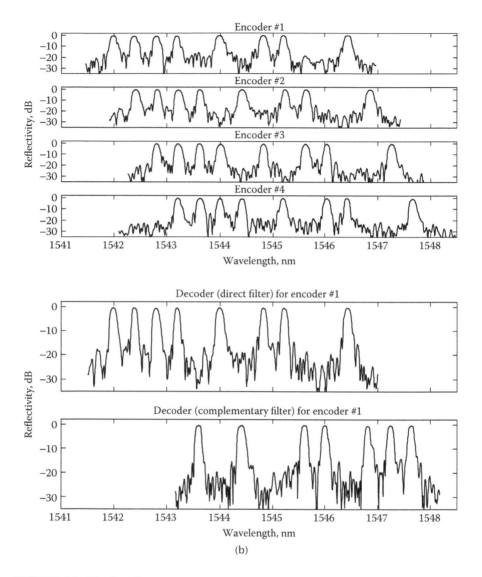

FIGURE 4.9 (Continued).

4.3.1.4 Discussion

In SAE FOCDMA, perfect cancellation of multiuser interference can be obtained if the power contained in each of the spectral bands of the different codes is the same. Thus, the reflection responses of the FBGs should be identical, i.e., have the same shape, bandwidth, etc., and the encoders and decoders need to be properly matched in terms of the reflection band wavelengths. This applies to whether a linear array or superimposed gratings are used. Moreover, the characteristics of the broadband optical source need to be accounted for. In particular, the grating reflectivities would be

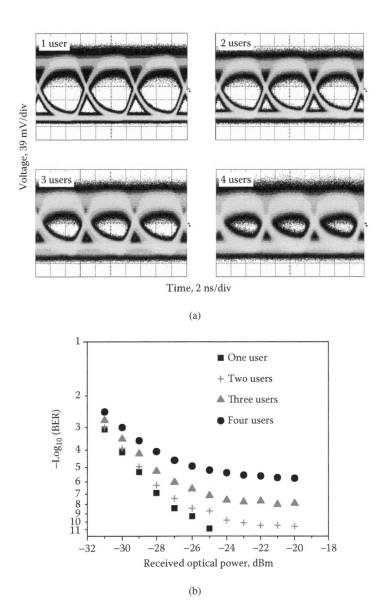

FIGURE 4.10 (a) Receiver eye diagrams and (b) BER measurements for different transmission scenarios.

identical in the case that the source is spectrally flattened. Otherwise, the grating reflectivities can be tailored to account for the source spectrum so that the power reflected by each wavelength band is the same [55]. With current grating fabrication technology, such requirements can be met within reasonably strict tolerances.

The primary limitation with SI-FBGs is the number of gratings that can be superimposed due to the finite photosensitive response of fiber. Moreover, all of the gratings require well-defined wavelengths which may be difficult to achieve in practice since the writing response is not necessarily linear due to saturation of the photoinduced index change, erasure of the grating index due to multiple exposures, and slight wavelength shifts due to an increase in the average effective mode index. Thus, a careful calibration process is required.

Finally, we note that it is also possible to design grating-based filters operating in *transmission* to implement SAE FOCDMA. In reflection, if a grating is weak to moderately reflecting and highly dispersive with an approximately linear group delay response, i.e., constant first-order dispersion $\ddot{\Phi}_v$ [ps^2] (such as can be obtained using a linearly chirped FBG), then its apodization profile $\upsilon(z)$ can be "mapped" directly into the magnitude of its corresponding complex reflection coefficient $|H^r(v)|$ [56]:

$$|H^r(v')| \propto \upsilon\left(z = \frac{c\ddot{\Phi}_v}{2n_{eff}}v'\right) \tag{4.24}$$

where $v' = v - v_0$ (with v_0 being the Bragg frequency at the input end of the grating and c being the speed of light in vacuum) is a scaled frequency. Equation 4.24 represents a space (z)-to-frequency(v) mapping and can be derived based on the observation that a linearly chirped FBG can be considered as a concatenation of many uniform gratings, each with a different period and reflecting a different spectral component according to the Bragg condition. In particular, due to the linear dependence of the grating period with the fiber length, the reflected (Bragg) frequency v also varies as a function of z according to the following relation:

$$v(z) = \frac{2n_{eff}}{c\ddot{\Phi}_v}z + v_0 \tag{4.25}$$

assuming that the waves are coupled exactly at the point where the phase-matching condition is verified. Going one step further, if the chirped FBG is also apodized, then each uniform grating segment has its own coupling coefficient $\kappa(z)$ as specified by the peak index modulation $\delta\overline{n}_{eff}$ and the apodization profile $\upsilon(z)$, i.e., $\kappa(z) \propto \delta\overline{n}_{eff}\upsilon(z)$. Physically, different spectral components incident upon the grating are reflected at different spatial positions with a reflection coefficient equal to the local refractive index modulation and the space-to-frequency mapping is ensured so as long as the dispersion $\ddot{\Phi}_v$ is sufficiently large.

At this point, we note that it is possible to define a more precise relationship between the grating apodization profile $\upsilon(z)$ and the magnitude of its reflection coefficient $|H_r(v)|$ as compared with the "direct" relation defined in Equation 4.24). To this end, we recall that the peak reflectivity for a uniform FBG at the resonant frequency v_0 is given by Equation 4.13. Following a development similar to the one presented above, we can write a new relationship between the grating apodization profile and the magnitude of its reflection coefficient:

$$|H_r(v')| \propto \tanh\left[mA\left(z = \frac{c\ddot{\Phi}_v}{2n_{eff}}v'\right)\right] \tag{4.26}$$

where m is a constant. A direct proportionality between the grating reflectivity and its coupling strength only occurs for very weak gratings, i.e., when $\kappa L \ll 1$ and as a result, Equation 4.13 can be approximated by $\left. |H_r(\nu = \nu_0)| \right|_{peak} \propto \kappa L$. Consequently, Equation 4.24 is simply a special case of Equation 4.26 that is only valid when the grating is weakly reflecting. However, Equation 4.26 is valid for weak as well as for strong gratings.

Based on this concept, we can design FBG-based spectral filters in a very simple fashion: the desired spectral response is spatially "recorded" in the grating apodization profile. However, since the space-to-frequency mapping occurs in reflection, the reflected signal can be chirped undesirably. To overcome this, we can operate the grating in transmission, which inherently induces nearly zero dispersion. We first determine the required filter transmission response $|H^t(\nu)|$ and then the corresponding reflection response $|H^r(\nu)| = 1 - |H^t(\nu)|$ is obtained from which the apodization profile $\upsilon(z)$ can be extracted using Equation 4.26. Note that both direct and complementary filters can be designed in this way.

For implementing SAE FOCDMA, linear arrays of FBGs, SI-FBGs, and transmission-based filters have their advantages and disadvantages. While transmission-based filters are simple in design and are nearly dispersionless (so there is no need for a second compensating grating structure), they cannot be tuned dynamically resulting in a static encoder/decoder and long structures may be required (smaller bandwidth features will require longer grating lengths). On the other hand, while linear arrays of FBGs or SI-FBGs can be tuned for providing reconfigurability, fabrication and packaging can be more complex and may require more careful control (especially for superimposed gratings). Ultimately, the choice grating structure will depend on the system requirements.

4.3.2 PHASE ENCODING

4.3.2.1 Overview

The 4-f pulse shaper shown in Figure 4.4 has been investigated extensively as a means for coherent optical signal processing [12,57,58]. In particular, the amplitude and/or phase of the spectral content of an ultrashort broadband optical input pulse is manipulated so that the output signal corresponds to the inverse Fourier transform of the signal appearing in the Fourier plane of the 4-f system. It has been used successfully for generating arbitrary pulse shapes and has developed into a mature technology, incorporating advanced functionality such as adaptive shaping, using programmable amplitude/phase filters based on liquid crystal modulator arrays.

In 1988, Salehi, Weiner, and Heritage recognized the potential of the 4-f pulse shaper system as a means for encoding and decoding femtosecond pulses [10,11]. In the encoder, a spatially patterned phase mask is used to scramble the phases of an input femtosecond pulse such that it becomes a pseudonoise burst several picoseconds in duration (the duration is defined by the spectral resolution of the filtering process that, in turn, is set by the spatial separation of the frequency content achieved with the grating and the resolution of the mask). In the decoder, a conjugate phase mask is used to unscramble the phases. When properly matched, the spectral phase shifts are removed and the original femtosecond pulse is reconstructed. Otherwise, the decoder simply respreads the encoded signal: the phases are rearranged but not unscrambled, and the pulse cannot be reconstructed. A thresholder, usually based on a nonlinear optical device, is used to distinguish auto-correlation peaks corresponding to properly decoded signals. Indeed, impressive encoding/ decoding results, as well as basic system trials have been demonstrated [59]. However, the approach may be limited practically due to lack of compactness and possible high cost.

With advances in grating fabrication technology, it is possible to realize very complex grating profiles to phase encode ultrashort pulses for pulse shaping and signal processing applications. Within the context of spectral phase encoded (SPE) FOCDMA, two main approaches have been considered using (1) segmented gratings [28,29,60] and (2) superstructure gratings (SS-FBGs) [35–37]. We first consider pulse shaping using FBGs operating within the weak-grating limit (or first-order Born approximation) and then discuss how these results can be applied to the generation of phase encoded signals.

4.3.2.2 Pulse Propagation in FBGs Under the Weak Grating Limit

We consider an FBG with a refractive index described by Equation 4.1. The grating reflection response can be characterized by the frequency transfer function (field reflection coefficient) $H_r(v) = |H_r(v)| \exp[j\Phi_r(v)]$ or alternatively, by the corresponding temporal impulse response $h_r(t) = \mathfrak{I}^{-1}\{H_r(v)\}$ where $\mathfrak{I}^{-1}$ denotes inverse Fourier transform. From a first-order analysis of the coupled-mode equations that govern wave propagation in a periodic medium perturbation, it can be shown that in the weak-grating limit, i.e., $|H_r(v)| \ll 1$,

$$\upsilon(z) \exp[j\varphi(z)] \propto \hat{h}_r \left(t = \frac{2}{c} \int_0^z n(z')dz' \right) \tag{4.27}$$

where $\hat{h}_r(t)$ is the complex envelope of the temporal impulse response $h_r(t)$ [61]. For a constant effective mode index n_{eff}, it is immediate from Equation 4.27 that

$$\hat{h}_r(t) \propto [\upsilon(z) \exp[j\varphi(z)]]_{z=\frac{c}{2n_{eff}}t} \tag{4.28}$$

for the interval $0 \le t \le 2n_{eff}L/c$ corresponding to the round-trip propagation time in the grating. This important result states that within the weak grating limit, the complex envelope of the reflection impulse response is proportional to the spatial index modulation profile of the grating in both amplitude and phase. Equation 4.28 effectively represents a direct space(z)-to-time(t) mapping, with a scale factor given by the speed of light in the grating medium, c/n_{eff}. Note that this mapping requires the reflected signal to closely resemble the grating impulse reflection response, which in turn requires the spectral bandwidth of the incident pulse to be larger than that of the grating reflection response $H_r(v)$.

4.3.2.3 General Phase Encoding/Decoding by Pulse Reflection From FBGs

By operating within the above conditions, the impulse reflection response (or ultrashort pulse reflection response) of a grating encoder has a temporal profile set by the corresponding complex refractive index profile. If $H_r^{enc}(v)$ represents the complex reflection response of the grating encoder and $E_i(v)$ is the input pulse spectrum, then the encoded signal is simply $E_{enc}(v) = E_i(v)H_r^{enc}(v)$. At the decoder, the encoded waveform is reflected from a second grating decoder with frequency response $H_r^{dec}(v)$ so that the decoded signal is given by $E_{dec}(v) = E_i(v)H_r^{enc}(v)H_r^{dec}(v)$ (this assumes either back-to-back transmission or that the transmission medium does not introduce any signal distortion). To properly recover the input signal, $H_r^{dec}(v) = [H_r^{enc}(v)]^*$ such that the decoded signal corresponds to the auto-correlation function of the grating. Physically, this requires that the grating profile of the decoder be the spatially reversed structure used at the encoder.

4.3.2.4 Segmented FBGs

The first demonstration of using FBGs for SPE FOCDMA was realized by Grunnet-Jepsen and colleagues [29,30]. The encoder structure is based on a segmented fiber grating, or essentially, a linear array of uniform FBGs with different peak refractive index modulations and spatial relative phases relative to a fixed coordinate system as shown in Figure 4.11(a). The main differences between the linear array of FBGs used to implement SAE and SPE are (1) the use of discrete phase shifts rather than effective phase changes created by the physical separation between gratings and (2) the peak index modulation is smaller for SPE. When a short input pulse is reflected from the grating, the reflected or encoded signal, comprises a train of time-delayed pulses having a duration equal to the chip window. The relative amplitudes, temporal phases, and bandwidths of the pulses in the encoded signal are determined by the peak refractive index modulation, phase, and length of the individual subgratings in the segmented structure, respectively. The length of the grating structure determines the round-trip propagation time, which in turn sets the bit window. The number of chips determines the code length and the chip rate. The matched grating decoder is the same grating structure as the encoder but with the light incident on the opposite end, and is used reconstruct the original short pulse. Again, modeling a segmented FBG is straightforward using the fundamental matrix formalism, with the addition of the following matrix between grating segments to represent the discrete phase shift ϕ:

$$[F]^{PS} = \begin{bmatrix} \exp\left(-j\dfrac{\phi}{2}\right) & 0 \\ 0 & \exp\left(j\dfrac{\phi}{2}\right) \end{bmatrix}. \tag{4.29}$$

In their proof-of-concept demonstration involving a sparse two-user system (the encoded waveforms did not overlap temporally), they used grating encoders based on a uniform FBG 2.4 cm in length that was divided into eight segments of specific

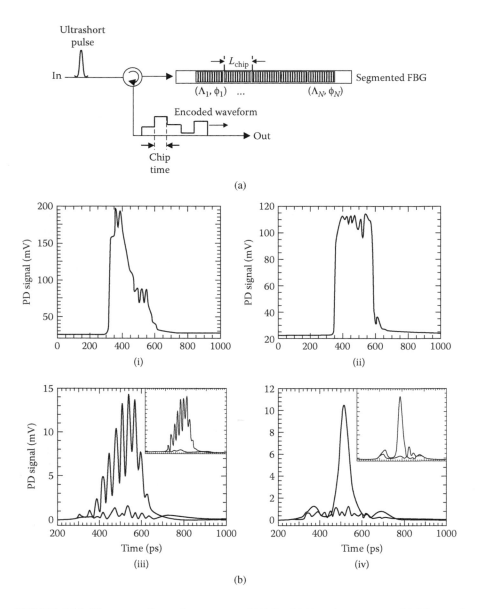

FIGURE 4.11 Phase-encoding using segmented FBGs. (a) Encoder structure. (b) Measured encoded and decoded waveforms, (i) and (ii), and decoded waveforms (following propagation through 20 km of SMF), (iii) and (iv), for both codes. For each code, auto-correlation and cross-correlation results are shown. The insets correspond to simulations. (Adapted from Grunnet-Jepsen, A. et al., 1999, *IEEE Photon. Technol. Lett.*, 11(10), 1283–1285, with permission from IEEE © 2004.)

phase and amplitude. The corresponding chip times were ≈30 psec and the encoded waveform duration was ≈240 psec. The codes were based on imparting discrete phase shifts of 0, $\frac{2\pi}{3}$, or $\frac{5\pi}{3}$ between grating segments and each segment had a reflectivity

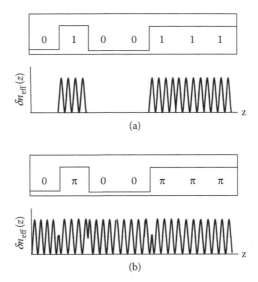

FIGURE 4.12 Schematic examples of temporal codes and corresponding refractive index modulation profiles along the SS-FBG: (a) unipolar code and (b) bipolar code.

of $\approx 1\%$. The specific phase codes used in the demonstration were $(0, \frac{2\pi}{3}, \frac{5\pi}{3}, 0, \frac{2\pi}{3}, \frac{5\pi}{3}, 0, \frac{2\pi}{3})$ for one code and $(0, 0, 0, \frac{2\pi}{3}, \frac{5\pi}{3}, 0, \frac{5\pi}{3}, \frac{5\pi}{3})$ for the second code. Figure 4.11(b) shows the measured encoded waveforms as well as the decoded signals following propagation through 20 km of SMF. The decoded signal shows a 15:1 contrast ratio between the auto-correlation peaks corresponding to the desired user and unwanted noise corresponding to the interfering user.

4.3.2.5 SS-FBGs

An SS-FBG is a conventional uniform period FBG onto which an additional, slowly varying refractive index modulation profile is imposed. This slowly varying profile can change the overall amplitude or add discrete phase shifts to the grating structure. By operating within the weak grating limit, the impulse response of the SS-FBG will have a temporal response determined by the slowly varying refractive index profile. In particular, when an ultrashort optical pulse is reflected from an SS-FBG with discrete phase shifts, the output comprises a series of coherent pulses whose relative phases are set by those in the grating structure.* An example of a 7 chip bipolar code and the corresponding SS-FBG refractive index modulation is shown in Figure 4.12(b). The grating has a uniform amplitude with discrete $\pm\pi$ phase shifts added at the boundaries of adjacent spatial chips (namely, a π phase shift is added if the two corresponding adjacent code elements are different).

---

* Note the similarity between an SS-FBG and the segmented structure discussed in section 3.2.4. Strictly speaking, the gratings used by Grunnet-Jepsen et al. are SS-FBGs. Segmented FBGs can be considered more general since they can comprise a concatenation of gratings having different periods.

As illustrated in Figure 4.12(a), it is also possible to define unipolar codes using SS-FBGs in which the amplitude of the refractive index varies in accordance with the code.

We now illustrate the principle of encoding and decoding using bipolar codes implemented with SS-FBGs. We consider two bipolar codes, $c_1 = (0,\pi,0,0,\ \pi,\pi,\pi)$ and $c_2 = (\pi,\ 0,\ \pi,\ 0,\ 0,\ \pi,\ \pi)$, and use 2 psec sech pulses as the ultrashort input pulses. The SS-FBG encoder comprises seven chips, each 1 mm in length (corresponding chip duration of 9.7 ps assuming $n_{eff} = 1.452$) and the total grating encoder is 7 mm long (encoded waveform duration ≈ 67.8 ps). The peak index modulation is 3×10^{-5}. Figure 4.13 shows the spectra of the grating encoders, the corresponding encoded

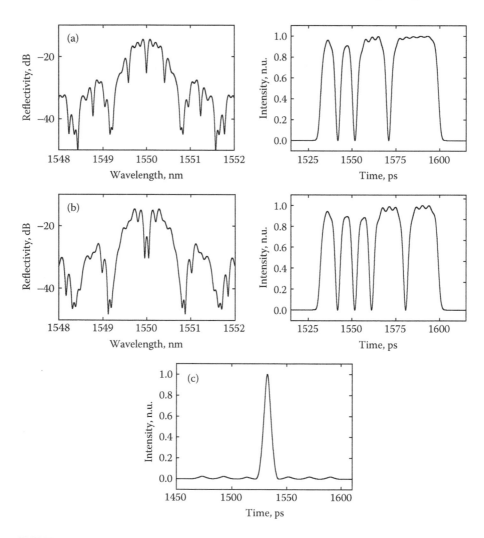

FIGURE 4.13 Calculated reflectivity of SS-FBG encoder and encoded waveform for (a) code c_1 and (b) code c_2. (c) Decoded waveform of code c_1.

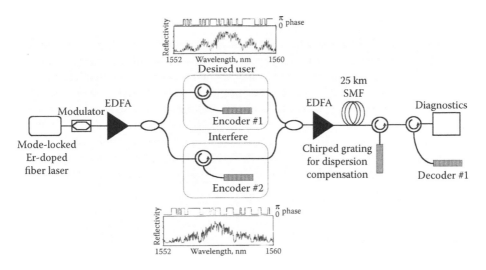

FIGURE 4.14 Experimental setup for demonstrating SPE using SS-FBGs.

waveforms, and the decoded waveform for c_1 (with no interfering user). The decoder comprises the physically reversed SS-FBG encoder structure and the response shows clearly a single distinct auto-correlation peak.

The use of SS-FBGs for implementing SPE FOCDMA with bipolar codes has been investigated extensively by researchers at Southampton University [35–37]. In particular, Teh and colleagues have performed a series of investigations using 7 chip and 64 chip bipolar Gold sequences. In these demonstrations, they use chip lengths of 0.66 mm, corresponding to chip times of 6.4 psec or chip rates of approximately 160 Gchip/sec. For 7 (64) chip codes, system operation at 10 Gb/sec (1.25 Gb/sec) is possible. The experimental setup for a two-user demonstration system is shown in Figure 4.14. The input pulses are generated from a mode-locked fiber laser producing transform-limited soliton pulses approximately 2.0 psec to 2.5 psec in duration. Figure 4.15(a) shows the measured decoded waveforms illustrating matched and unmatched filtering of two different 64 chip bipolar codes. The actual codes and corresponding grating responses for the two users are shown in the inset of Figure 4.14. Matched filtering results in a clear auto-correlation peak and no distinct cross-correlation spikes are discernible. To quantify system performance, data from the two users were transmitted over 25 km of SMF. A linearly chirped FBG was used before the decoders and receiver in order to compensate for chromatic dispersion. BER measurements at 1.25 Gb/sec are shown in Fig. 15(b). When only the desired user was transmitting, error-free transmission was possible and a negligible power penalty of 0.3 dB was incurred following transmission through the fiber. When the interfering user was added, the power penalty increased to 2.6 dB due to multiuser interference (this can be seen from the cross-correlation results shown in Figure 4.15).

4.3.2.6 Discussion

With SPE, since the phase is being manipulated, it is important to properly compensate transmission impairments, especially those due to chromatic fiber dispersion. If transmission impairments associated with dispersion are properly accounted for, the focus shifts to reducing multiuser interference in order to improve SNR and enhance BER performance. This can be done using several approaches. For example, a nonlinear optical loop mirror can be used to remove the low-level pedestal associated with the presence of the interfering channels and the finite background seen even for properly decoded pulses obtained by matched filtering [62,63]. Time gating can suppress additional sidelobes in the decoded signal and also provides additional

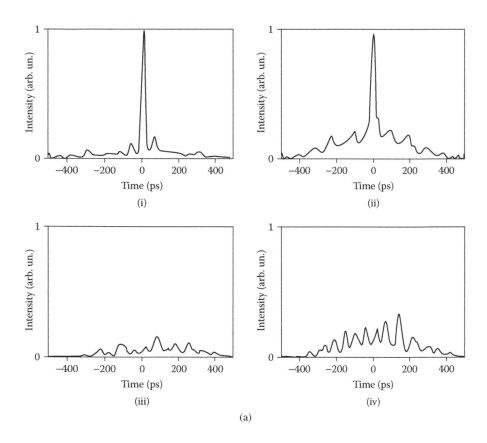

FIGURE 4.15 Results of experiment. (a) Measured signals after decoding for 2.5-psec soliton input pulses for the grating combinations: (i) encoder 1: decoder 1, (ii) encoder 2: decoder 2, (iii) encoder 2: decoder 1, and (iv) encoder 1: decoder 2. (b) BER measurements for the encoder 1: decoder 1 combination (closed circles: laser back-to-back; closed squares: no transmission; open squares: after transmission; closed triangles: with second channel present and no transmission; open triangles: with second channel present and transmission). (Reprinted from Teh, P.C. et al., 2001, *IEEE/OSA J. Lightwave Technol.* 19(9), 1352–1365, with permission from IEEE © 2004.)

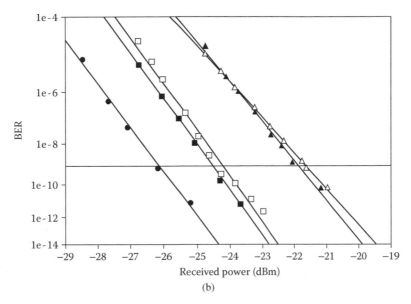

FIGURE 4.15 (Continued).

resilience to degradation arising from transmission impairments [64]. Finally, multiuser interference can also be reduced further using quaternary phase encoding, i.e., using discrete phase shifts of 0, 0.5π, π, and 1.5π as opposed to 0 and π as in the binary case. Recently, a four-user system employing 255 chip quaternary phase codes at a chip rate of 320 Gchip/sec was demonstrated [65].

In order to support reconfigurable operation, it is necessary to tune the discrete phase shifts of the SS-FBGs. Since phase shifts induced at the time of grating fabrication are permanent, alternate means are required to obtain tunable phase shifts. Effective phase shifts can be created using highly localized temperature effects [66]. The localized temperature will cause a change in the propagation constant creating a confined chirp, which has a similar effect to a discrete phase shift. Such localized temperature changes can be implemented by passing a fine electric conductor over the FBG and passing a current to obtain resistive heating. By varying the current, and hence temperature, the effective phase shifts can be varied (these changes are impermanent so as long as the temperature does not exceed the grating erasure temperature which is about 150° C). Thus, a phase encoder with multiple tunable phase shifts can be formed from a single uniform FBG. The local grating phase is changed by varying the temperature along the grating according to desired code sequence as shown schematically in Figure 4.16. Such a structure has been used as a reconfigurable encoder for 16 chip quaternary phase codes at a chip rate of 20 Gchip/sec [67]. In a simple system measurement, the use of the reconfigurable encoder caused only a 2.5 dB power penalty compared to the use of a static encoder. Indeed, impressive results have been demonstrated and illustrate the potential for using phase-shifted SS-FBGs for SPE FOCDMA.

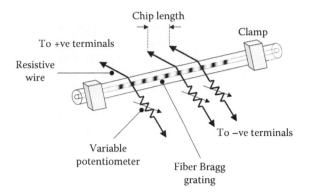

FIGURE 4.16 Schematic of a reconfigurable FBG phase encoder/decoder based on localized thermal effects.

4.3.3 WAVELENGTH-TIME ENCODING

4.3.3.1 Overview

While results have demonstrated the potential for SAE and SPE FOCDMA systems, performance can be limited ultimately by the one-dimensional nature of the codes. On the other hand, the use of multidimensional codes allows for fundamental improvements in the performance of FOCDMA systems. There are many types of hybrid multidimensional coding schemes; those that have attracted recent interest are based on two-dimensional (2D) wavelength-time codes [17–20,26,28,42,43,45].

Figure 4.17 illustrates the encoding process of a unipolar 2D wavelength-time code. A broadband pulse representing a "1" bit is wavelength encoded using optical filters that select the desired wavelengths as defined by the code sequence. These wavelengths are then encoded in the temporal domain (temporally spread) using optical time delays. For matched filtering, the decoder filters the same wavelengths

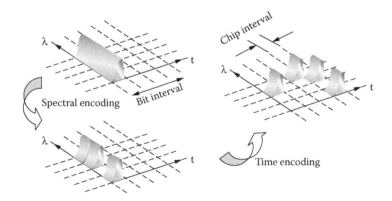

FIGURE 4.17 (**Color Figure 4.17 follows page 240.**) Schematic of encoding of 2D wavelength-time code.

as the encoder and the time delays for the corresponding wavelengths are reversed in order to temporally despread the signal, i.e., the wavelengths are realigned into the same time chip.

2D wavelength-time codes have several advantages for implementing incoherent FOCDMA. First, active code chips are defined not only by their temporal position in the sequence but also by their wavelength. Thus, since cross-correlation is defined in the time domain, multiuser interference is reduced. Second, by exploiting the additional degree of freedom in using multiple wavelengths together with temporal spreading, there is increased flexibility in code design, especially in terms of achieving specific auto-correlation or cross-correlation constraints. Third, compared to DS FOCDMA, a higher bit rate can be accommodated for the same code dimension since in 2D, the aggregate code dimension is given by the product of the number of wavelength slices and time chips. Alternatively, to maintain a reasonable bit rate, very short chip times are not required which eases matched filtering and detection processes. Taken together, these features support code families with increased cardinality and with improved system performance in that a larger number of simultaneous users can operate at a specified BER.

Many different code families have been developed for 2D wavelength-time FOCDMA. These include frequency-hopping sequences from wireless CDMA technology, such as Fast Frequency Hopping (FFH) [28,29] or prime-hop (PH) sequences [17], codes based on mapping algorithms to produce 2D spreading sequences from 1D prime sequences [20], 2D spatial optical orthogonal codes [13–15], and newly developed wavelength-time codes [18,20,42].

In terms of technology, research has focused on developing suitable encoder/decoder structures that implement the codes, such as the use array waveguide gratings or multiwavelength lasers combined with fiber optic delay lines [70–74]. The wavelength selective and dispersive (group delay) properties of FBGs also make them suitable for implementing wavelength-time codes. Advantages of the FBG approach include all-fiber encoding/decoding for increased efficiency and the potential to support reconfigurable systems using tunable devices. In this section, we discuss two grating structures that have been considered for implementing 2D wavelength-time FO-CDMA: linear arrays of FBGs and chirped gratings.

4.3.3.2 Single and Parallel Linear FBG Arrays

As discussed in Section 3.1, a linear array of FBGs can be used to implement the codes in an SAE FOCDMA system. One advantage is that the use of tunable gratings allows for a reconfigurable encoder/decoder structure. However, in SAE FOCDMA applications, the disadvantage of the linear array was the need of an additional structure comprising the same gratings but in reverse order in order to compensate for round-trip delays so that all of the reflected wavelengths have the same delay. On the other hand, it is precisely this delay between reflected wavelengths that allow a linear array of FBGs to implement wavelength-time codes since the array effectively decomposes a broadband input pulse simultaneously in wavelength and time [25].

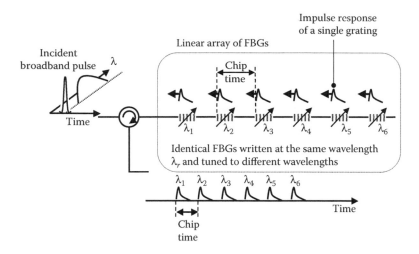

FIGURE 4.18 2D wavelength-encoding/time-spreading encoder/decoder using serial FBG array. The decoder is the same structure as the encoder with the wavelengths in the reverse order.

Figure 4.18 illustrates the principle of generating a 2D wavelength-time code using a linear FBG array. Consider a broadband input occupying the wavelength range $[\lambda_{min}, \lambda_{max}]$ that is incident on the encoder structure. Each FBG reflects a specific wavelength as defined by the wavelength-time code. Based on the *first-in, first-out principle*, the corresponding positions of the FBGs in the array define the time chips of the reflected wavelengths. More concretely, if the ith FBG in the linear array is set at $\lambda_i \in [\lambda_{min}, \lambda_{max}]$, then the reflection at λ_i will occupy the ith time slot. Note that the bit window $T = 1/B = nT_c$ is determined by the round-trip propagation time through the complete linear array (B is the data rate, n is the number of FBGs in the linear array, and T_c is the duration of a code chip). The weight of the code defines the number of FBGs used in the linear array. Moreover, their physical separation (grating to grating) corresponds to the chip time. For a given code weight, higher data rates can be accommodated using shorter chip times, or shorter physical separations between gratings.

For proper decoding, the peak wavelengths of the FBGs must be arranged in the reverse order to temporally despread the wavelengths (of course, the FBGs in the decoder should filter the same wavelengths as those in the encoder). If the FBGs in the decoder do not physically occupy the reverse positions as in the encoder, then the wavelengths are simply spread further temporally. Thus, the encoder and decoder to implement a given code are identical structures, with encoding/decoding performed by bi-directional reflections (i.e., from opposite ends) from the linear array of FBGs. As mentioned earlier, the response of a linear array of FBGs is straightforward to compute using the fundamental matrix.

There have been several recent demonstrations of using linear arrays of FBGs for 2D wavelength-time FOCDMA. For example, Kitayama and colleagues used linear arrays of FBGs to implement PH sequences [31,43,45]. PH sequences are obtained by using a combination of prime sequences to define the temporal spreading and wavelength hopping patterns. For a given prime number p, the code weight and length are, respectively, p and p^2; a total of $p(p - 1)$ codes can be generated. The codes have an auto-correlation peak intensity of p with no sidelobes and the cross-correlation is at most 1. The only difference in the grating encoder/decoder structure used in these demonstrations is that the gratings are chirped as opposed to having a uniform period. As before, the total encoder length is set nominally by the bit rate (i.e., the round-trip propagation time of the grating encoder is approximately equal to the bit window). The bit window is divided in p^2 chips of equal duration (the encoder length is divided into p^2 segments of equal length) and the p FBGs are suitably located so that each reflected wavelength occupies its designated time chip in the code. In a simple two-user experiment (see Figure 4.19), PH sequences with $p = 3$ were considered. The chip times were 35 psec and each FBG had a reflection bandwidth of 3 nm. A supercontinuum source that generates 4 psec pulses with over 170 nm bandwidth (centered at 1550 nm) at 2.5 Gb/sec was used as the input. Since the gratings are chirped, each reflected wavelength occupies a duration longer than the chip time; however; the inverse chirp of the decoder (which is the same structure as the encoder only operated from the opposite end) compensates for the elongation of the reflected pulses. The group delay characteristics of the two grating encoders and decoders are shown in the insets of Figure 4.19. Figure 4.20 shows the codes and corresponding encoded waveforms, as well as typical auto-correlation and cross-correlation traces. The properly decoded waveform consists of a sharp pulse with no sidelobes and has a peak approximately three times larger than the cross-correlation (as expected based on the properties of the codes used). Error free transmission at 2.5 Gb/sec over 33 km of dispersion-shifted fiber was also obtained with negligible power penalty compared to back-to-back transmission. Finally, data rate enhancement was also demonstrated by allowing the bit window to be less than the round-trip propagation time through the encoder.

In another series of demonstrations, researchers at the University of Laval demonstrated a 16-user system at 1.25 Gb/sec using FFH codes based on a bin of 30 frequencies spaced by 50 GHz and a weight of 8 [75]. Each FFH code is implemented with a linear array of 8 FBGs, each 14 mm in length and separated by 1 mm for a total length of 11.9 cm. Four series of five identical linear arrays are used and the different codes are obtained by strain-tuning the appropriate grating array. The codes and spectral response of the five identical linear arrays are shown in Figure 4.21(b). The chip time is $\approx$145 psec and the data rate is $\approx$860 Mb/sec, as dictated by the round-trip propagation time in the encoder ($\approx$1.15 nsec). Note that higher transmission rates are possible if two data pulses are allowed to be simultaneously present in the encoder (as the results show, this does not degrade the matched filtering process). The reflectivity and 3-dB bandwidths of the FBGs are 13 dB and 20 GHz (0.16 nm). The experimental setup is shown in Figure 4.21(a). An electro-optic modulator generates pulses (from an incoherent broadband source) approximately 100 psec in duration (the pulses do not fill an entire chip time) in the data pattern "1 0 0 0 1 1 0". The modulated data is directed to 16 different encoders and the encoded signals are transmitted through 77 km of standard single-mode fiber and a suitable length of dispersion compensating fiber. The transmitted signals are then decoded

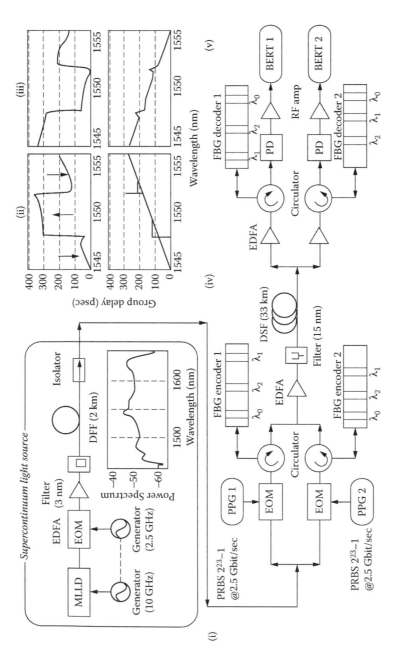

FIGURE 4.19 Experimental set-up for 2-user 2D wavelength-time FOCDMA system based on PH sequences. (i) power spectrum of supercontinuum source; group delay response for encoders (ii), (iii), and decoders (iv), (v). (Reprinted from Wada, N. et al., 2000, *IEE Electron. Lett.*, 36(9), 815–817, with permission from IEE © 2004 and N. Wada.)

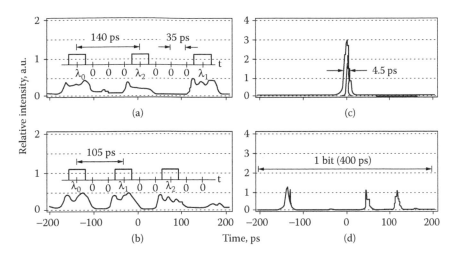

FIGURE 4.20 Measured waveforms: (a), (b) encoded waveforms corresponding to the two codes shown in the inset, (c) auto-correlation for code 1, and (d) cross-correlation of code 1 using decoder 2. (Reprinted from Wada, N. et al., 2000, *IEE Electron. Lett.*, 36(9), 815–817, with permission from IEE © 2004 and N. Wada.)

and observed using an optical sampling module with 30-GHz bandwidth. Figure 4.21(a) shows the optical spectrum and temporal waveform of the signal comprising data from the 16 users before propagation in the fiber span. Figure 4.22 compares the decoded signal for transmission back-to-back and through the fiber span with the output pulses from the modulator. The decoded signal clearly exhibits auto-correlation peaks (corresponding to the desired user), which can be distinguished from the interference of the other 15 transmitting users and there was little degradation following transmission through the fiber span. Similar results were also obtained when the data rate was increased to 2.5 Gb/s. Thus one other important conclusion from these experiments is that the bit rate may be higher than what is initially set by the round-trip propagation time in the encoder/decoder structures.

Yet another code family that can be used for wavelength-time FOCDMA systems are 2D spatial codes [13–15]. A 2D spatial code consists of a 2D array ($m \times n$) of "1"s and "0"s to indicate which spatial elements are active in defining a user code (the total number of "1"s in the array gives the code weight w). Spatial codes can be used to represent wavelength-time codes where the "1"s in the array define the wavelengths and their corresponding temporal positions. In this case, the m rows represent the wavelengths and the n columns represent the time chips (slots) in each code. Spatial codes can support multiple weight, multiple pulse per row (MMPR), which allows for wavelength reuse, or multiple pulse per column (MPPC) in which two or more wavelengths occupy the same time slot. These properties, which are unique to spatial codes, may offer increased flexibility in code design and other potential advantages such as varying levels of QoS [15]. While a single linear array of FBGs is well suited for implementing wavelength-time codes based on frequency-hopping sequences, they are not necessarily adaptable for 2D spatial codes.

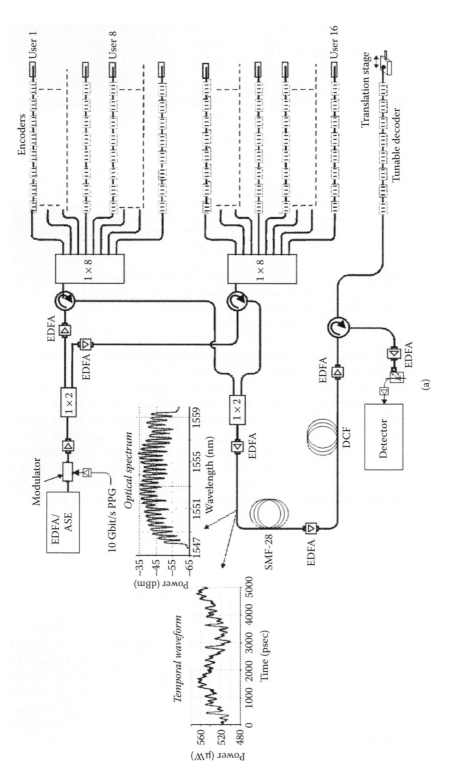

FIGURE 4.21 (a) 16-User wavelength-time FOCDMA demonstrator system at 1.25 Gb/sec based on FFH codes.

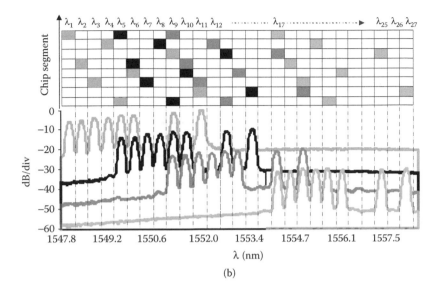

FIGURE 4.21 (Continued). (**Color Figure 4.21b follows page 240.**) (b) Codes used in demonstrator system and spectral response of grating encoders.

A logical extension of using a single linear array of FBGs to implement wavelength-time codes is to use parallel linear arrays [38]. Consider a wavelength-time FOCDMA system based on 2D spatial codes of size $m \times n$, weight w, and the possibility of allowing $\tilde{w}$ ($\tilde{w} \leq w$) "1"'s per column (i.e., up to $\tilde{w}$ wavelengths can occupy the same time slot). Furthermore, assume that the m wavelengths occupy the window [λ_{min}, λ_{max}] and are equally spaced apart. Figure 4.23 shows a parallel linear array of FBGs that can be used for implementing such codes. The encoder comprises a $1 \times \tilde{w}$ power divider/combiner where each arm is connected to a linear array of n FBGs, each separated by a constant length L_{sep} (this is the separation between the midpoints of consecutive gratings). In total, there are $\tilde{w} \times n$ FBGs. If we use tunable gratings, each can be set at the same wavelength λ_r where $\lambda_r \notin [\lambda_{min}, \lambda_{max}]$.

The FBGs in each array are tuned as necessary, for example by applying a compressive or tensile strain, to reflect the required wavelengths defined by the spatial code (in order to have a small L_{sep} to ensure compactness or to support a higher data rate, the FBGs can be arranged on a segmented piezoelectric stack where each segment can be individually addressed to independently strain the FBGs as required). The positions of the tuned FBGs in the array define the time slots of the reflected wavelengths, i.e., the *first-in, first-out principle*. Tuning FBGs having the same position in each array (for example, the jth FBG in multiple arrays) so that $\lambda_r \in [\lambda_{min}, \lambda_{max}]$ will result in multiple wavelengths occupying the same time slot (the jth one) thereby realizing MPPC spatial codes. On the other hand, tuning FBGs occupying different physical positions in the arrays to the same wavelength allows MPPR spatial codes to be realized (to minimize the effects of spectral shadowing, the FBGs tuned to the same wavelength can come from different arrays). If an FBG is not tuned from its nominal wavelength λ_r, then it represents a "0" in the 2D matrix.

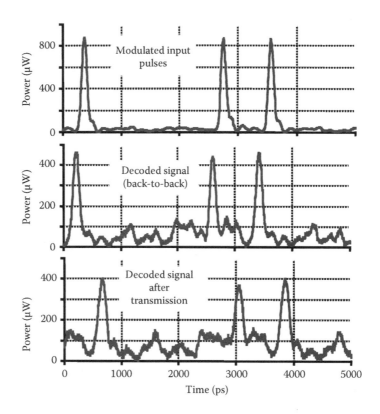

FIGURE 4.22 Results of demonstrator experiments: top, broadband data pulses at 1.25 Gb/sec; middle, decoded signal for desired user (back-to-back); and bottom, decoded signal for desired user after transmission through fiber span. [OFC paper TuV3, 2001.]

Finally, the code weight w depends on the total number of FBGs that are detuned with $\lambda_r \in [\lambda_{min}, \lambda_{max}]$ — this can be easily varied so that codes with different weights can be generated. Note that it is also possible to use static FBGs in the parallel arrays. In fact, we would only require as many FBGs as the code weight. However, we do not have the flexibility of reconfiguration so a trade-off will need to be made between encoder/decoder reconfigurability and the required resources.

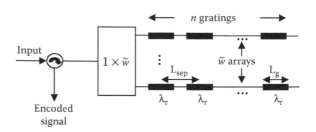

FIGURE 4.23 FBG-based encoder for implementing 2D spatial codes.

To demonstrate the principle of such an encoder structure, we consider spatial codes (SCs) comprising five wavelengths and five time chips and with a weight of 3. Each linear array comprises five identical FBGs and we use two such arrays in parallel. Each FBG is Gaussian apodized and has a peak refractive index $\delta\overline{n}_{eff} = 3.7 \times 10^{-4}$, length $L = 1.2$ cm, and $L_{sep} = 1.8$ cm. The 3-dB bandwidth of the gratings is 50 GHz; the wavelengths are separated by 100 GHz and the central wavelength is 1550 nm. Figure 4.24(a) shows the grating responses of encoders corresponding to the three spatial codes (denoted SC1, SC2, and SC3) shown in the inset (in a physical implementation, λ_r would be set outside the range of interest [1548.4 nm, 1551.6 nm] and the FBGs would be tuned accordingly to select the appropriate wavelengths). The amplitude response (solid lines) and the corresponding group delays (dashed lines) are plotted together to illustrate the time slot of each reflected wavelength. Clearly, the combined amplitude and group delay response represent the appropriate wavelength-time relationship defined by the spatial codes. Specifically, for the SC2, λ_3 is reflected in two different time slots (the two group delay curves, which have the same reference point, are for the two arrays in the encoder) whereas for SC3, we see that λ_1 and λ_3 are reflected in the same time slot.

We then reflect 1-psec Gaussian pulses centered at 1550 nm from the different encoders. In these simulations, the short pulse is used simply to provide large bandwidth. In practical realizations, any incoherent broadband input signal can be used. Figure 4.24(b) shows the corresponding encoded signal. The encoded waveforms are as expected from the encoder amplitude/group delay response. For example, in the encoded signal of SC3, there are only two pulses. However, the latter pulse contains two wavelengths. This can be observed from its larger amplitude and by comparison with the corresponding inset, which shows the encoded waveform from only one of the two serial arrays in the encoder. Note that since the input power spectrum is not flat, there is a slight variation in the peak of the signals in the encoded waveform. Ideally, this variation would not exist. The same structure used for encoding can be used for decoding. In this case, the tuned FBGs need to provide the inverse group-delay response (the physical positions are in reverse order compared to the encoder) so that the temporally separated wavelengths in the encoded signal can be despread. As an example, we also show in Figure 4.24(b) the properly decoded signal associated with spatial code SC1.

Finally, we note that such a 2D grating encoder can be used to implement a newly developed family of multiple-length, constant weight wavelength-time codes. With these codes, short code patterns perform better than long patterns in the same code set. This property allows faster rate services to have better performance (or higher priority).

4.3.3.3 Chirped Moiré Gratings

An alternative FBG-based implementation of wavelength-time codes is to use chirped grating structures, one of which is the chirped moiré grating (CMG) [28]. A CMG consists of two superimposed linearly chirped Bragg gratings [76,77]. The refractive index modulation representing a CMG can be written as

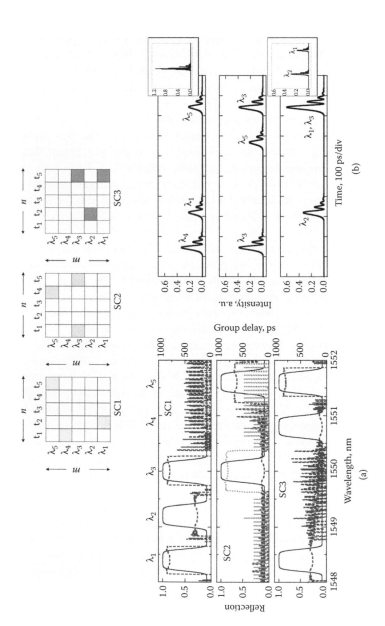

FIGURE 4.24 (Color Figure 4.24 follows page 240.) (a) Reflection (solid line) and group delay (dotted/dashed lines) of encoder structures (comprising two parallel linear arrays of FBGs) for the spatial codes shown in the inset. (b) Corresponding encoded signals obtained by reflecting a 1-psec Gaussian pulse from the grating encoders. The inset in the top figure shows the corresponding decoded signal of SC1 and the inset in the bottom figure shows the encoded signal corresponding to only one of the two arrays in the encoder structure. For the insets, the time scale is 100 psec/div. The slight variation in the waveform peaks is due to the nonuniform power distribution of the broadband input.

$$\delta n_{eff}(z) = \overline{\delta n}_{eff,1} \cos\left(\frac{2\pi z}{\Lambda_1}\right) + \overline{\delta n}_{neff,2} \cos\left(\frac{2\pi z}{\Lambda_2}\right) = \overline{\delta n}_{eff}\upsilon(z)\cos\left(\frac{2\pi z}{\Lambda_s}\right) \qquad (4.30)$$

where $\overline{\delta n}_{eff,i}$ are the peak refractive index modulations of the two superimposed gratings (assumed to be equal, $\overline{\delta n}_{eff,1} = \overline{\delta n}_{eff,2} = \overline{\delta n}_{eff}$); $\upsilon(z) = \cos(2\pi z/\Lambda_c)$; $\Lambda_c = 2[|1/\Lambda_1 - 1/\Lambda_2|]^{-1}$ and $\Lambda_s = 2[1/\Lambda_1 + 1/\Lambda_2]^{-1}$ are respectively the period of the slowly varying envelope and rapidly varying component of the grating; and Λ_1 and Λ_2 are the periods of the superimposed gratings, each varying linearly as a function of position: $\Lambda_i = \Lambda_{0,i} + (\delta\Lambda_i/\Lambda_{0,i})z$, $i = 1, 2$. There are crossover points of the beat in the grating fringe pattern where the phase of the grating changes by π, each producing a passband in the transmission response. The multiple passbands can also be interpreted as Fabry-Pérot resonances of the two mirrors formed by superimposing the two gratings. Obviously, these passbands have the effect of creating spectrally separate stopbands in reflection. The grating chirp in part determines the stop bandwidth of the individual gratings and the combined stop bandwidth of their superposition. The difference in the central wavelengths of the two gratings, given by $\Delta\lambda = 2n_{eff}|\Lambda_{0,1} - \Lambda_{0,2}|$, determines the number of and spacing between passbands in the transmission response.

In Figure 4.25(a), we show the calculated spectral response of a 3-cm CMG with $\overline{\delta n}_{eff} = 8.0 \times 10^{-4}$, $\lambda_1 = 2n_{eff}\Lambda_{0,1} = 1550.0$ nm, $\lambda_2 = 2n_{eff}\Lambda_{0,2} = 1550.2$ nm, and equal chirp parameters for both gratings $\delta\Lambda_1 = \delta\Lambda_2 = 5.0 \times 10^{-14}$ m²/m. This CMG has seven reflection bands. Of greater interest is the corresponding group delay in

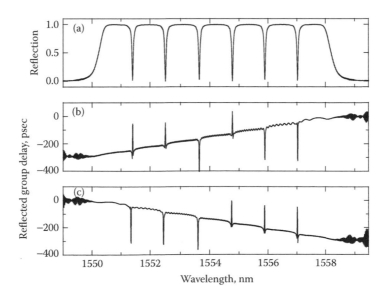

FIGURE 4.25 (a) Reflection response of a 3-cm-long CMG with $\overline{\delta n}_{eff} = 4.0 \times 10^{-4}$, $\lambda_1 = 1550.0$ nm, $\lambda_2 = 1550.2$ nm, and equal chirp parameters for both gratings $\delta\Lambda_1 = \delta\Lambda_2 = d\Lambda = 5.0 \times 10^{-14}$ m²/m; (b) corresponding reflected group delay of the CMG; and (c) its physically reversed structure.

reflection, shown in Figure 4.25(b). Since the relationship between reflected wavelengths and time is linear (due to the linear chirp in the gratings), each reflection band is reflected in its own time slot. A CMG in reflection can then be used to decompose a short broadband pulse simultaneously in both wavelength and time domains and has similar functionality to a linear array of FBGs. The physically reversed CMG structure has the same reflectivity (spectral amplitude response) and, to first order, an opposite reflected group delay, see Figure 4.25(c). We can thus envision encoding and decoding broadband pulses by successive reflections from a CMG (encoding) and its physically reversed structure (decoding).

A CMG has N spectrally separate reflection bands centered at λ_i ($i = 1, \ldots, N$) which occupy N time chips each having a duration $T_c \approx T/N$ where T is the round-trip propagation time through the grating and could correspond to the bit window. This results in a code length equal to N. Since the reflection bands are reflected sequentially in time due to the linear relationship between reflected wavelengths and time, a CMG gives rise to only one code with a weight $w = N$, where the weight corresponds to the number of pulses in the coded waveform (and equivalently, the number of wavelengths). Suppose that from a given CMG we choose only q of the available N reflection bands (wavelengths) for one code so that $w = q < N$. This can be accomplished by selectively eliminating stopbands from the CMG response as described in [78,79]. The code length is still N as defined by the number of time slots corresponding to the original CMG; however, $N - w$ of the entries in the code are 0 (no signal) and the w nonzero entries correspond to the wavelengths of retained reflection bands. We can then define a maximum of $\tilde{m} = [N/Q] = [N/w]$ codes that are strictly orthogonal (the gratings corresponding to these codes have no spectral overlap so that the cross-correlation is identically 0), where [•] denotes the integer part of the argument. To generate additional codes, the $m \leq \tilde{m}$ strictly orthogonal codes that we define can be simply wavelength shifted. These new codes are only quasi-orthogonal since there may be some spectral overlap between the original and shifted codes.

A detailed discussion of code design that accounts for the properties of the CMG encoders is given in [28]. However, some details are provided here, especially in order to understand the limitations that the grating structures can impose. Assume that the N reflection bands of a given CMG occupy the range $[\lambda_{min}, \lambda_{max}]$ which spans the entire bandwidth of the broadband input pulse, i.e., $[\lambda_{min}, \lambda_{max}] = [\lambda_{min}^{pulse}, \lambda_{max}^{pulse}]$. In the following, we will use the notation where the wavelengths represent the central wavelengths of the reflection bands. We can then write $\lambda_{max} = \lambda_{min} + (N - 1)\Delta\lambda$ where $\Delta\lambda$ is the bandwidth of a reflection band. If we measure wavelength in units of $\Delta\lambda$ so that in these normalized units $\Delta\lambda = 1$, then $\lambda_{max} = \lambda_{min} + (N - 1)$. As mentioned earlier, we can choose to define m ($\leq \tilde{m} = [N/w]$) codes that use only w of the available N reflection bands, (i.e., a code of length N having a weight w). These strictly orthogonal codes define m time-spreading patterns coupled to m wavelength-encoding patterns. We will refer to them as the original m codes. Note that the combined wavelength selective and dispersive natures of the CMGs are responsible for coupling the time-spreading and wavelength-encoding patterns. They are not independent. Additional codes can be obtained by wavelength-shifting the original m codes: all the wavelengths in each of the m time-spreading patterns are simply shifted by $n\Delta\lambda = n$, an integer multiple of the bandwidth of a reflection band.

This can be accomplished physically by strain or compression tuning the m CMG structures corresponding to the original m codes. Since we have assumed that the N reflection bands span the entire pulse bandwidth, the new wavelengths in the shifted codes must still lie in the range $[\lambda_{min}, \lambda_{max}]$. Let the w wavelengths for a given code c_j occupy $[\lambda_{min}^j, \lambda_{max}^j]$. This code allows lower and upper wavelength shifts of ($\lambda_{min}^j - \lambda_{min}$) and ($\lambda_{max} - \lambda_{max}^j$), respectively. From the time-spreading pattern corresponding to c_j, there are then a total of $N - (\lambda_{max}^j - \lambda_{min}^j)$ wavelength-encoding patterns, all having the same time-spreading pattern, that can be generated. Thus, from a given N, w, and set of original m time-spreading patterns, we can obtain the following maximum number of codes:

$$\text{Max. number of codes} = Nm - \left(\sum_{j=1}^{m} \left[\lambda_{max}^j - \lambda_{min}^j \right] \right) \qquad (4.31)$$

The original m time-spreading patterns (or codes) are then chosen to satisfy the following: (1) a single auto-correlation peak (no sidelobes) and (2) a maximum cross-correlation peak of 1.

We now consider a simple example that illustrates encoding and both proper and improper decoding processes. For simplicity, let $N = 7$ and $w = 3$. The codes are represented by $c_j = [c_j^1, c_j^2, ..., c_j^N]$, where the index of the array defines the time slot (for example, c_j^1 corresponds to time slot 1, and so on) and the values of the $c_j^1, c_j^2, ..., c_j^N$ elements represent the central wavelengths in these time slots. We can define at most two strictly orthogonal codes, for example $c_1 = [0,2,3,0,0,6,0]$ and $c_2 = [1,0,0,4,0,0,7]$. We can also define quasi-orthogonal codes that are wavelength-shifted versions of c_1 and c_2, one of which is $c_3 = [0,1,2,0,0,5,0]$. The encoded and decoded waveforms for a user with code c_j are denoted as $c_j^{enc}(t)$ and $c_j^{dec}(t)$, respectively. In Figure 4.26(a), we show the spectral responses of three CMGs that generate the three codes c_1, c_2, and c_3 (in the codes, λ_1 corresponds to the reflection band centered at ≈ 1550.9 nm and so on). These three CMGs have the same parameters (length, peak index modulation, difference in central periods or wavelengths of the two superimposed gratings) as in Figure 4.25 except that they incorporate regions of no refractive index modulation within the grating structure to suppress selected reflection bands in order to obtain the desired spectral response [78,79]. Codes c_1 and c_3 have the same time-spreading pattern but different wavelength-encoding patterns. Thus the encoded signals $c_1^{enc}(t)$ and $c_3^{enc}(t)$ will have pulses in the same time slots but the pulses will be at different wavelengths. In Figure 4.26(b), we show a spectrogram for the encoded waveform $c_1^{enc}(t)$, which is obtained by reflecting a transform-limited 0.5-psec Gaussian pulse at $\lambda = 1554.25$ nm from the corresponding grating encoder structure. The spectrogram is a joint time-frequency representation [80], which illustrates the intensity of a signal as a function of both frequency (wavelength) and time simultaneously (darker regions in the image correspond to higher intensities) and is a convenient way to highlight further the encoding process.

In Figure 4.27(a) through (c), we show the result of decoding $c_1^{enc}(t)$, $c_2^{enc}(t)$, and $c_3^{enc}(t)$ using the decoder for code c_1. As expected, $c_1^{dec}(t)$ has the largest signal due to proper decoding. Although c_1 and c_2 are strictly orthogonal codes,

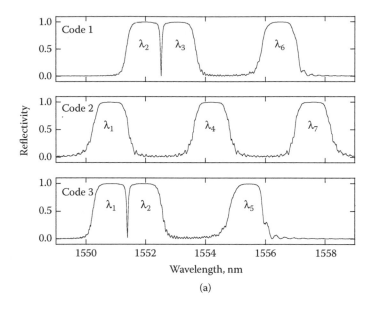

(a)

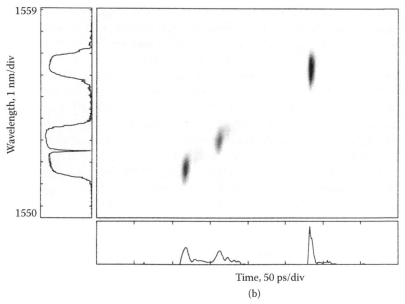

Time, 50 ps/div

(b)

FIGURE 4.26 Simulation of encoding a 0.5-psec transform-limited Gaussian pulse. (a) spectral response of the CMGs corresponding to the three codes c_1, c_2, and c_3. (b) Spectrogram for encoded waveform $c_1^{enc}(t)$.

there is some spectral overlap between the corresponding CMG spectral responses (see Figure 4.26 [a]) so that strictly $c_2^{dec}(t) \neq 0$; nevertheless the energy of $c_1^{dec}(t)$ is clearly much greater than that of $c_2^{dec}(t)$. Also, codes c_1 and c_3 have one overlapping reflection band at $\lambda_2 \approx 1552$ nm so that part of $c_3^{enc}(t)$ will be decoded by the decoder

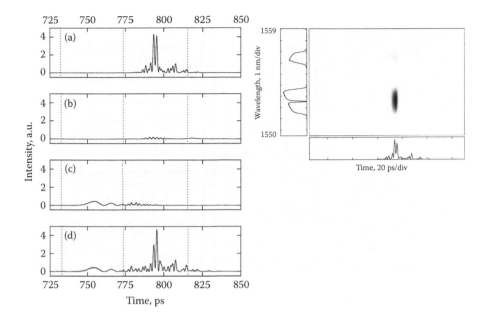

FIGURE 4.27 Simulation of proper and improper decoding of encoded waveforms. Decoded waveforms: (a) $c_1^{dec}(t)$, (b) $c_2^{dec}(t)$, and (c) $c_3^{dec}(t)$; (d) the decoded waveform of $c_1^{enc}(t) + c_2^{enc}(t) + c_3^{enc}(t)$. Note the similarity with $c_1^{dec}(t)$. All decoding was performed with the decoder for code c_1. The dotted lines show the time slots (chips). The inset shows the spectrogram of $c_1^{dec}(t)$.

for code c_1. We also consider decoding the input signal $c_1^{enc}(t) + c_2^{enc}(t) + c_3^{enc}(t)$ with the decoder for code c_1. As shown in Figure 4.27(d), the desired signal $c_1^{dec}(t)$ can still be recovered even in the presence of undesired signals from c_2 and c_3 by using the appropriate decoder. The inset in Figure 4.27 shows the spectrogram of the decoded signal $c_1^{dec}(t)$ and illustrates how the wavelengths are temporally despread after proper decoding.

The use of CMGs for implementing 2D wavelength-time FOCDMA has also been investigated experimentally [81]. A simple system has been assembled to demonstrate the principle of encoding/decoding and verify the rejection of multiuser interference. The system comprises four users, one desired and three interfering, each having its own CMG encoder. The user codes are as follows: $c_{des} = [0,2,0,4,0,0,7]$ and the three interferers are $c_{int}^1 = [0,2,3,0,5,0,0]$, $c_{int}^2 = [1,0,3,0,5,0,0]$, and $c_{int}^3 = [0,3,0,5,0,0,8]$. The decoder is the same grating structure as c_{des} except operated from the opposite end. The grating encoders, each 3.5 cm long with a grating period chirp of 0.5 nm/cm, were fabricated using the technique described in [79] and were tailored to the code specifications. Subpicosecond pulses from a mode-locked fiber laser were then encoded and decoded with desired user grating encoders/decoders. Figure 4.28(a) shows the corresponding waveforms measured with a detector having a response time of ≈100 psec. The encoded signal has three pulses, each at a different wavelength as defined by the wavelength-encoding pattern of the grating encoder. This was confirmed by tuning the central wavelength of the output pulses from the fiber

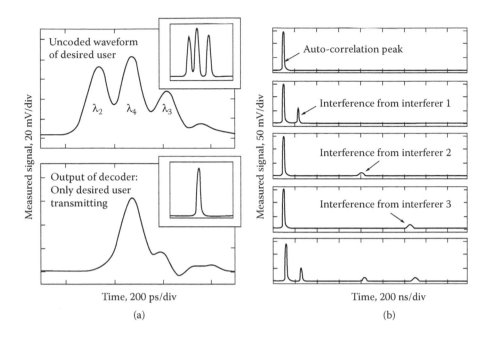

FIGURE 4.28 (a) Measured encoded and decoded waveforms for the desired user. Simulations are shown in the inset. (b) Decoder output when different users are transmitting.

laser and observing an increase or decrease in the relative peaks of the three reflected pulses. The decoded signal consists of a single pulse and is clearly shorter in duration than the encoded signal. This shows that, within the detector response time, the signal has been properly decoded. Simulations are shown in the inset and there is good agreement with the measured results. Figure 4.28(b) illustrates the decoder output when different users are transmitting. In all cases, an auto-correlation peak can be easily distinguished from the multiuser interference. Note that although interferers two and three have strictly orthogonal codes relative to the desired user, an output peak is nevertheless observed. This is due to the nonideal spectral characteristics of the fabricated CMG encoders and decoders (i.e., some spectral overlap between adjacent reflection bands).

4.3.3.4 Discussion

To assess the full potential of FBGs for implementing 2D wavelength-time FOCDMA, several important issues need to be considered. These include, among others, reconfigurability, the number of users that can be supported, and the impact of chromatic fiber dispersion.

The use of linear arrays of tunable uniform FBGs provides maximum reconfigurability: based on the first-in, first-out principle, wavelength-time codes from any code family can be implemented simply by tuning the appropriate FBG in the encoder array. These include PH sequences; various FFH sequences; the multilength, constant weight wavelength-time codes discussed in [42]; and even the codes developed specifically for the CMGs. Reconfigurability is limited by the tuning range of the FBGs. As mentioned earlier, the change in peak reflection wavelength $\Delta\lambda_{shift}$ in terms of an applied strain (or fiber stretching ΔL) is given by Equation 4.18. While tunable gratings over 90 nm have been demonstrated, for a typical FBG operating at 1550 nm, the peak shift is about 6 to 12 nm, corresponding to $\frac{\Delta L}{L} = 0.005 - 0.01$. Thus, if we want each encoder to be able to tune to any arbitrary code, then the codes need to be defined within the tunable wavelength range of FBGs. On the other hand, we can allow the codes to be defined over a larger wavelength range at the expense of limited, rather than full, reconfigurability.

The number of codes or users that can be supported depends on the code family. For example, while PH sequences are simple and straightforward to generate, they are only able to support very few users. Recall that with PH sequences, for a prime number p, the number of codes that are supported is $p(p-1)$ and the spreading length (number of time chips) is p^2. Thus, increasing the number of users requires longer spreading sequences and can only be achieved at the expense of a lower data rate or higher chip rates which may be more difficult to process. FFH codes and specially designed wavelength-time codes can support more users, but these codes generally involve more time chips than frequencies, which again may result in lower data rates. With SCs having m wavelength, n time chips, and a weight w, the upper bound on the number of simultaneous codes (Φ) has been derived to be [14]:

$$\Phi(mn, w, \lambda_a, \lambda_c) \leq \frac{(mn-1)\ldots(mn-\hat{\lambda})}{w(w-1)\ldots(w-\hat{\lambda})} \tag{4.32}$$

for the case $\lambda_a = \lambda_c = \hat{\lambda}$ where λ_a and λ_c are, respectively, the auto- and cross-correlation constraints of the codes (typically, $\hat{\lambda} = 1$). Again if we assume full reconfigurability and a nominal tuning range for the FBGs of 12 nm, then with a wavelength spacing of 100 GHz, this implies $m = 15$. The number of time chips (which are equal in duration) is set by the number of FBGs in a single array — this depends on L, L_{sep}, and the data rate $B = 1/T$. If we assume that the bit window should correspond to the round-trip propagation time through the FBG array, then a small L_{sep} is required. For the previous example considered where $L = 1.2$ cm, $L_{sep} = 1.8$ cm, and a fiber refractive index of 1.45, this gives $n = 9$ (or 36) for a data rate at 622.08 Mb/sec (or 155.52 Mb/sec), i.e., OC-12 (or OC-3) transmission. For a weight $w = 3$, $\Phi \leq 22$ or 90 for $n = 9$ or 36, respectively. For a fixed tuning range, additional wavelengths can be obtained by reducing the wavelength spacing, say to 50 GHz. However, there will be a tradeoff in terms of the number of time chips that can be supported as longer gratings are required to obtain the narrower bandwidth and to minimize channel crosstalk.

The final issue to deal with is chromatic fiber dispersion. Dispersion can be problematic, even over the shorter transmission distances typical of local access networks where FOCDMA is likely to be deployed, especially when the codes are defined over a very large bandwidth [82–84]. Dispersion induces additional relative temporal shifts among the pulses at the different wavelengths thereby resulting in temporal skew. This skew can reduce the auto-correlation peak and increase the impact of multiuser interference. Although dispersion compensating fiber, dispersion shifted fiber, or operation near the zero-dispersion wavelength at 1300 nm in standard single mode fiber are techniques that can be used to manage or avoid dispersion, they may be less cost-effective. An alternate approach is to compensate for dispersion directly in the grating encoder/decoder structures, either through preskewing at the encoder (similar to prechirping or precompensation), at the decoder (similar to postcompensation), or both [45,84]. This is achieved by setting the physical separation between the gratings in the arrays so as to compensate for the relative delays induced by dispersion. Tamai and colleagues recently demonstrated the use of postcompensation in the grating decoder with PH sequences [45]. However, the use of pre- and postcompensation in the encoders/decoders will not work with reconfigurable devices since the compensation spacings are tailored for a given wavelength-time code.

4.4 SUMMARY AND CONCLUSION

In this chapter, we have provided a brief overview of FBG technology and described their application for encoding/decoding optical signals in various FOCDMA coding strategies.

First, FBGs can be used as simple optical filters for SAE FOCDMA. Several different grating encoder/decoder structures are possible, including a linear array of FBGs, SI-FBGs, and filters operating in transmission based on a space-to-frequency mapping. Each of these structures can be designed and used to implement the different codes families suitable for SAE; however, issues such as simplicity in implementation or fabrication and flexibility for reconfigurable operation may favor the use of a particular structure. For example, while a linear array of tunable uniform FBGs provides the most flexibility in terms of reconfigurable operation, the encoder/decoder actually requires two arrays in order to compensate the round-trip delay of different spectral components (this group delay variation is inherent in the linear array structure due to the first-in, first-out principle), which makes the implementation more complex and sensitive to mismatch in the grating responses. On the other hand, SI-FBGs and filters operating in transmission based on space-to-frequency mapping avoid the problem of group delay variation and present a simpler implementation, albeit at the expense of reconfigurability (with SI-FBGs or gratings operating in transmission, limited reconfigurability is possible). We have also discussed possible limitations that the grating structures can impose in terms of code size and hence the maximum number of users supported.

Second, segmented FBGs and SS-FBGs can be used to generate phase-encoded signals for SPE FOCDMA. When operating under the weak grating limit, the impulse reflection response of an FBG corresponds directly to its complex refractive index profile. Thus, discrete phase shifts that are incorporated in the refractive index profile of segmented FBGs or SS-FBGs are directly translated in the reflected (encoded) signal. SS-FBGs have been studied extensively and impressive experimental results have been achieved, including high data rates (up to 10 Gb/sec), high chip rates (up to 320 Gchip/sec), and even reconfigurability using tunable phase shifts generated using highly localized thermal effects.

Finally, the combined wavelength selective and dispersive properties of single or multiple linear arrays of FBGs and chirped grating structures make them suitable for implementing 2D wavelength-time codes. Results from simple experiments that verify the principle of encoding/decoding as well as multiuser system demonstrations show the feasibility and potential of this approach. Again, we have discussed the potential limitations the gratings may impose in terms of system performance, such as the number of codes that can be generated or the data rates that can be supported.

Grating fabrication technology has advanced considerably in the past years and it is now possible to make gratings reliably and with nearly identical characteristics. Gratings with very complex profiles, such as those to obtain boxlike spectral responses with low in-band dispersion or SS-FBGs, as well as tunable devices are becoming increasingly available. If transmission impairments, primarily chromatic fiber dispersion, are managed properly, then one primary issue focuses on reducing multiuser interference. Time-gating and nonlinear approaches have been considered; but these increase system complexity or may have practical limitations. More complex schemes, such as combining 2D wavelength-time bipolar codes, have also been proposed [85–87]. Finally, several studies show that the performance of FOCDMA systems can be limited by beat noise rather than multiuser interference [88], and further detailed investigations are required.

OCDMA still remains largely a research topic, the development of low-cost enabling technologies such as FBG encoders and decoders and their successful demonstrations in system experiments, may eventually bring the commercial deployment of FOCDMA systems.

ACKNOWLEDGMENTS

I would like to thank my colleagues and students for numerous stimulating discussions; J. Magné, S. LaRochelle, L. A. Rusch, and B. Cauro for providing several of the figures used in the text; and J. Azaña for critical reading of this manuscript.

REFERENCES

[1]   Stok, A., Sargent, E. H. (2000). Lighting the local area: Optical code-division multiple access and quality of service provisioning. *IEEE Network.* 14(6):42–46.

[2]   Stok, A., Sargent, E. H. (2002). The role of optical CDMA in access networks. IEEE Commun. Mag. 40(9):83–87.

[3]   Shah, J. (2003). Optical CDMA. *Opt. and Photon. News* 42(4):42–47.

[4]   Prucnal, P. R., Santoro, M. A., Fan, T. R. (1986). Spread spectrum fiber-optic local area network using optical processing. *IEEE/OSA J. Lightwave Technol.* 4(5):547–554.

[5]   Salehi, J. A. (1989). Code division multiple-access techniques in optical fiber networks: Part 1: Fundamental principles. *IEEE Trans. on Commun.* 37(8):824–833.

[6]   Zaccarin, D., Kavehrad, M. (1993). An optical CDMA system based on spectral encoding of LED. *IEEE Photon. Technol. Lett.* 4(4):479–482.

[7]   Kavehrad, M., Zaccarin, D. (1995). Optical code-division-multiplexed systems based
      on spectral encoding of noncoherent sources. *IEEE/OSA J. Lightwave Technol.*
      13(3):534–545.

[8]   Nguyen, L., Aazhang, B., Young, J. F. (1995). All-optical CDMA with bipolar codes.
      Electron. Lett. 31(6):469–370.

[9]   Lam, C. F., Tong, D. T. K., Wu, M. C., Yablonovitch, E. (1998). Experimental
      demonstration of bipolar optical CDMA system using a balanced transmitter and
      complementary spectral encoder. *IEEE Photon. Technol. Lett.* 10(10):1504–1506.

[10] Weiner, A. M., Heritage, J. P., Salehi, J. A. (1988). Encoding and decoding of
      femtosecond pulses. *Opt. Lett.* 13(4):300–302.

[11] Salehi, J. A., Weiner, A. M., Heritage, J. P. (1990). Coherent ultrashort light pulse
      code-division multiple access communication systems. *IEEE/OSA J. Lightwave
      Technol.* 8(3):478–491.

[12] Weiner, A. M. (1995). Femtosecond optical pulse shaping and processing. *Prog.
      Quantum. Electron.* 19:161–237.

[13] Kitayama, K. (1994). Novel spatial spread spectrum based fiber optic CDMA
      networks for image transmission. *IEEE. J. Sel. Areas Commun.* 12(5):762–772.

[14] Yang, G. -C., Kwong, W. C. (1996). Two-dimensional spatial signature patterns. *IEEE
      Trans. on Commun.* 44(2):184–191.

[15] Kwong, W. C., Yang, G. -C. (1998). Image transmission in multicore-fiber code-
      division multiple-access networks. *IEEE Commun. Lett.* 2(10):285–287.

[16] Park, E., Mendez, A. J., Garmire, E. M. (1992). Temporal/spatial optical CDMA
      networks-design, demonstration, and comparison with temporal networks. *IEEE Pho-
      ton. Technol. Lett.* 4(10):1160–1162.

[17] Tančevski, L., Andonovic, I. (1994). Wavelength hopping/time spreading code divi-
      sion multiple access systems. *Electron. Lett.* 30(17):1388–1390.

[18] Jugl, E., Kuhwald, T., Iversen, K. (1997). Algorithm for construction of (0,1)-matrix
      codes. *Electron. Lett.* 33(3):227–229.

[19] Kim, S., Yu, K., Park, N. (2000). A new family of space/wavelength/time spread three
      dimensional optical code for OCDMA networks. *IEEE/OSA J. Lightwave Technol.*
      18(4):502–511.

[20] Mendez, A. J., Gagliardi, R. M., Feng, H. X. C., Heritage, J. P., Morookian, J. -P. (2000).
      Strategies for realizing optical CDMA for dense, high-speed, long span, optical
      network applications. *IEEE/OSA J. Lightwave Technol.* 18(12):1685–1696.

[21] Yu, K., Park, N. (1999). Design of new family of two-dimensional wavelength-time
      spreading codes for optical code division multiple access networks. *Electron. Lett.*
      35(10):830–831.

[22] Hill, K. O., Meltz, G. (1997). Fiber Bragg grating technology fundamentals and
      overview. *IEEE/OSA J. Lightwave Technol.* 15(8):1263–1276.

[23] Kashyap, R. (1999). *Fiber Bragg Gratings*. San Diego: Academic Press.

[24] Hunter, D. B., Minasian, R. (1999). Programmable high-speed optical code recognition
      using fiber Bragg grating arrays. *Electron. Lett.* 35(5):412–414.

[25] Chen, L. R., Benjamin, S. D., Smith, P. W. E., Sipe, J. E. (1998). Applications of
      ultrashort pulse propagation in Bragg gratings for wavelength-division multiplexing
      and code-division multiple access. *IEEE J. Quantum Electron.* 34(11):2117–2129.

[26] Fathallah, H., Rusch, L. A., LaRochelle, S. (1999). Passive optical fast frequency-hop
      CDMA communications system. *IEEE/OSA J. Lightwave Technol.* 17(3):397–405.

[27] Town, G. E., Chan, K., Yoffe, G. (1999). Design and performance of high-speed
      optical pulse-code generators using optical fiber Bragg gratings. *IEEE J. Sel. Topics
      in Quantum Electron.* 5(5):1325–1331.

[28]   Chen, L. R., Smith, P. W. E., de Sterke, C. M. (1999). Wavelength-encoding/temporal-spreading optical code division multiple-access system with in-fiber chirped moiré gratings. *Appl. Opt.* 38(21):4500–4508.

[29]   Grunnet-Jepsen, A., Johnson, A. E., Maniloff, E. S., Mossberg, T. W., Munroe, M. J., Sweetser, J. N. (1999). Demonstration of all-fiber sparse lightwave CDMA based on temporal phase encoding. *IEEE Photon. Technol. Lett.* 11(10):1283–1285.

[30]   Grunnet-Jepsen, Johnson, A. E., Maniloff, E. S., Mossberg, T. W., Munroe, M. J., Sweetser, J. N. (1999). Fiber Bragg grating based spectral encoder/decoder for light-wave CDMA. *Electron. Lett.* 35(13):1096–1097.

[31]   Wada, N., Sotobayashi, H., Kitayama, K. (2000). 2.5 Gbit/s time-spread/wavelength-hop optical code division multiplexing using fiber Bragg grating with supercontinuum light source. *Electron. Lett.* 36(9):815–817.

[32]   Huang, J. -F., Hsu, D. -Z. (2000). Fiber-grating-based optical CDMA spectral coding with nearly orthogonal M-sequence codes. *IEEE Photon. Technol. Lett.* 12(9):1252–1254.

[33]   Wei, Z., Shalaby, H. M. H., Ghafouri-Shiraz, H. (2001). Modified quadratic congruence codes for fiber-Bragg-grating based spectral-amplitude-coding optical CDMA systems. *IEEE/OSA J. Lightwave Technol.* 19(9):1274–1281.

[34]   Wei, Z, Ghafouri-Shiraz, H., Shalaby, H. M. H. (2001). New code families for fiber-Bragg-grating-based spectral-amplitude-coding optical CDMA systems. *IEEE Photon. Technol. Lett.* 13(8):890–892.

[35]   Teh, P. C., Petropoulos, P., Ibsen, M., Richardson, D. J. (2001). Generation, recognition and recoding of 64-chip bipolar optical code sequences using superstructured fiber Bragg gratings. *Electron. Lett.* 37(3):190–191.

[36]   Teh, P. C., Petropoulos, P., Ibsen, M., Richardson, D. J. (2001). Phase encoding and decoding of short pulses at 10 Gb/s using superstructured fiber Bragg gratings. *IEEE Photon Technol. Lett.* 13(2):154–156.

[37]   Teh, P. C., Petropoulos, P., Ibsen, M., Richardson, D. J. (2001). A comparative study of the performance of seven- and 63-chip optical code-division multiple-access encoders and decoders based on superstructured fiber Bragg gratings. *IEEE/OSA J. Lightwave Technol.* 19(9):1352–1365.

[38]   Chen, L. R. (2001). Flexible fiber Bragg grating encoder/decoder for hybrid wavelength-time optical CDMA. *IEEE Photon. Technol. Lett.* 13(11):1233–1235.

[39]   Torres, P., Valente, P. C. G., Carvalho, M. C. R. (2002). Security system for optical communication signals with fiber Bragg gratings. *IEEE Trans. on Micro. Theory and Tech.* 50(1):13–16.

[40]   Inaty, E., Shalaby, H. M. H., Fortier, P., Rusch, L. A. (2002). Multirate optical fast frequency hopping CDMA system using power control. *IEEE/OSA J. Lightwave Technol.* 20(2):166–177.

[41]   Huang, J. -F., Yang, C. -C. (2002). Reductions of multiple-access interference in fiber-grating-based optical CDMA network. *IEEE Trans. on Commun.* 50(10):1680–1687.

[42]   Kwong, W. C., Yang, G. -C. (2002). Wavelength-time codes for multimedia optical CDMA systems with fiber Bragg grating arrays. In: Proceedings IEEE Global Communications Conference.

[43]   Kutsuzawa, S., Minato, N., Oshiba, S., Nishiki, A., Kitayama, K. (2003). 10 Gb/s × 2 ch signal unrepeated transmission over 100 km of data rate enhanced time-spread/wavelength-hopping OCDM using 2.3-Gb/s-FBG en/decoder. *IEEE Photon. Technol. Lett.* 15(2):317–319.

[44]   Magné, J., Wei, D.-P., Ayotte, S., Rusch, L. A., LaRochelle, S. (2003). Experimental dem-
       onstration of frequency-encoded optical CDMA using superimposed fiber Bragg gratings.
       In: Proceedings OSA Topical Meeting on Bragg Gratings, Photosensitivity, and Poling.

[45]   Tamai, H., Iwamura, H., Minato, N., Oshiba, S. (2004). Experimental study on
       time-spread/wavelength-hop optical code division multiplexing with group delay
       compensating en/decoder. *IEEE Photon. Technol. Lett.* 16(1):335–337.

[46]   Erdogan, T. (1997). Fiber grating spectra. *IEEE/OSA J. Lightwave Technol.*
       15(8):1277–1294.

[47]   Othonos, A., Lee, X., Measures, R. M. (1994). Superimposed multiple Bragg gratings.
       Electron. Lett. 30(23):1972–1974.

[48]   Goh, C. S., Mokhtar, M. R., Butler, S. A., Set, S. Y., Kikuchi, K., Ibsen, M. (2003).
       Wavelength tuning of fiber Bragg gratings over 90 nm using a simple tuning package.
       IEEE Photon. Technol. Lett. 15(4):557–559.

[49]   Mokhtar, M. R., Goh, C. S., Butler, S. A., Set, S. Y., Kikuchi, K., Richardson, D. J.,
       Ibsen, M. (2003). Fiber Bragg grating compression-tuned over 110 nm. *Electron.
       Lett.* 39(6):509–511.

[50]   Skaar, J., Sahlgren, B., Fonjallaz, P. -Y., Storøy, H., Stubbe, R. (1998). High-reflec-
       tivity fiber-optic bandpass filter designed by use of the iterative solution to the
       Gel'fand-Levitan-Marchenko equations. *Opt. Lett.* 23(12):933–935.

[51]   Feced, R., Zervas, M. N., Muriel, M. A. (1999). An efficient inverse scattering
       algorithm for the design of nonuniform fiber Bragg gratings. *IEEE J. Quantum
       Electron.* 35(8):1105–1115.

[52]   Kersey, A. D., Davis, M. A., Patrick, H. J., LeBlanc, M., Koo, K. P., Askins, C. G.,
       Putnam, M. A., Friebele, E. J. (1997). Fiber grating sensors. *IEEE/OSA J. Lightwave
       Technol.* 15(8):1442–1463.

[53]   Meghavoryan, D. M., Daryan, A. V. (2003). Superimposed fiber Bragg grating sim-
       ulation by the method of single expression for optical CDMA systems. *IEEE Photon.
       Technol. Lett.* 15(11):1546–1548.

[54]   Smith, E. D. J., Gough, P. T., Taylor, D. P. (1995). Noise limits of optical spectral-
       encoding CDMA systems. *Electron. Lett.* 31(17):1469–1470.

[55]   Huang, J. -F., Tsai, C. -M., Lo, Y. -L. (2004). Compensating fiber gratings for source
       flatness to reduce multiple-access interferences in optical CDMA network
       coder/decoders. *IEEE/OSA J. Lightwave Technol.* 22(3):739–745.

[56]   Azaña, J., Chen, L. R. (2002). Synthesis of temporal optical waveforms by fiber
       Bragg gratings: A new approach based on space-to-frequency-to-time mapping. *J.
       Opt. Soc. Am. B.* 19(11):2758–2769.

[57]   Weiner, A. M. (2000). Femtosecond pulse shaping using spatial light modulators.
       Rev. of Scientific Inst. 71:1929–1960.

[58]   Weiner, A. M. (2000). Femtosecond pulse processing. *Opt. and Quantum Electron.*
       32:473–487.

[59]   Sardesai, H. P., Chang, C. -C., Weiner, A. M. (1998). A femtosecond code-division
       multiple-access communication system test bed. *IEEE/OSA J. Lightwave Technol.*
       16(11):1953–1964.

[60]   Fang, X. Wang, D.-N., Li. S. (2003). Fiber Bragg grating for spectral phase optical code-
       division multiple-access encoding and decoding. *J. Opt. Soc. Am. B.* 20(8):1603–1610.

[61]   Kogelnik, H. (1976). Filter response of nonuniform and almost-periodic structures.
       Bell Syst. Tech. J. 55:109–126.

[62]   Lee, J. H., Teh, P. C., Petropoulos, P., Ibsen, M., Richardson, D. J. (2001). Reduction
       of interchannel interference noise in a two-channel grating-based OCDMA system
       using a nonlinear optical loop mirror. *IEEE Photon. Technol. Lett.* 13(5):529–531.

[63] Lee, J. H., Teh, P. C., Petropoulos, P., Ibsen, M., Richardson, D. J. (2002). A grating-based OCDMA coding-decoding system incorporating a nonlinear optical loop mirror for improved code recognition and noise reduction. *IEEE/OSA J. Lightwave Technol.* 20(1):36–46.

[64] Petropoulos, P., Wada, N., The, P. C., Ibsen, M., Chujo, W., Kitayama, K. -I., Richardson, D. J. (2001). Demonstration of a 64-chip OCDMA system using superstructured fiber gratings and time-gating detection. *IEEE Photon. Technol. Lett.* 13(11): 1239–1241.

[65] Teh, P. C., Ibsen, M. Lee, J. H., Petropoulos, P., Richardson, D. J. (2002). Demonstration of a four-channel WDM/OCDMA system using 255-chip 320-Gchip/s quaternary phase coding gratings. *IEEE Photon. Technol. Lett.* 14(2):227–229.

[66] Uttamchandani, D., Othonos, A. (1996). Phase-shifted Bragg gratings in optical fibers by post-fabrication thermal processing. *Opt. Commun.* 127:200–204.

[67] Mokhtar, M. R., Ibsen, M., Teh, P. C., Richardson, D. J. (2003). Reconfigurable multilevel phase-shift keying encoder-decoder for all-optical networks. *IEEE Photon. Technol. Lett.* 15(3):431–433.

[68] Kwong, W. C., Yang, G. -C. (1999). Frequency-hopping codes for multimedia services in mobile telecommunications. *IEEE Trans. on Vehicular Technol.* 48(6):1906–1915.

[69] Kim, J., Lee, C. -K., Seo, S. -W., and Lee, B. (2002). Frequency-hopping optical orthogonal codes with arbitrary time-blank patterns. *Appl. Opt.* 41(20):4070–4077.

[70] Yegnanarayanan, S., Bhushan, A. S., Jalali, B. (2000). Fast wavelength-hopping time-spreading encoding/decoding for optical CDMA. *IEEE Photon. Technol. Lett.* 12(5):573–575.

[71] Yu, K. Shin, J., Park, N. (2000). Wavelength-time spreading optical CDMA system using wavelength multiplexers and mirrored fiber delay lines. *IEEE Photon. Technol. Lett.* 12(9):1278–1280.

[72] Kim, S. (2000). Cyclic optical encoders/decoders for compact optical CDMA networks. *IEEE Photon. Technol. Lett.* 12(4):428–430.

[73] Takiguchi, K., Shibata, T., Itoh, M. (2002). Encoder/decoder on planar lightwave circuit for time-spreading/wavelength-hopping optical CDMA. *Electron. Lett.* 38(10):469–470.

[74] Mendez, A. J., Gagliardi, R. M., Hernandez, V. J., Bennett, C. V., Lennon, W. J. (2003). Design and performance analysis of wavelength/time (W/T) matrix codes for optical CDMA. *IEEE/OSA J. Lightwave Technol.* 21(11):2524–2533.

[75] Ben Jaafar, H., LaRochelle, S., Cortès, P. -Y., Fathallah, H. (2001). 1.25 Gbit/s transmission of optical FFH-OCDMA signals over 80 km with 16 users. In: Proceedings Conference on Optical Fiber Communication. 2:TuV3-1-TuV3-2.

[76] Town, G. E., Sugden, K, Williams, J. A. R., Bennion, I., Poole, S. B. (1995). Wideband Fabry-Pérot-like filters in optical fiber. *IEEE Photon. Technol. Lett.* 7(1):78–80.

[77] Everall, L. A., Sugden, K., Williams, J. A. R., Bennion, I., Liu, X., Aitchison, J. S., Thoms, S., De La Rue, R. (1997). Fabrication of multipassband moiré resonators in fibers by the dual-phase-mask exposure method. *Opt. Lett.* 22(19):1473–1475.

[78] Chen, L. R., Cooper, D. J. F., Smith, P. W. E. (1998). Transmission filters with multiple flattened passbands based on chirped moiré gratings. *IEEE Photon. Technol. Lett.* 10(9):1283–1285.

[79] Chen, L. R., Loka, H. S., Copper, D. J. F., Smith, P. W. E., Tam, R., Gu, X. (1999). Fabrication of transmission filters with single and multiple flattened passbands based on chirped Moiré gratings. *Electron. Lett.* 35(7):584–585.

[80] Cohen, L (1989). Time-frequency representations—A review. *Proc. IEEE.* 77:941–981.

[81]   Chen, L. R., Smith, P. W. E. (2000). Demonstration of incoherent wavelength-encoding/time-spreading optical CDMA using chirped moiré gratings. *IEEE Photon. Technol. Lett.* 12(9):1281–1283.

[82]   Zuo, C., Ma, W., Pu, H., Lin, J. (2001). The impact of group velocity on frequency-hopping optical code division multiple access system. *IEEE/OSA J. Lightwave Technol.* 19(10):1416–1419.

[83]   Wang, X., Chan, K. T. (2002). The effect of grating-position deviation and fiber dispersion in the fiber-optic-CDMA network with the FBG encoder/decoder for incoherent 2D coding. *Microwave and Opt. Technol. Lett.* 35(1):16–19.

[84]   Ng, E. K. H., Weichenberg, G. E., Sargent, E. H. (2002). Dispersion in multiwavelength optical code-division multiple-access systems: impact and remedies. *IEEE Trans. Commun.* 50(11):1811–1816.

[85]   Wan, S. P., Hu, Y. (2001). Two-dimensional optical CDMA differential system with prime/OOC codes. *IEEE Photon. Technol. Lett.* 13(12):1373–1375.

[86]   Yim, R. M. H., Bajcsy, J., Chen, L. R. (2003). A new family of 2D wavelength-time codes for optical CDMA with differential detection. *IEEE Photon. Technol. Lett.* 15(1):165–167.

[87]   Heo, H., Min, S., Won, Y. H., Yeon, Y., Kim, B. K., Kim, B. W. (2004). A new family of 2-D wavelength-time spreading code for optical code-division multiple-access system with balanced detection. *IEEE Photon. Technol. Lett.* 16(9):2189–2191.

[88]   Tančevski, L., Rusch, L. A. (2000). Impact of the beat noise on the performance of 2-D optical CDMA systems. *IEEE Commun. Lett.* 4(8):264–266.

5 Coherent Optical CDMA Systems

Paul Toliver, Shahab Etemad, and Ron Menendez

CONTENTS

5.1 INTRODUCTION

Optical CDMA systems can be divided into two broad categories based on the way in which a particular user's code is applied to the optical signal. These classifications include *coherent optical CDMA* and *incoherent optical CDMA* approaches. In a coherent OCDMA system, a given user's code is generally applied via phase coding of the optical signal field, which is often derived from a highly coherent wideband source, such as a mode-locked laser. The receiver for a coherent OCDMA system relies on a coherent reconstruction of the signal field to recover the decoded user's data. In contrast, an incoherent OCDMA system typically relies on amplitude-modulated codes rather than directly manipulating the optical phase. Also, the receiver is based upon an incoherent decoding and recovery process. A number of incoherent OCDMA system architectures utilize wideband incoherent sources, such as a broadband amplified spontaneous emission (ASE) source, while other incoherent architectures utilize coherent laser sources as part of their implementation. This chapter focuses on coherent optical CDMA, while incoherent approaches are discussed in greater detail in Chapter 6.

The chapter begins with a description of different coherent OCDMA approaches based on spectral and temporal phase coding. We then discuss the subsystem technologies used to implement these types of OCDMA systems. A discussion of code selection specifically for narrowband spectral phase coded system is presented next. Finally, we discuss networking architectures compatible with coherent OCDMA systems.

5.2   COHERENT OCDMA APPROACHES

Within the category of coherent systems, it is useful to further classify OCDMA systems based on the way in which phase coding is applied to the optical signal field. Since optical phase can be manipulated in either the frequency domain or the time domain, two types of coherent OCDMA systems are possible:

- Spectral Phase Coded Optical CDMA (SPC-OCDMA)
- Temporal Phase Coded Optical CDMA (TPC-OCDMA)

5.2.1   SPECTRAL PHASE CODED OCDMA (SPC-OCDMA)

A block diagram of the spectral phase coded OCDMA system architecture is illustrated in Figure 5.1. In addition, Figure 5.2 provides an illustration of the signals at various stages in the system. The SPC-OCDMA system requires a broadband multiwavelength

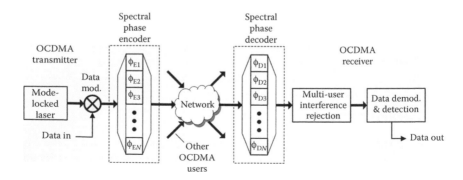

FIGURE 5.1 Block diagram of the spectral phase coded OCDMA (SPC-OCDMA) system architecture.

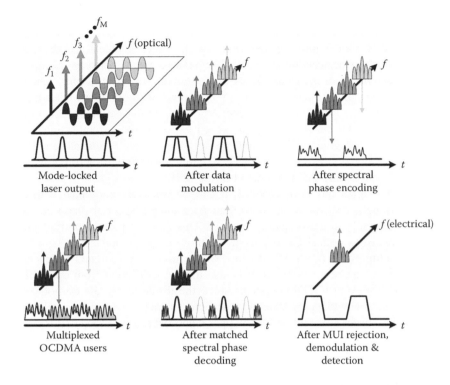

FIGURE 5.2 Representation of signals at various stages of the SPC-OCDMA system, with conceptual illustrations both in time domain (bottom) and frequency domain (top).

source of light that is highly coherent from a frequency domain perspective, such as is available from a mode-locked laser. The mode-locked laser, which is discussed further in Section 5.3.1, produces at its output a stream of short optical pulses in the time domain. The pulsed signal can also be shown to be equivalent to a comb of phase-locked continuous wave optical frequencies equally spaced on a frequency grid determined by the laser repetition rate. These two signal representations are illustrated conceptually in upper and lower part of Figure 5.2. In the frequency representation, phase locked relationship between the different frequency components is represented as discrete spectral components, all of which are aligned in phase at a particular instant.

The stream of mode-locked laser pulses is then modulated with the user's data. The data modulation, which occurs in the time domain, could be either simple on-off-keying (OOK), where optical pulses are simply turned on or off depending up whether the user's data bit is a "1" or "0," respectively, or one of the more advanced modulation techniques described in Section 5.3.2. In the frequency representation shown Figure 5.2, data modulation is equivalent to spectral broadening of the discrete spectral components generated by the mode-locked laser.

After data modulation, the signal is sent into a spectral phase encoder, which applies a particular OCDMA phase code to the spectrum. Each user is assigned one of a set of N-element spectral phase codes. The spectral phase encoders, the technology of which is described in detail in Section 5.3.3, separate the modulated spectrum into spectral bins and apply a distinct phase shift to each bin. The phase components could be simple binary codes, such as 0 or π, or more advanced multilevel phase codes. As shown in Figure 5.2, the effect of the spectral phase encoder from a frequency domain perspective is to alter the relative phase between the different spectral components of the signal. In the time domain, the spectral phase encoding results in temporal broadening of the original narrow mode-locked laser pulse, making the signal appear more noise-like. As discussed in Section 5.3.3, a number of proposed SPC-OCDMA systems have been experimentally demonstrated using static or slowly programmable spectral phase codes at each user, although more secure implementation may require highly dynamic reconfiguration of phase codes, potentially approaching the user's data bit rate [22].

Once the signal has been encoded, it can be passively combined with other OCDMA signals, each of which have their own unique spectral phase codes but overlap completely in the frequency domain. This form of passive multiplexing is what distinguishes optical CDMA from the more conventional dense wavelength division multiplexing (DWDM) systems where users are assigned independent, nonoverlapping spectral passbands. It is also uniquely different from optical time-division multiplexing (OTDM) systems since the different signals do not occupy pre-assigned, nonoverlapping time slots and can in fact overlap temporally. Note, however, that for the coherent OCDMA approaches that are described in this chapter, it is quite common to synchronize transmission from the different users to minimize the multichannel coherent interference that would result in the case of completely asynchronous timing.

In order to recover a particular OCDMA user's signal at the receive end of the system, a spectral phase decoder is first employed. Physically it is nearly identical to the spectral phase encoder located at the transmitter, but it has a conjugate spectral phase mask. As shown in Figure 5.2, the process of matched spectral phase decoding realigns the individual spectral components, which also results in restoring the original mode-locked laser pulse shape, but for the desired channel only. On the other hand, waveforms from other OCDMA transmitters are not restored since the spectral phase decoder does not properly realign their spectral components and, therefore, these signals continue to remain noiselike.

After spectral phase decoding, it is necessary to remove the multiuser interference (MUI) noise resulting from undesired OCDMA users. As discussed in Section 5.3.4, two possible implementations for this function include optical time gating, which optically extracts the desired pulse only, and optical thresholding, which acts as a limiter to extinguish low level MUI noise.

After the MUI noise has been eliminated, the desired user's data signal can be recovered through data demodulation and detection. Various approaches are possible, a number of which are discussed in Section 5.3.2. As an example, the most straightforward approach would be the use of direct detection with a simple photoreceiver, which can be used when a simple intensity modulation format such as OOK is used at the transmitter.

5.2.1.1 Wideband vs. Narrowband SPC-OCDMA

Depending upon the frequency resolution at which individual spectral phase components are applied relative to the mode-locked laser comb spacing, spectral phase coded OCDMA systems can further classified into the following two categories:

- Wideband SPC-OCDMA
- Narrowband SPC-OCDMA

Wideband SPC-OCDMA systems, also referred to as ultrashort or femtosecond pulse OCDMA [1,2], utilize spectral phase encoders and decoders whose phase bin spacing, Δf_{PHASE_MASK}, is much larger than the laser source comb spacing, f_{MLL}. That is, a given phase component of the spectral phase mask is applied to multiple components of the modulated laser comb. In contrast, narrowband SPC-OCDMA systems [3] are designed with a phase bin spacing equal to the laser source comb spacing so that different phase comb components are applied to each laser comb component. The discussions in this chapter are focused primarily on narrowband SPC-OCDMA components, systems, code selection, and architectures.

5.2.2 TEMPORAL PHASE CODED OCDMA (TPC-OCDMA)

As mentioned earlier in the chapter, it is also possible to manipulate optical phase in the time domain rather than the spectral domain. This approach is taken in the temporal phase coded OCDMA system [13, 16, 23], which is illustrated in Figure 5.3. Figure 5.4 also provides an illustration of the signal waveforms at various stages through the system. For the TPC-OCDMA system, it is only necessary to consider the evolution of signal waveforms in the time domain to understand its operation.

The TPC-OCDMA system architecture exhibits a number of similarities with the SPC-OCDMA approach described in the previous subsection. For example, the light source in the TPC-OCDMA system is also frequently a mode-locked laser. However, in this case, it is not the multiwavelength spectral characteristics of the mode-locked

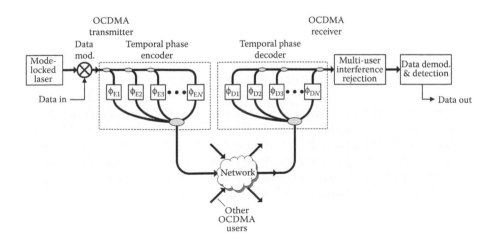

FIGURE 5.3 Block diagram of the temporal phase coded OCDMA (TPC-OCDMA) system architecture.

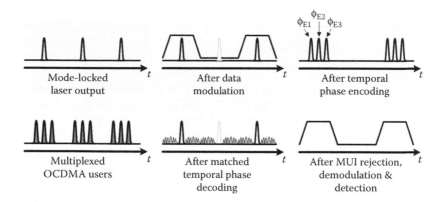

FIGURE 5.4 Time-domain representations of signals at various stages of the TPC-OCDMA system.

laser that are exploited as was the case for SPC-OCDMA, but rather its short pulse capabilities. The mode-locked laser pulses are modulated with the user data stream. This could be done with either simple on-off keyed intensity modulation or the more complex modulation formats also available in SPC-OCDMA systems.

After modulation, a temporal phase encoder is used to create N pulse copies, each of which delayed so that they lie on an equally spaced time grid. The spacing between pulses is defined as the *temporal chip interval*. In addition to the coarse time delay between pulse copies, there is also a fine relative phase shift. Each pulse copy is set to a specific relative phase shift based on the particular user's assigned OCDMA code. Similar to SPC-OCDMA systems, the individual phase components could be constructed from simple binary codes, such as 0 or π, or more advanced multilevel phase codes.

After temporal phase encoding, the signals from the various OCDMA users can now be passively combined into a single transmission medium, such as a common single-mode optical fiber. As was the case for SPC-OCDMA, the signals, which are synchronously timed, overlap completely in both the time as well as the frequency domain, but now it is their temporal phase codes that uniquely identify them rather than a spectral phase code.

To recover a given TPC-OCDMA user's data stream, a matched filtering process is employed. However, in contrast to the SPC-OCDMA matched filtering, which is best understood from a frequency domain perspective, the matched filtering for TPC-OCDMA can be best described as a time domain operation. At the receiver, a temporal phase decoder similar to the transmitter's temporal phase encoder is used. To recover a particular OCDMA user's data stream, the decoder's phase elements are set to the conjugate of the desired transmitter's encoder. The decoder essentially performs temporal correlation of the received signal by making N copies of the received signal, applying the specific decoder phase element, ϕ_{Di}, to each, and delaying the copies by the temporal chip interval before passively recombining them. As a result, when the receiver's temporal phase decoder is properly set to correspond to a particular transmitted code, a strong autocorrelation peak is observed in the waveform at the output of the decoder. On the other hand, signals from unmatched OCDMA transmitters do not result in a sharp autocorrelation peak, but continue to remain noiselike.

After matched decoding, it is necessary to remove multiuser interference (MUI) resulting from the unmatched users in the TPC-OCDMA system. Similar technologies that are utilized in the SPC-OCDMA systems, such as optical time gating and optical thresholding, could be used for this function. Once the MUI has been eliminated, the desired user's data signal can be recovered through data demodulation and detection, again using the same approaches available in SPC-OCDMA systems.

5.3 SUBSYSTEM TECHNOLOGIES

In this section, we will discuss some of the key technologies for implementing coherent OCDMA systems.

5.3.1 MODE-LOCKED LASERS

Mode-locked lasers are commonly used as the primary optical source in both SPC-OCDMA and TPC-OCDMA systems. For SPC-OCDMA, it is the frequency characteristics that are exploited, while its time domain properties are utilized in TPC-OCDMA systems.

A normal continuous wave (CW) laser, such as the distributed feedback (DFB) laser used in most DWDM systems, operates with a single dominant center wavelength defined by a narrowband filter within the laser cavity, resulting in a very narrow optical spectrum. This mode of operation is referred to as single mode. In contrast, a mode-locked laser operates with a broad spectrum of wavelengths, each corresponding to a different laser cavity mode. The width of the spectrum is defined both by the bandwidth of the laser gain medium as well as any wideband filtering that may be present inside the laser cavity.

By modulating the round trip gain or loss of the laser cavity with a periodic signal, it is possible to "lock" all of the cavity modes so that they maintain a well-defined phase relationship with one another, hence the name mode-locked laser. The equivalent time domain signal can be determined by taking the Fourier transform of this phase-locked, multiwavelength comb. These concepts are illustrated in Figure 5.5.

The mode-locked laser signal in the time domain consists of a continuous stream of short optical pulses separated by time $1/f_{MLL}$, where f_{MLL} is the pulse repetition

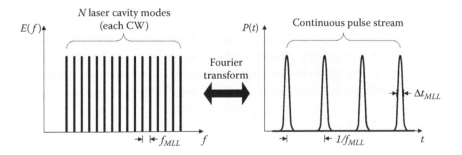

FIGURE 5.5 Representation of the output signal from a mode-locked laser, both the frequency domain optical field (left) and time domain intensity (right).

frequency.

$$E(t) = \sum_{n=1}^{N} E_0 e^{j2\pi n f_{MLL} t} \cdot e^{j2\pi f_0 t} \tag{5.1}$$

$$P(t) \propto |E(t)|^2 \tag{5.2}$$

The temporal width of each pulse, Δt_{MLL}, is proportional to $1/(N \cdot f_{MLL})$, where N is the number of laser cavity modes available.

Techniques for mode-locking can be divided into both active and passive techniques. Active techniques include amplitude modulation, frequency modulation, and synchronous pump mode-locking, while passive techniques typically rely on a saturable absorber mechanism. We only briefly discuss active mode-locking based on amplitude modulation here, since it is the most widely used approach used in coherent OCDMA systems. Typically, mode-locked lasers operating at telecom wavelengths are constructed using either semiconductor optical amplifiers (SOA) or erbium-doped fiber amplifiers (EDFA) for the primary gain medium. For SOA-based lasers, the cavity modulation required to set up the mode-locking condition can be performed either using current injection or a separate external amplitude modulator within the cavity. In contrast, EDFA-based mode-locked lasers use external modulation almost exclusively.

5.3.2 DATA MODULATION AND DEMODULATION

At the transmit end of an OCDMA system, each user must convert their digital data from the electronic domain into an appropriate optical signal before OCDMA coding is performed. This operation, which was illustrated in Figure 5.1 for the case of SPC-OCDMA and in Figure 5.3 for TPC-OCDMA systems, is referred to as data modulation. A variety of *optical modulation formats* are possible. Depending upon the particular modulation format chosen, it may be necessary to perform a complementary *optical demodulation* operation before the signal finally reaches a photodetector, which is the device that does the conversion of the optical signal back to the electrical domain.

A few of the more common formats applicable to the coherent OCDMA systems discussed in this chapter are presented here. Note that since both the SPC-OCDMA and TPC-OCDMA rely on pulsed laser sources, the modulation formats tend to fall in the category of short-pulse *return-to-zero* (RZ) modulation formats. In contrast, conventional DWDM transmission systems (with bit rates up to 10 Gb/sec) tend to use *non-return-to-zero* (NRZ) formats, where the optical power stays at a constant level between two adjacent "1" bits, although RZ formats are becoming more common in high bit rate systems at 40 Gb/sec and above.

5.3.2.1 RZ-OOK

One of the most straightforward pulsed laser modulation formats is return-to-zero on-off-keying (RZ-OOK). In this modulation scheme, a "1" data bit is encoded as the presence of the optical pulse, while a "0" data bit is encoded as the absence of the optical pulse. As shown in transmitter block diagram in Figure 5.6, laser pulses are

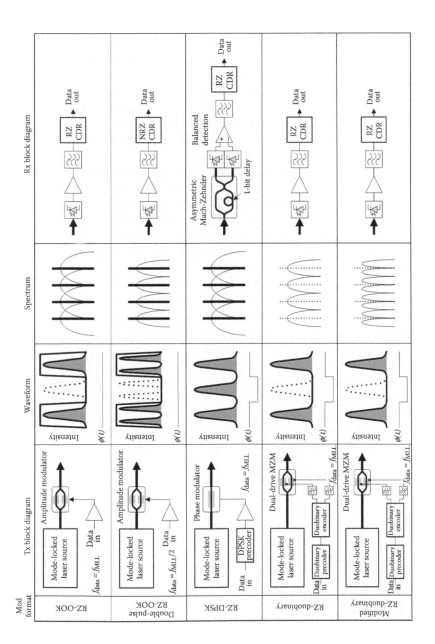

FIGURE 5.6 Table summarizing the common RZ optical modulation formats available for pulsed coherent OCDMA systems.

turned on and off through the use of an external amplitude modulator, which is electronically driven with the user's data stream. For data modulation rates up to 40 Gb/sec, the amplitude modulator is often a lithium niobate ($LiNbO_3$) electro-optic modulator arranged in a Mach-Zehnder configuration.

As shown in Figure 5.6, the receiver for RZ-OOK modulation is similar to that required for conventional NRZ-OOK. The front end consists of a photodetector, transimpedance amplifier/preamplifier, and appropriate electrical filtering. It is followed by a clock and data recovery (CDR) circuit, which extracts a clock signal for timing purposes and has a decision circuit for deciding when the signal is a digital "1" or a digital "0." Since the incoming optical signal is pulsed, the CDR circuit should be optimized for RZ formats. Fortunately, clock signals are relatively easy to recover from a RZ waveform as compared to NRZ since the signal spectrum has a component at the data modulation bit rate.

The optical spectrum of the RZ-OOK modulation format is illustrated in Figure 5.6. Since the modulation process is essentially the product of the mode-locked laser signal and the time-domain modulation envelope, one can compute the spectrum by convolving the laser comb with the equivalent modulation spectrum. The convolution is done using optical *field* amplitudes, and the intensity spectrum (which is what is observed on a typical optical spectrum analyzer) can be computed by squaring the result. One way to view the RZ-OOK optical spectrum is as an equivalent NRZ-OOK spectrum, but repeated for each mode-locked laser cavity mode. The NRZ spectrum has nulls in its spectrum at $\pm f_{data}$, the modulation data rate. As are result, when the data rate is equal to the mode-locked laser repetition rate, f_{MLL}, the data modulation sidebands overlap as shown in Figure 5.6.

For an SPC-OCDMA system, one would like to eliminate the modulation sideband overlap shown in Figure 5.6 for RZ-OOK so that each element of the phase code is only applied to a single frequency component of the mode-locked laser and its associated data modulation sidebands. Therefore, techniques for spectral shaping need to be considered. One variation to the RZ-OOK that limits the spectral extent of the modulation sidebands is the use of double-pulse RZ-OOK. As illustrated in Figure 5.6, double-pulse RZ-OOK uses two laser pulses for each data bit. In other words, $f_{data} = f_{MLL}/2$. In this case, the modulation sideband spectral nulls for the different laser cavity modes overlap and the spectrum illustrated in Figure 5.6 results. One other advantage to the double pulse RZ-OOK is that conventional NRZ clock and data recovery can be used, since the electrically filtered receiver signal results in a more NRZ-like waveform when the bandwidth is restricted to be approximately equal to the data bit rate. However, the technique does have the disadvantage that the mode-locked laser needs to operate at twice the data modulation rate.

5.3.2.2 RZ-DPSK

A more advanced optical modulation technique that offers some advantages in terms of certain transmission impairments is differential phase shift keying (DPSK). As illustrated in Figure 5.6, rather than modulating the intensity of laser pulses by turning them on or off, RZ-DPSK modulates the optical phase. However, unlike simple phase

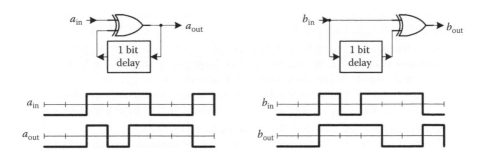

FIGURE 5.7 Equivalent logic block diagrams (consisting of an exclusive OR gate and 1-bit delay) and example timing diagrams for (a) differential encoding and (b) differential decoding.

shift keying (PSK), which would require a long-term stabilized optical phase relationship (over the entire data bit stream) between the laser source at the transmitter and the local oscillator used for coherent detection at the receiver, DPSK requires high optical coherency in the transmitter laser source only between adjacent bits, and a slow drift (occurring over many data bits) in the absolute optical phase alignment can be tolerated. In addition, relatively simple differentially coherent demodulation and intensity detection approaches can be applied.

The transmitter for RZ-DPSK modulation consists of a phase modulator rather than an intensity modulator. Similar to RZ-OOK, lithium niobate ($LiNbO_3$) electro-optic modulator can be used, but they are in a much simpler waveguide phase modulator configuration. A second key component in the transmitter is the DPSK digital precoder. This component is required since the receiver recovers data by differentially examining the incoming waveform; therefore, the transmitted signal must be first encoded in a differential manner.

The logic required to implement the DPSK differential encoding is illustrated in Figure 5.7(a). It consists of an exclusive-OR gate and a 1-bit feedback delay. As shown in the timing diagram, the output remains at its previous value if the current input bit is a logical 0. If the current bit is a logical 1, the output toggles.

At the receiver, complementary differential DPSK decoding is required. A logical block diagram is shown in Figure 5.7(b), where it can be seen to consist of the same logical components as the encoding. As shown in the timing diagram, the output is a logical 1 when the current input bit is different from the previous bit, while it is a logical 0 if the current bit is the same as the previous.

In terms of an actual implementation, differential DPSK encoding can be performed using an electronic precoder based on digital electronic components with the same functions as shown in the block diagram. The resulting digital signal can then be used to drive the electro-optic phase modulator.

Differential DPSK decoding can be performed using an all-optical DPSK demodulator followed immediately by photodetection to convert the signal to the electronic domain. This can be seen in the RZ-DPSK receiver block diagram shown in Figure 5.6. Here, an optical interferometer, such as Mach-Zehnder interferometer with a 1-bit delay in one of its arm, can be used as the differential demodulator/decoder. Through the use of balanced detection at the output of the interferometer, one can achieve an inherent 3 dB receiver sensitivity advantage over RZ-OOK. A simplified explanation for the 3 dB performance advantage of DPSK over OOK is that the separation between two DPSK symbols when viewed on a complex phase constellation diagram of the optical *field* is larger by a factor of $\sqrt{2}$. In terms of optical power, this translates to a factor of 2 or 3 dB [4].

Since RZ-DPSK generates modulation sidebands comparable in width to RZ-OOK, spectral shaping techniques would be needed for a SPC-OCDMA system if one desires to have the laser repetition rate equal to the data bit rate. Similar to RZ-OOK, one possible approach would be to use a double-pulse approach but use it in conjunction with RZ-DPSK. In the following section, we discuss a completely different approach to limiting the spectral extent of data modulation sidebands.

5.3.2.3   RZ-Duobinary

Duobinary modulation is a bandwidth efficient modulation scheme that allows one to transmit f_{data} bit/sec within a bandwidth less than $f_{data}/2$ Hz. In order to accomplish this based on Nyquist's results, it is inevitable that the signal will suffer from some amount of intersymbol interference (ISI). However, for the case of duobinary modulation, the ISI is introduced in a carefully controlled manner [5].

As illustrated in Figure 5.6, the duobinary transmitter consists of a duobinary precoder, a duobinary encoder, and an optical modulator. The duobinary precoder is similar in construction to the DPSK digital precoder and is shown in Figure 5.8, consisting of an exclusive-OR gate and feedback delay, although the delay may not necessarily be a single bit but depends upon the particular version of duobinary used. The duobinary encoder is also illustrated in Figure 5.8 and consists of second k-bit delay line, an adder or subtraction circuit, and finally a low pass filter. The low pass filter is an ideal rectangular filter with a cutoff frequency at one half of the data bit rate. The resulting waveform, which can be seen to be a 3-level signal at the output of the encoder, finally drives the optical modulator.

Different duobinary modulation modes are possible based on the following parameters: (1) the value chosen for k in delay elements in Figure 5.8, (2) whether an add or subtract operation is chosen in the duobinary encoder, and (3) how the optical Mach-Zehnder modulator interference condition is biased when the electrical modulation signal is zero as well as the amplitude of the driving signal. Depending upon the selection of these parameters, different modulation spectrums will result. We will briefly describe two of the more standard configurations next.

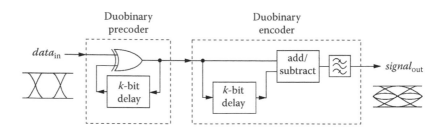

FIGURE 5.8 Equivalent logical block diagram of duobinary precoder and duobinary encoder.

Figure 5.6 shows the resulting spectrum for two specific duobinary configurations. In the first configuration, (1) the delay, k, is set to a value of 1, (2) the duobinary encoder is set to perform an add operation, and (3) the optical modulator is biased at the null operating point while the peak-to-peak modulation amplitude is equal to $2V_{pi}$, where $2V_{pi}$ is the voltage required to cause a relative optical phase shift of π radians between the two arms of the Mach-Zehnder interferometer. For this case, as shown in the spectrum of Figure 5.6, the modulation sidebands of the different laser modes no longer overlap when the modulation data rate, f_{data}, is equal to the mode-locked laser repetition rate, f_{MLL}. Note also that in this particular duobinary configuration, the center carrier frequencies are suppressed as indicated by the dashed lines in Figure 5.6. The carrier suppression is a result of the alternating π –phase shift that is imposed on the optical field between consecutive ones in the symbol sequence, similar to the carrier-suppressed RZ modulation format [6].

A second duobinary configuration, referred to as modified duobinary, is shown in Figure 5.6. The primary difference between the modified configuration and the standard duobinary is that the duobinary encoder is set to a subtract operation rather than add. In this case, not only is the carrier suppressed, but in addition, there is a null in the modulation spectrum where the carrier once existed. Note, that the modulation sidebands are constrained to $\pm f_{data}$, limiting the sideband overlap; therefore, it is still possible to operate the mode-locked laser at the same frequency as the data modulation rate.

Table 5.1 summarizes the key differences between the various modulation formats discussed in this chapter. Column 1 of Table 5.1 provides the name of the modulation format, while column 2 provides the ratio of the mode-locked laser repetition frequency to the data bit rate. Column 3 identifies if the modulation sidebands surrounding the mode-locked laser carrier frequencies overlap one another, preventing each spectral phase code element from being applied to a single frequency component of the mode-locked laser and its associated data modulation sidebands. Also, if the mode-locked laser carrier frequencies are suppressed as a result of the modulation format, this is listed in column 5 of Table 5.1. Finally, column 5 identifies whether a null is created in the modulated spectrum where the carrier was once present.

TABLE 5.1
Table Summarizing Relevant Differences between Different Optical Modulation Formats

Modulation Format	f_{MLL}/f_{data}	Modulation Sideband Overlap?	Carriers Present?	Modulation Null?
RZ-OOK	1	Yes	Yes	No
Double-pulse RZ-OOK	2	No	Yes	No
RZ-DPSK	1	Yes	No?	No
Duobinary	1	No	No	No
Modified duobinary	1	No	No	Yes

5.3.3 Spectral Phase Encoder/Decoder Technologies

The ability to access and modify uniformly the phase of the frequency bins is fundamental to an OCDMA technology that is based on spectral phase coding. The essential operation of such encoder/decoders consists of three functions, demultiplexing of the WDM channels into addressable frequency bins, imparting a prescribed phase change uniformly across on the spectral content of the bin, and multiplexing the phase encoded frequency bins back into a single fiber. It is clear that to maintain the integrity of the phase encoding process, the phases of all frequency bins have to be modified through the use of optically stable paths. Demultiplexing and multiplexing processes can be done in parallel or in series depending on the particular technology used. In this section we describe three technologies that have been used in spectral phase encoder/decoders, namely diffraction gratings, virtually imaged phased array (VIPA), and cascaded micro-ring resonators. We note filtering technologies such as Fabry-Perot etalons, fiber Bragg grating, and Mach-Zehnder interferometer that may be attractive for other OCDMA technologies such as fast frequency hopping are not considered here because of loss due to cascading or difficulty in maintaining phase integrity due physical separation of individual filters.

5.3.3.1 Diffraction Grating–Based OCDMA

The first demonstration of an OCDMA system based on spectral phase encoding modulated the spectral content of an ultrashort 75 fsec pulse [2]. The encoder/decoder shown in Figure 5.9 used a pair of bulk free-space gratings and a confocal lens pair optics with unit magnification to spread the spectral content in space, while a phase mask was used to impart the prescribed phase change for each spectral component.

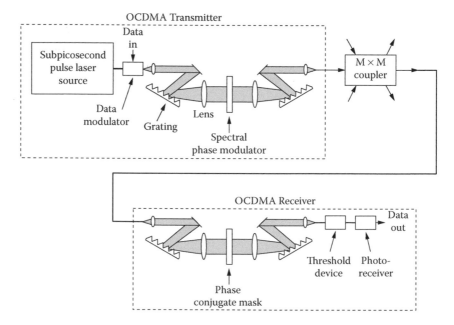

FIGURE 5.9 Encoder and decoder for ultra short pulse spectral phase OCDMA.

The phase mask consisted of a linear arrangement of frequency bins with phase modulations of 0 and π, and is placed at the focal plane midway between the lenses at the point where the optical spectral components experience maximum separation. Due to the large frequency spread of the ultrashort pulse, the spectral resolution of simple grating technology was shown to be adequate for addressing and modifying the phases for up to 127 frequency bins with pseudorandom codes of length 127.

The operation of this OCDMA system starts with a coherent ultra short pulse source shared between different transmitters, each having a distinct spectral phase mask. Because of the coherent nature of the ultra short pulse, the action of the encoder is to spread the short pulse over approximately 100 psec, thus reducing the peak intensity below a predetermined threshold. The encoded signal from different transmitters are passively combined and broadcast to all receivers that share the same optical channel. The receiver consists of a spectral phase decoder and an optical threshold device. The optical decoder has a phase mask that is the complex conjugate of the mask of its proper transmitter. For the case of binary spectral phase codes ([+1, −1] codes assigned [0,π] phase shifts), the encoder and decoder phase masks are identical. When the incoming signal passes through a matched decoder, the spectral phase shifts are removed and the original pulse is reconstructed. On the other hand, when the encoder/ decoder pair is not matched, the spectral phase shifts are rearranged but not removed. As a result the output of the decoder remains noise like. The threshold device is set to detect data corresponding to the intense decoded pulses and to reject the improperly decoded pseudo noise bursts from other transmitters. In Reference [2], where only a single user was present, a contrast ratio of ~25:1 between properly and improperly decoded pulses was demonstrated. In a realistic OCDMA system, where there is more than one transmitter, the background noise will be overwhelmed by the multiuser interference (MUI) noise, requiring additional noise suppression steps such as use of orthogonal coding, optical time gating, and/or optical thresholding. Some of these technologies are discussed further in Section 5.3.4. More recent efforts on diffraction-grating based OCDMA can be found in [18, 26].

5.3.3.2 VIPA-Based OCDMA

The desire to have an OCDMA system compatible with existing DWDM optical networking compels one to consider the possibility of a system with spectral range limited to a single optical passband, where the spectral width could be on the order of 100 GHz. Taking into account various coding requirements, the challenge is to be able to impart phase modulations on distinct frequency bins of width <10 GHz, much smaller than what is practical with a bulk grating–based device. Such high resolution filtering is possible through the use of either a virtually imaged phased array (VIPA) device or a micro ring resonator device discussed in the following section. Figure 5.10 shows the detailed operation of a VIPA structure as was first proposed and developed by Shirasaki [7]. The device consists of two parallel surfaces, one with 100% reflection coating and the other with >95% reflection coating. The beam is initially focused on the partially reflecting surface where, in this case, 5% of the light leaks out at the first bounce and starts propagating as a wave front. The remaining 95% bounces twice and on the third bounce 0.95 × 5% leaks as another

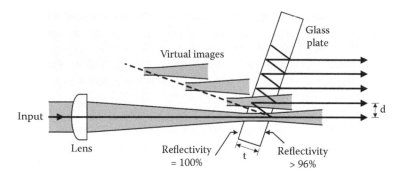

FIGURE 5.10 Illustration of virtually imaged phased array (VIPA) filter.

wave front but displaced by d and time delayed. For a small incident angle, θ, a number of wavefronts propagate from virtual images of the same beam waste, but are separated by $2t$, where t is the distance between the two parallel surfaces. The positions of the beam waists in the virtual images are self-aligned and there is no need to adjust their individual positions. The interference of these wavefronts leads to a collimated beam, where for a central wavelength λ_0, light propagates in the direction set by the angle θ. For other wavelengths, however, light propagates at a different angle depending on the wavelength relative to λ_0, with the red-shifted and blue-shifted wavelengths propagating on the opposite side of λ_0. The VIPA has the following advantages over diffraction gratings: 10 to 20 times larger angular dispersion, polarization independent operation, simple structure and alignment, and compactness.

Using a configuration similar to the VIPA structure, the Essex Corporation has developed a Hyperfine optical demultiplexer, which has a gradually changing reflectivity to compensate for the leakage of light and produces equal intensity wave fronts [8]. This device has been modified for application to spectral phase encoding in a spectrally efficient OCDMA system [3,9]. Figure 5.11 shows the basic operation of the spectral phase encoder/decoder. Collimated wave fronts are focused onto a phase mask that imparts $[0, \pi]$ phase shifts on frequency bins by making light travel $[0, \lambda/2]$ roundtrip distance with respect to a reference plane. The phase encoded beam is

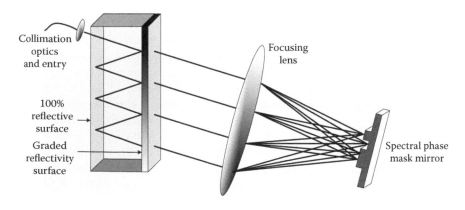

FIGURE 5.11 VIPA-based spectral encoder-decoder.

reflected back from the phase mask, travels back through the glass substrate and is focused into an output fiber using a slight tilt of the phase mask. Successful operation of such a device has been demonstrated for SPC-OCDMA with a resolution of ~1 GHz. The free spectral range of the device, which is defined as the span of frequencies over which spectral features can be unambiguously identified before the filter transfer function repeats, is determined by the thickness of the glass substrate and was set to 100 GHz. This value sets the maximum frequency extent over which spectral phase coding can be performed with independent phase values applied to each frequency bin.

The detection of reconstructed pulses after decoder is similar to those pointed out for the case of diffraction grating. However, as discussed in earlier sections, the operation of such a system with narrow frequency code bins of 5 GHz requires careful design of the data modulation subsystem as well as optimized code selection.

5.3.3.3 Micro Ring Resonator–Based OCDMA

Recent developments in planar lightwave devices have propelled micro-ring resonator (MRR) devices to a competitive position for ultra-high resolution filtering [10]. The advantages of MRR filters are: compatibility with optical integration, compactness, phase stability, and mass producible. Narrowband MRR filters capable of wavelength tuning through thermal means have recently been demonstrated. These devices have performance similar to VIPA filters in terms of spectral resolution, quality of the spectral shape, and insertion loss. However, for MRR-based coders, care must be taken to account for the group velocity delay variations or dispersion at the bin edges when designing an OCDMA system. For example, the bandwidth-efficient modulation techniques discussed previously in Section 5.3.2 can aid in limiting the spectral extent of the data modulation sidebands, thereby reducing the impact of dispersion.

Figure 5.12(a) shows a MRR filter fabricated in a SiO2/Si platform by Little Optics, Inc. [11]. The device waveguides are defined by a doping process that increases the index of refraction by ~0.2 with respect to glass. Light from the input is coupled to the first micro ring resonator whose resonant frequency is defined by its radius. The wavelength for maximum coupling can be tuned by changing the ring radius lightly with an on-chip heater. The larger the number of rings used in series, the more control that is available on the spectrum of coupled light. The direction of

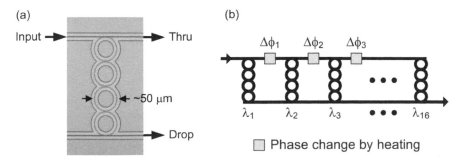

FIGURE 5.12 (Color Figure 5.12 follows page 240.) (a) A three-ring MRR tunable filter, and (b) optical circuit for a MRR-based 16-channel coder.

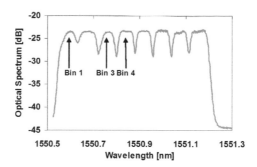

FIGURE 5.13 Experimental spectral amplitude response of four frequency bin MRR-based coder.

coupling is such that for an even number of micro rings, the coupled or dropped light emerges from the device opposite the input fiber.

Figure 5.13 shows a MRR-based coder for a DWDM compatible OCDMA system [24]. In contrast to the VIPA-based coder, the frequency bins of the MMR-based coder are tuned individually. Note that in order to maintain the same phase relation between different frequency bins, coupled or dropped light will need to exit from the opposite side of the circuit from the input, translating to requiring an even number of MRRs. The [0, π] phase shift on a frequency bin is imparted by making light travel [0, λ/2] extra distance by heating and lengthening a given path.

Figure 5.13 shows the spectral response of a prototype 8-frequency bin MRR-based coder. The coder is set to code 2 of the Hadamard-8 code set, resulting in a pi phase shift between each of the two neighboring frequency bins. The measurements were obtained using ASE source and a 0.01 nm resolution optical spectrum analyzer, which results in the finite drop in the insertion loss at the bin boundaries. Each of the 10 GHz-spaced frequency bins provides flat-topped passbands combined with sharp roll-offs at the bin edges. Unfortunately, the sharp drop at the bin edges also results in relatively large dispersion, limiting the useable width of the frequency bin to approximately 60% of the nominal 10 GHz passband.

5.3.4 MultiUser Interference Rejection

As mentioned earlier in the high level descriptions of both the spectral phase coded OCDMA and temporal phase coded ODCMA systems, multiuser interference noise from undesired users remains even after the signals have passed through the matched OCDMA decoder. In addition, since the optical signal energy present in both the decoded and undesired channels are similar in magnitude, both will appear essentially identical from the perspective of a typical photoreceiver that is band-limited to the data bit rate, preventing the desired signal from being recovered correctly. Therefore, further processing techniques are necessary in order to eliminate the interference. In principle, an ultrawide bandwidth optical photoreceiver could be used at the OCDMA receiver front end, and the removal of interference could be performed in the electrical domain. However, given the bandwidth requirements of an OCDMA system, which is typically on the order of many tens or even hundreds

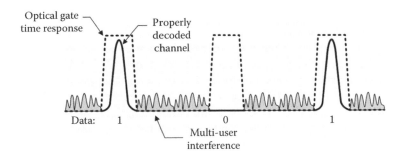

FIGURE 5.14 Illustration of optical time gating for multiuser interference rejection.

of GHz, this is likely to be impractical due to the need for ultrafast electronics. As a result, multiuser interference rejection is most commonly performed in the optical domain, where wideband optical processing can be accomplished relatively easily. Two categories of optical processing techniques for multiuser interference rejection, including optical time gating and optical thresholding, are discussed here.

5.3.4.1   Optical Time Gating

The basic concept behind the application of optical time gating to extract the decoded OCDMA signal is illustrated in Figure 5.14. Through the proper selection of an appropriate code set for a synchronous coherent OCDMA system (discussed further in Section 5.4), it is possible to design the system such that the multiuser interference energy falls outside a time interval where the properly decoded signal pulse resides. Therefore, by optically gating the composite signal in order to provide low loss within the desired time window while at the same time providing for high extinction outside that window, one can extract only the properly decoded signal bit stream.

For the purposes of application to coherent optical CDMA system, some of the more important performance metrics for optical gating technologies include:

- Gate width (typically on the order of 10 psec or less)
- Gate repetition rate (comparable to data rate, typically >1 GHz or higher)
- Gate extinction rate (depends upon number of users but typically 10–20 dB)
- Data pulse energy levels and dynamic range
- Gating control/clock pulse energy levels and dynamic range

As a result of these high performance requirements, relatively high-speed optical processing techniques must be employed, such as those used for all-optical demultiplexing. Although there are a wide variety of options, some of the technologies that have been demonstrated specifically for coherent OCDMA systems include:

- Nonlinear interferometers
- Four-wave mixing (FWM) techniques

These are the technologies that we will focus on in the following subsections. Both optical fiber-based and semiconductor optical amplifier (SOA)-based approaches to these techniques are possible.

5.3.4.1.1 Fiber-Based Nonlinear Interferometers for Optical Time Gating

By using an optical clock pulse in order to alter the effective phase shift through one arm of an interferometer, one can construct an all-optical gate. The phase changed can be accomplished through a distributed medium, such as nonlinear propagation through a length of optical fiber, or it can be a concentrated nonlinearity, such as can be accomplished with a semiconductor optical amplifier (SOA). Since interferometers generally require stabilization for proper operation, a common approach for fiber-based nonlinear interferometers, which generally require long lengths of fiber and therefore path lengths can drift with environmental conditions, is the nonlinear optical loop mirror (NOLM) [12]. Illustrated in Figure 5.15, the NOLM is built in a Sagnac interferometer configuration, which by its construction is self-stabilizing.

The operation of the NOLM can be described as follows, assuming for simplicity that the input signal is a single optical pulse. The incoming signal, input at Port A, is split into two counterpropagating pulse replicas at a 50:50 fiber coupler. When the clock pulse, which can be injected at Port C, is not present, the low amplitude data pulses simply counterpropagate around the loop and recombine at the coupler. The interference condition is such that signals interfere destructively at the output port B, but interfere constructively at the original input port A, thereby reflecting the data pulse and hence the name "loop mirror." On the other hand, by injecting a large amplitude clock signal that is of a close but distinguishable wavelength relative to the data pulse wavelength, it is possible to overlap the clock pulse with the clockwise propagating data pulse and introduce a nonlinear phase shift of π. In this case, the interference condition is altered such that the data pulse now exits at port B. An optical bandpass filter at port B suppresses the remaining clock signal, leaving only the desired gated data pulse. The width of the time gating window for the NOLM is defined by the overlap between the clock and co-propagating data pulse. The first NOLMs that were constructed required very long dispersion-shifted fibers (>1 km) to obtain the required nonlinear phase shift; however, recent developments in highly nonlinear fibers have allowed for a reduction in fiber length to approximately 100 m or less. In conjunction with optical thresholding, the NOLM optical time gate has been successfully applied to an implementation of phase-coded OCDMA system [13].

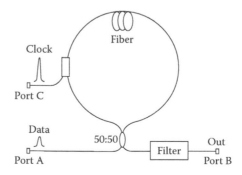

FIGURE 5.15 Illustration of nonlinear optical loop mirror (NOLM) time gate.

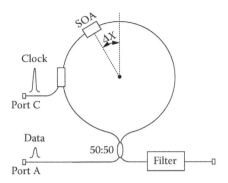

FIGURE 5.16 Illustration of the terahertz optical asymmetric (TOAD) time gate.

5.3.4.1.2 SOA-Based Nonlinear Interferometers for Optical Time Gating

The nonlinear phase change required in the NOLM can be accomplished by other means, such as the use of as a semiconductor optical amplifier (SOA), allowing for the potential of device integration. An interferometric architecture similar to the NOLM can be used, as shown in Figure 5.16. When the SOA is offset from the center of loop by Δx, the device is referred to as the terahertz optical asymmetric demultiplexer or TOAD [14].

Similar to the NOLM, when the clock pulse is not present, incoming data pulses reflect from the TOAD. By injecting a clock pulse, which is typically chosen to be on the order of 10 dB larger in amplitude than the data pulse intensity, gating can occur. The clock pulse saturates the SOA, thereby changing its effective index. The clock pulse, which travels only in the clockwise direction, is injected following the clockwise propagating data pulse to give the clockwise data pulse the opportunity to propagate through the SOA before the clock pulse saturates the SOA index. Since the SOA slowly recovers on the time scale of hundreds of picoseconds, counter-propagating data pulses that arrive immediately after the clock pulse event has occurred see the SOA in approximately the same relative state and do not experience a differential phase shift. The temporal duration of the gating window is set by the offset of the SOA, Δx, from the center of the loop. As the offset is reduced, the gating window width decreases until the actual length of the SOA needs to be taken into account. The nominal gate width is related to the offset by

$$\Delta t_{gate} = 2\Delta x / c_{fiber} \tag{5.3}$$

where c_{fiber} is the speed of light in fiber. Gating windows as short as 1.6 picoseconds have been demonstrated experimentally using a TOAD [15]. The TOAD optical time gate has been successfully applied to an implementation of SPC-OCDMA [3].

5.3.4.1.3 Four Wave Mixing (FWM) Optical Time Gating

Another approach to optical time gating is through the use of four wave mixing (FWM). FWM is a third-order nonlinearity, similar to intermodulation distortion in the electrical domain. In FWM, the nonlinear beating between the data signal and

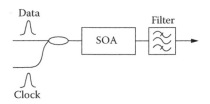

FIGURE 5.17 Illustration of the four-wave mixing (FWM) time gate.

a control signal at a different wavelength generates new optical tones as sidebands. FWM-based gating can be accomplished in optical fiber or in waveguide devices such as semiconductor optical amplifiers (SOAs).

In SOAs, FWM occurs through carrier density modulation. Two copolarized optical signals are coupled into the SOA. One is the control signal at frequency f_c and typically has a much higher intensity than the other input signal (data) to be wavelength converted, which is at frequency f_d. The two copropagating signals mix and, through carrier density modulation, form an index grating off which signals can be scattered. The scattering of the control signal from this grating generates two waves, one at the data frequency and one at a new frequency, $f_{converted} = 2f_c - f_d$. This is the useful wavelength-converted signal. In addition, data signal scattering also generates two much weaker waves, one at the control frequency and one at a new frequency, $f_{satellite} = 2f_c - f_d$. This is called the satellite wave and is generally not used.

By injecting a short optical control pulse along with the incoming OCDMA signal into the SOA as shown in Figure 5.17, it is possible to create an optical time gate by filtering out the resulting wavelength converted signal. The clock pulse is temporally aligned to the correct position relative to the desired OCDMA pulse, and an optical bandpass filter is placed at the output of the SOA in order to extract the FWM signal only. The FWM optical time gate has been successfully applied to an implementation of TPC-OCDMA [16].

5.3.4.2   Optical Thresholding

Another approach to OCDMA multiuser interference rejection is nonlinear optical thresholding. The basic concept, illustrated in Figure 5.18, relies on a device that has a nonlinear power-transfer response as indicated. The pulses for the properly decoded pulses are of large peak amplitude and therefore pass through the device with low relative loss. On the other hand, the undesired channels, which are of low peak intensity since they have been spread in time, are effectively suppressed. One of the advantages of optical thresholding over optical time gating is in eliminating the need for a synchronized clock pulse.

As was the case for optical time gating, there are various alternatives to implementing an all-optical thresholding device. Two approaches that have been specifically applied to OCDMA systems include:

- Nonlinear optical loop mirrors (NOLM)
- Second harmonic generation (SHG) in periodically poled lithium niobate (PPLN)

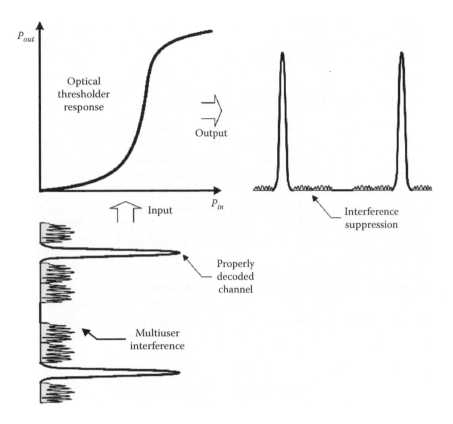

FIGURE 5.18 Illustration of optical thresholding for multiuser interference rejection.

5.3.4.2.1 NOLM-Based Thresholding

The NOLM used for all-optical nonlinear thresholding are very similar in construction to the NOLMs used for optical time gating with the exception of the clock pulse and its associated input port. In addition, rather than using a 50:50 coupler to create the loop, an asymmetric coupling ratio is used, such as 70:30. By introducing this asymmetry, the nonlinear phase shift experienced by the two counterpropagating pulses of large amplitude is also different, and can result in the incoming pulse to be sent to the output port rather than get reflected. On the other hand, for low amplitude pulses, the nonlinear phase shift is not large for either and the pulses are primarily reflected as in the standard loop mirror configuration. NOLM-based thresholding has been used in phase-coded OCDMA systems both as a stand-alone device [17], as well as in conjunction with optical time gating [13].

5.3.4.2.2 PPLN-Based Thresholding

Periodically poled lithium niobate (PPLN) devices can be used for nonlinear second harmonic generation. Since the output power is strongly dependent on the input signal intensity, they can be used as nonlinear thresholding devices. PPLN-based optical thresholding has been successfully applied in wideband SPC-OCDMA systems [18], using waveguide-based devices.

5.4   CODE SELECTION FOR SPC-OCDMA

Codes are, of course, central to any OCDMA system since it is on the basis of the code signatures applied to the multiple transmissions that one is able to discriminate multiple transmissions occupying the same optical passband and overlapping in time. The choice of the code influences everything from how many users may be simultaneously active, the total number of available codes, the expected performance of the system (multiaccess interference levels), to the need (or lack thereof) for global network synchronization. However, before addressing the different code types in use, it is helpful to review the role that codes play in OCDMA networks.

The discussion here will focus on code selection for narrowband spectral phase coded optical CDMA systems (SPC-OCDMA). In these applications, we begin with the set of N well-defined, phase-locked frequencies that are the output of a mode-locked laser (MLL). The temporal electric field is a pulse train that can be represented as

$$E(t) = \frac{1}{N} e^{j2\pi f_0 t} \sum_{n=1}^{N} e^{j2\pi n f_{MLL} t} \tag{5.5}$$

where f_{MLL} is the frequency spacing between the MLL lines and is of the order of 5–10 GHz, f_0 is the optical carrier frequency on the order of 10^{14} Hz, and the term in the summation represents the envelope of the pulse train. The pulse repetition rate is equal to $(1/f_{MLL})$ and the pulse widths are equal to $(1/N f_{MLL})$. $|E(t)| = 1$ at integer multiples of the period $T = (1/f_{MLL})$.

The pulse train is "spectrally phase encoded" by applying a spectral phase mask $\Phi_{n,code}$ that multiplies frequency n by a phase term corresponding to a given code (in some cases, the amplitude of the spectral phase component also changes, but in the nominal case the phase function is of unit amplitude). In the absence of data modulation, the effects of encoding the MLL signal with code E and decoding with code D can be expressed as:

$$E(t) = \frac{1}{N} e^{j2\pi f_0 t} \sum_{n=1}^{N} \Phi_{n,D} \Phi_{n,E} e^{j2\pi n f_{MLL} t} \tag{5.6}$$

Spectral-phase codes can be categorized as either orthogonal or nonorthogonal codes. Both code categories are well known in wireless/cellular CDMA technology, where orthogonal codes are widely used for the downstream links (from the headend to the end stations) while nonorthogonal codes are used for the upstream links. Orthogonal codes have the advantage of limited multiuser interference and high spectral efficiency but must be operated in synchronous fashion to preserve orthogonality. The synchronization requirement is more easily satisfied for the downstream links because all of the transmitters are co-located at the cell tower.

Nonorthogonal codes have the great advantage of not requiring synchronization between transmitters, but this advantage comes at the cost of greater multiuser interference and relatively lower spectral efficiency.

5.4.1 ORTHOGONAL CODES: BINARY WALSH-HADAMARD

The primary example of orthogonal codes is based on using well-known Hadamard matrices H_N to determine the sequence of 1's and −1s (or equivalently relative phase changes of $[0,\pi]$) across the frequency spectrum. We refer to the columns (or rows) of these matrices as Hadamard codes. In general, all of the elements of Hadamard matrices are complex roots of unity and H satisfies the orthogonality condition

$$H_N^* H_N = N \cdot I_N \tag{5.7}$$

where * indicates conjugate transpose, and H_N and I_N represent Hadamard and identity matrices of order N, respectively. An important subset of the Hadamard matrices in which all the code elements are either +1 or −1 can be generated recursively from the definitions $H_1 = 1$ and

$$H_{2N} = \begin{bmatrix} H_N & H_N \\ H_N & -H_N \end{bmatrix} \tag{5.8}$$

In the absence of data modulation, the effects of encoding with code E and decoding with code D can be expressed as (here we represent H_N as simply H with element $H_{i,j}$)

$$E(t) = \frac{1}{N} e^{j2\pi f_0 t} \sum_{n=1}^{N} H_{n,D} H_{n,E} e^{j2\pi n f_{MLL} t} \tag{5.9}$$

As defined in Equation 5.8, the Hadamard codes obey $H = H*$ and

$$\frac{1}{N} \sum_{n=1}^{N} H_{n,i} H_{n,j} = \delta_{i,j} \text{ where } \delta_{i,j} = 1 \text{ iff } i = j, 0 \text{ otherwise} \tag{5.10}$$

As a consequence, when $D = E$ (matched encoder and decoder), (4) reduces to (1) and again $|E(t)| = 1$ at integer multiples of the period T. When $D\ E$ (mismatched encoder and decoder), $|E(t)| = 0$ at integer multiples of the period T. The effect of the spectral phase coding is that matched codes recover the original mode-locked laser pulses while mismatched coders result in a null at the time of the peak of the desired signal. With appropriate synchronization between coded transmitters, a receiver can discriminate its matching coded signal from the $(N-1)$ other signals by sampling the decoded signal at integer multiples of T where the desired signal is maximized

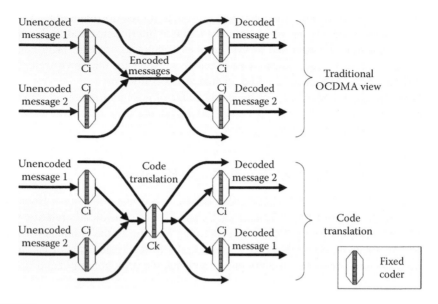

FIGURE 5.19 A conventional OCDMA configuration is shown at the top in which messages are exchanged between matched encoder/decoder pairs. The addition of an appropriate intermediate code translation stage, as shown at bottom, can interchange the recipients of the two input signals.

and the interfering signals are nulled. Essentially the same process operates when all transmitters are sending signals that are data modulated.

Cascaded application of two such coders is equivalent to element-by-element multiplication of their respective phase codes to yield the output code. The interesting observation is that *the set of Hadamard codes is closed under such element-by-element multiplication*: cascading two Hadamard codes results in another code in the set [25]. Thus, as indicated in the code translation table in Figure 5.20, a signal encoded with code i, and then passed through a code-j coder emerges in code k, where code k is another element of the Hadamard set. For example, an incoming signal in code 9 arriving at a code-12 coder, emerges in code 4. One consequence of the closure of the Hadamard codes is that *any* of the Hadamard codes can be passively converted to *any* other code in the set by means of passing it through an appropriately chosen phase coder. As shown in the lower part of Figure 5.19, the addition of an appropriate code translation stage [25, 26, 27] in a conventional OCDMA network can passively route communication between mismatched encoders and decoder.

5.5   OCDMA NETWORK ARCHITECTURES FOR SPC-OCDMA

5.5.1   POINT-TO-POINT MULTIPLEXING TRANSMISSION LINKS

Optical code division multiple access networks typically operate in a configuration in which communication is established between matching encoders and decoders (Figure 5.19[a]). To establish arbitrary connectivity amongst all users, the encoders

Incoming code

Coder =		1	2	3	4	5	6	7	8	9	10	11	12	13	14	15	16
	1	1	2	3	4	5	6	7	8	9	10	11	12	13	14	15	16
	2	2	1	4	3	6	5	8	7	10	9	12	11	14	13	16	15
	3	3	4	1	2	7	8	5	6	11	12	9	10	15	16	13	14
	4	4	3	2	1	8	7	6	5	12	11	10	9	16	15	14	13
	5	5	6	7	8	1	2	3	4	13	14	15	16	9	10	11	12
	6	6	5	8	7	2	1	4	3	14	13	16	15	10	9	12	11
	7	7	8	5	6	3	4	1	2	15	16	13	14	11	12	9	10
	8	8	7	6	5	4	3	2	1	16	15	14	13	12	11	10	9
	9	9	10	11	12	13	14	15	16	1	2	3	4	5	6	7	8
	10	10	9	12	11	14	13	16	15	2	1	4	3	6	5	8	7
	11	11	12	9	10	15	16	13	14	3	4	1	2	7	8	5	6
	12	12	11	10	9	16	15	14	13	4	3	2	1	8	7	6	5
	13	13	14	15	16	9	10	11	12	5	6	7	8	1	2	3	4
	14	14	13	16	15	10	9	12	11	6	5	8	7	2	1	4	3
	15	15	16	13	14	11	12	9	10	7	8	5	6	3	4	1	2
	16	16	15	14	13	12	11	10	9	8	7	6	5	4	3	2	1

FIGURE 5.20 The code translation table for Hadamard 16 codes — an incoming code at the top of the matrix which encounters a coder as indicated at the left, emerges from the coder in the code indicated in the body of the matrix.

and/or decoders at the edges of the network must be tunable. However, since these networks operate in a broadcast-and-select mode, tunable decoders make eavesdropping relatively easy, but fixed decoders obviate simple multicast operation.

5.5.2 Star Networks, Single Cascaded

Given the ability to perform optical code translations, a number of physical-star networks can be composed in which each end user is assigned a unique fixed code to use to transmit to the central hub. The hub is equipped with adjustable coders (see Figure 5.21 below). (For simplicity, Figure 5.21 only shows a one-fiber configuration in which a single physical coder functions simultaneously as an encoder and a decoder and uses direction-division multiplexing to separate upstream and downstream signals. Two-fiber network configurations, not shown, would function similarly.)

Absent the adjustable coders at the hub, no two users can communicate since no two share matching coders. However, using adjustable coders suitably arranged at the entrance and exits ports of the central broadcast optical star, arbitrary connections among the four users can be established. At the hub, messages from a given user could be converted to the code used by the destination user and distributed via a broadcast star. Note that in this configuration, a message between any two users would traverse a cascade of four encoders (in a two-fiber configuration, message need

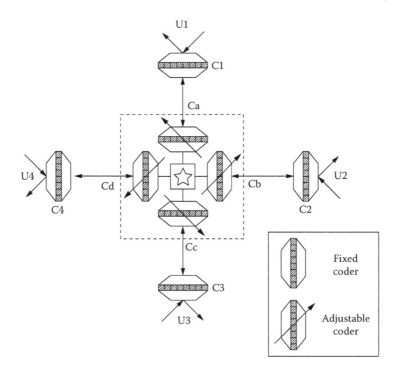

FIGURE 5.21 A star network using code translation to route connections between users.

only traverse a cascade of three encoders). While there would have to be some active elements to switch the adjustable coder, once the coder is set, the code conversion process itself is passive and all-optical. As described below, such a network could be configured at the hub to provide bi-directional point-to-point connectivity between any pairs of end users and/or unidirectional point-to-multipoint connectivity, all by means of changing the adjustable coders at the hub.

For this network, the settings of the adjustable hub coders are not unique. For example, suppose we wish to set up two-way links between U1 (user 1) and U2 and between U3 and U4. Then adjustable coder Ca could be set to any code and would thereby translate incoming code C1 to some code Cα, and adjustable coder Cb must be set to whatever code converts Cα to C2. For the U3/U4 connection, adjustable coder Cc can be set to any code Cβ except that which would convert incoming code C3 to Cα, to produce code Cχ. Then adjustable coder Cd must be set to whatever code converts Cχ to C4. The counter propagating signals would simply undergo the code translations in reverse order. To set up a multicast such that U1 can broadcast to the other three stations, one approach would be set Ca to C1 (effectively decoding U1's signal), and Cb, Cc, and Cd to C2, C3 and C4, respectively. Again, this combination of adjustable coder settings is not unique.

The above configuration provides robust centralized control at the hub. The network operator managing the hub can control what signals reach which end user in decodeable form and prevent eavesdropping or user-to-user interference due to code collision (again assuming each user is assigned a unique code). In comparison to a network with the adjustable coders placed at the user ends of the network, eavesdropping and interference would be more easily managed in this "hub-controlled" network than in the "edge-controlled" network. If only the adjustable coders were placed at the core of the network adjacent to the broadcast star and no coders were placed at the end user, network control of eavesdropping and interference would still be possible, but both the downstream and upstream signals to and from the end user would be essentially unencoded.

5.5.3 RING NETWORKS AND ADD/DROP ARCHITECTURES

The ability to successfully translate encoded signals enables ring architectures be constructed in which a combination of passive code translation and fast optical switching permits a code-based equivalent of wavelength add/drop multiplexing to be performed. With this capability, both peer-to-peer and hub-to-subnode networks are feasible. Here we will focus on the latter, in which a hub wishes to communicate bidirectionally with several subnodes by sending M spectrally encoded downstream signals on one fiber and receiving upstream signals from each node. At the hub, the outputs of M parallel synchronized spectral encoders would be passively combined and launched together on the ring fiber (for simplicity, we ignore ring protection). A code add/drop multiplex, as shown in Figure 5.22, would be required at each subnode.

The operation of the add/drop unit is as follows: at the left, an ensemble of suitably encoded signals arrives on the ring fiber as input to a decoder matched to code n. The decoder reconstructs the original pulse train sent on code n while mismatched codes are shifted into other code states according to the translation table in Figure 5.20 and are all nulled at the sampling time. Following the decoder, a fast optical cross-bar switch, such as the TOAD-based cross-bar switch demonstrated in [19], is gated into the cross state at the sampling instant effectively extracting the one downstream signal from the ring for local detection and allowing the other

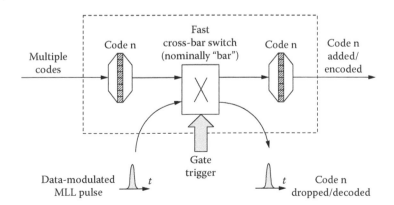

FIGURE 5.22 A code-based add-drop multiplexer permits one code to be extracted from several codes circulating on a ring network and at the same time, new data inserted/added to the other signals circulating on the ring.

signals to pass the gate during its bar-state. For upstream transmission from the subnode to the hub, a local mode-locked laser signal is data modulated and launched into the cross-bar switch at the moment that the switch is in the cross state, thereby replacing the downstream signal. Following the gate, a code-n coder reverses the effects of the code-n decoder and the composite signal leaving the subnode contains the new upstream data in code n and the $(M - 1)$ other signals back in their original code states. At the hub, the received signal would be split to M parallel decoders, one for each signal for reception of the several upstream signals.

One consequence of the closure of the Hadamard code set is that whenever there are two coders back to back, as would be the case when one considers the output coder of one subnode and the input decoder at the next subnode, the two coders can effectively be replaced by one coder. This means a code add/drop unit would not need two coders as shown in Figure 5.22, but the ring could function with a single coder and fast optical gate at each subnode.

Peer-to-peer networking on a ring brings an additional challenge not found on the hubbed ring. In principle, on the hubbed ring, a single code is launched from hub and dropped and replaced just once at the intended node and the replaced code is then optically terminated back at the hub. There is no possibility of a code (or a fraction thereof) circulating endlessly around the ring because the ring is not optically closed. But on a peer-to-peer ring, such endless circulation may be an issue because the path is optically closed and imperfections in the coding process and optical gating process will inevitably lead to some feed-through of the "nominally dropped" signal component. This feed-through component represents a source of coherent crosstalk with the added signal (recall that the mode-locked laser sources are high coherent optically) and may be a limiting factor for peer-to-peer ring applications.

5.5.4 SHARED CODE SCRAMBLING

Thus far we have focused on conventional Hadamard codes generated according to Equation 5.8. For a given order N of Hadamard codes the number of possible orthogonal code states so generated is N (actually, a limited number of non-isomorphic configurations of H_N exist for $N > 8$ but the number of code states is still quite modest). An eavesdropper equipped with an adjustable decoder would have to guess only on the order of N possible code settings in order to tune in on any given transmission. For increased data obscurity/scrambling, it would be useful if the eavesdropper were required to search a far larger number of possible codes before guessing the settings for the decoder.

Following the work in [20], we recognize that the set of possible orthogonal codes can be far larger than N. In [20], the authors point out that a large number of equivalently orthogonal matrices W_N can be generated from H_N by premultiplying by a diagonal matrix D_N of order N with all of the on-diagonal elements being arbitrarily chosen complex roots of unity ($e^{j2\pi/m}$ where m is an integer)

$$W_N = D_N H_N \tag{5.11}$$

It is straightforward to demonstrate that

$$W_N^* W_N = N \cdot I_N \tag{5.12}$$

and that the columns of the W_N matrix would therefore be suitable to use as codes in this type of system.

There are two important points to make about these modified Hadamard codes. First, while there are still only N distinct orthogonal codes available at any one time, the number of possible W_N code configurations is governed by the number of different arbitrarily chosen D_N matrices that can exist and that number can be quite large. With m possible phase states at each of the N diagonal elements, the number of distinct D_N matrices is m^N (consider for example, a case with $N = 16$ and $m = 4$ possible phase states, the number of distinct D_N matrices is 2^{32}, while $N = 32$ and $m = 8$ possible phase states yield 2^{96} distinct D_N). While some of these states are degenerate in that they would yield essentially the same fields (i.e., when two D_N's differ only by a scalar multiple), it is clear that the search space for guessing the possible code settings in order to eavesdrop on a given connection can be made far larger than N possible Hadamard codes. It is an open question whether or not one would be reduced to brute force guessing the code or if some systematic means of zeroing in on the code in use might exist. Consequently, we do not contend that this approach provides cryptographic security comparable to digital encryption of data, but might provide useful obscuration of signals in transit [21].

Second, it is important to recognize that the random diagonal matrix D_N can be implemented as a separate encoder similar to the same sort used to apply Hadamard codes to the MLL signal. This means one can physically separate the Hadamard coding stages (used for directing communication between end users) and the diagonal matrix "scrambling" stages in a network. Since it would be desirable to change the scrambling code with some regularity, the scrambling coders should be dynamically adjustable in synchrony and there is therefore some advantage to sharing these units to keep their number small.

As shown in Figure 5.23, one can imagine a network scenario consisting of multiple secure islands within which Hadamard coding would be used for signal

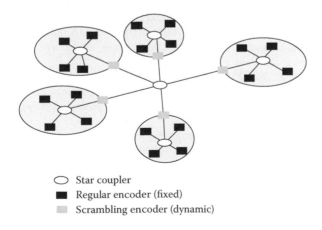

◯ Star coupler
■ Regular encoder (fixed)
▨ Scrambling encoder (dynamic)

FIGURE 5.23 (Color Figure 5.23 follows page 240.) Secure islands with conventional Hadamard coders interconnected with dynamic "scrambling" encoders for increased data obscurity between islands.

routing purposes, but before these signals leave the secure islands they would be scrambled by passing through a shared scrambling stage, a shared descrambling stage and finally a Hadamard decoder for signal selection.

5.6 SUMMARY

This chapter has provided a description of coherent optical CDMA systems. In particular, the focus has been on OCDMA systems utilizing short-pulse mode-locked lasers as the source of coherent light and employing phase coding in either the temporal or spectral domain. A description of a variety of key subsystems needed to construct a coherent optical CDMA system, including laser source technology, data modulation techniques, spectral phase encoder/decoder components, and multiuser interference rejection subsystems has also been provided. In addition to hardware requirements, a brief description of OCDMA codes has been provided, focusing in particular on orthogonal code sets for spectral phase coded OCDMA. Through the use of orthogonal coding and synchronous operation, coherent OCDMA systems have the potential to support the largest number of simultaneous users in a relatively narrowband spectral window with minimal multiuser interference compared to incoherent OCDMA approaches. Finally, a variety of system applications for coherent OCDMA systems have been described, including point-to-point links, star network architectures, ring networks, and security-enhanced networks enabled through the use of dynamic code scrambling.

REFERENCES

[1]   Weiner, A. M., Heritage, J. P., Salehi, J. A. (1988). Encoding and decoding of femtosecond pulses. *Opt. Lett.* 13:300–302.

[2]   Salehi, J. A., Weiner, A. M., Heritage, J. P. (1990). Coherent ultrashort light pulse code-division multiple access communication systems. *J. Lightwave Technol.* 8:478–491.

[3]   Etemad, S., Banwell, T., Galli, S., Jackel, J., Menendez, R., Toliver, P., Young, J., Delfyett, P., Price, C., Turpin, T. (2004), Optical-CDMA incorporating phase coding of coherent frequency bins: Concept, simulation, experiment. *Proc. Optical Fiber Communications Conf.* FG5.

[4]   Xu, C., Liu X., Wei, X. (2004). Modulation formats for high spectral efficiency fiber optic communications. *IEEE J. of Select Topics in Quantum Electron.* 10:281–293.

[5]   Sklar, B. (2001). *Digital Communications: Fundamentals and Applications*, 2nd ed. New York: Prentice Hall.

[6]   Miyamoto, Y., Hirano, A., Yonenaga, K., Sano, A., Toba, H., Murata, K., Mitomi, O. (1999). 320 Gbit/s (8 × 40 Gbit/s) WDM transmission over 367 km with 120 km repeater spacing using carrier-suppressed return-to-zero format. *Elect. Lett.* 35:2041–2042.

[7]   Shirasaki, M. (1996). Large angular dispersion by a virtually imaged phased array and its application to a wavelength demultiplexer. *Optics Lett.* 21:366–368.

[8]   Turpin, T. M., Froehlich, F. F., Nichols, B. D. Optical tapped delay line. United States Patent 6,608,721.

[9]   Toliver, P., Young, J., Jackel, J., Banwell, T., Menendez, R., Galli, S., Etemad, S. (2004). Optical network compatibility demonstration of O-CDMA based on hyperfine spectral phase coding. Proc. IEEE LEOS '04, Puerto Rico.

[10] Little, B. E., Chu, S. T., Haus, H. A., Foresi, J., Laine, J.-P. (1997). Micro ring resonator channel dropping filters. *J. Lightwave Technol.* 15:998–1005.

[11] www.littleoptics.com.

[12] Agrawal, G. P. (1995) *Nonlinear Fiber Optics*, 2nd ed. New York: McGraw Hill.

[13] Sotobayashi, H., Chujo, W., Kitayama, K. (2002). 1.6-b/s/Hz 6.4 Tb/s QPSK-OCDM/WDM (4OCDM × 40WDM × 40Gb/s) transmission experiment using optical hard thresholding. *IEEE Photon, Technol. Lett.* 14:555–557.

[14] Sokoloff, J. P., Prucnal, P. R., Glesk, I., Kane, M. (1993). A terahertz optical aymmetric demultiplexer (TOAD). *IEEE Photon. Technol. Lett.* 5:787–790.

[15] Runser, R. J., Zhou, D., Coldwell, C., Wang, B. C., Toliver, P., Deng, K. L., Glesk, I., Prucnal, P. R. (2001). Interferometric ultrafast SOA based optical switches: From devices to application. *Optical and Quant. Electron.* 33:841–874.

[16] Kitayama, K., Sotobayashi, H., Wada, N. (1999). Optical code division multiplexing (OCDM) and its applications to photonic networks. *IEICE Trans. Fundamentals* E82:2616–2626.

[17] Lee, J. H., Teh, P. C., Petropoulos, P., Ibsen, M., Richardson, D. J. (2001). Reduction of interchannel interference noise in a two-channel grating-based OCDMA system using a nonlinear optical loop mirror. *IEEE Photon. Technol. Lett.* 13:529–531.

[18] Jiang, Z., Seo, D. S., Yang, S. -D., Leaird, D. E., Weiner, A. M., Roussev, R. V., Langrock, C., Fejer, M. M. (2004). Four user, 2.5 Gb/s, spectrally coded O-CDMA system demonstration using low power nonlinear processing. Proc. Optical Fiber Communication Conference, Los Angeles, CA, PDP29.

[19] Glesk, I., Kang, K. I., Prucnal, P. R. (1997). Demonstration of ultrafast all-optical packet routing. *Electron Lett.* 33:794–795.

[20] Wysocki, B. J., Wysocki, T. (2002). Modified Walsh–Hadamard sequences for DS-CDMA wireless systems. *Int. J. Adapt. Control Signal Proc.* 16:589–602.

[21] Shake, T. H. (2005). Security performance of optical CDMA against eavesdropping. *J. Lightwave Technol.* 23:655–670.

[22] Shake, T. H. (2005). Confidentiality performance of spectral phase encoded optical CDMA. *J. Lightwave Technol.* 23:1652–1663.

[23] Sotobayashi, H., Chujo, W., Kitayama, K. (2004). Highly spectral efficient optical code division multiplexing transmission system (invited). *IEEE J. Sel. Top. Quant. Electron.* 10(2):250–258.

[24] Agrawal, A., Toliver, P., Menendez, R., Etemad, S., Jackel, J., Young, J., Banwell, T., Little, B. E., Chu, S.T., Delfyett, P. (2005). Fully Programmable Ring Resonator Based Integrated Photonic Circuit for Phase Coherent Applications. *Proc. Optical Fiber Communications Conf.* PDP6.

[25] Seo, D. S., Jiang, Z., Leaird, D. E., Weiner, A. M. (2004). Pulse shaper in a loop: demonstration of cascadable ultrafast all-optical code translation. *Optics Letters.* 29:1864–1866.

[26] Jiang, Z., Seo, D.S., Leaird, D.E., Roussev, R.V., Langrock, C., Fejer, M.M., Weiner, A.M. (2005). Reconfigurable all-optical code translation in spectrally phase-coded O-CDMA. *J. Lightwave Technol.* 23:1979–1990.

[27] Kitayama, K. (1998). Code division multiplexing lightwave networks based upon optical code conversion. *IEEE J. Select. Areas Commun.* 16(9):1309–1319.

6 Incoherent Optical CDMA Systems

Varghese Baby, Darren Rand, Camille-Sophie Brès, Lei Xu, Ivan Glesk, and Paul R. Prucnal

CONTENTS

6.1 INTRODUCTION

Since the initial efforts (Prucnal, Santoro, Fan [1986]; Salehi [1989]; Hui [1985]) toward CDMA in the optical domain (OCDMA), different approaches have been proposed for its implementation. The demonstrated systems can be classified into two main categories: coherent and incoherent. These two categories differ primarily in the signal modulation and detection. Incoherent schemes use the simpler, more standard techniques of intensity modulation with direct detection while coherent schemes are based on the modulation and detection of optical phase (Lam [2000]). Coherent approaches have been reviewed in Chapter 5 and are also the topic for discussion in many review papers (Salehi, Weiner, Heritage [1990]; Marhic [1993]; Griffin, Sampson, Jackson [1995]; Sardesai, Chang, Weiner [1998]; Huang, Nizam, Andonovic, Tur [2000]). The most common approaches to incoherent OCDMA are based on spectral-amplitude coding, spatial coding, temporal (time) spreading, and two-dimensional (2D) wavelength-hopping time-spreading (WHTS). This chapter will give a brief overview of all these approaches and will focus in particular on WHTS.

In most incoherent OCDMA systems, each user is assigned a specific code sequence: a coded transmission is sent to represent a data bit "1," and a null is used to represent a bit "0." Due to the incoherent nature of the system, there are no negative signal components and the signals are unipolar. To avoid loss of code confidentiality using simple energy level detectors, these schemes can be modified to assign two codes per user; a 1 being represented by a code and a 0 being represented by another (Shake [2005]). M-ary techniques have also been proposed (Hui [1985]; Narimanov, Kwong, Yang, Prucnal [2005]), which can be used for higher spectral efficiency.

Spectral-amplitude coding is implemented by spectrally decomposing a broadband light source followed by an optical element that can modulate the intensity of the different spectral components before again combining them. Both grating elements with spatial-amplitude masks (Zaccarin, Kavehard [1993], see Figure 6.1), and filters with periodic spectral transfer functions (Pfeiffer, Deppisch, Witte, Heidemann [1999]) have been used for blocking/passing the different spectral components. Typical broadband sources used include light emitting diodes (LEDs), superluminescence diodes (SLDs) and erbium doped fiber sources. Demonstrations have shown the capability to provide high-speed (155 Mb/sec) access network connections without wavelength stabilization and spectral control (Pfeiffer, Deppisch, Witte, Heidemann [1999]).

The use of the spatial domain for coding has been proposed in different forms—for parallel transmission and simultaneous access of 2D images using "multicore" fibers (Kitayama [1994]), in multiple-fiber systems using fiber tapped delay lines for decoding (Hui [1985]), spatial techniques using a 2D spatial mask (Hassan, Hershey, Riza [1995]) for encoding specific speckle patterns as code sequences, and

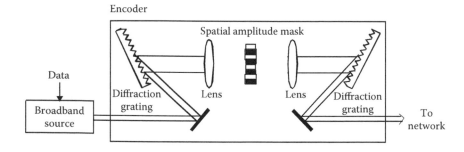

FIGURE 6.1 Spectral-amplitude encoding using diffraction gratings and spatial masks.

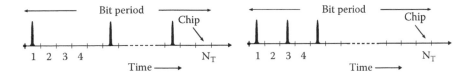

FIGURE 6.2 Example of two code sequences for temporal spreading OCDMA. The bit period is divided into N_T chips (N_T = code length) and codes are formed by placing w (w = code weight) pulses in these chips.

temporal/spatial (T/S) schemes (Park, Mendez, Garmire [1992]; Mendez, Lambert, Morookian, Gagliardi [1994]) using multimode fibers. Fiber optic implementations of these schemes (Park, Mendez, Garmire [1992]) have been in general limited by the requirement of multiple star couplers (one per spatial channel) and the requirement of equal optical path lengths from each distribution star to the en/decoder (usually solved by using fiber ribbons; Kim, Yu, Park [2000]).

A temporal spreading approach was historically one of the first OCDMA schemes to be implemented (Prucnal, Santoro, Fan [1985]). In this scheme, each bit period is divided into N_T smaller time intervals (N_T is code length), called chips. Optical codes are formed by placing short optical pulses (the number of short pulses used per bit is called the code weight) at different temporal positions. Implementations of temporal spreading OCDMA has employed different code families such as prime codes, optical orthogonal codes (OOCs), gold codes, etc. Figure 6.2 shows the representations of two temporal-spreading OCDMA code sequences. However, this approach has been limited due to its requirement of short optical pulses and long code lengths for good correlation properties.

Wavelength-hopping time-spreading (WHTS) systems is a 2D coding approach that spreads the codes in both the time and wavelength domains simultaneously (Tancevski, Andonovic [1994]), achieving increased code design flexibility as well as code performance. Zero autocorrelation sidelobes can be obtained with low cross correlations and higher cardinality at reduced code lengths, in comparison to temporal OCDMA. As in temporal spreading OCDMA, pulses are placed in different chips across the bit period. But WHTS differs from temporal spreading OCDMA in that the pulses in different chips also are of different wavelengths, thus following a wavelength-hopping pattern. Thus, WHTS codes can be represented as code matrices with time and wavelength as its two axes—the wavelength domain is divided into N_l wavelength channels and the time domain is divided into N_T chips. The 2D representation of two code sequences is given in Figure 6.3. The w pulses are then positioned within that matrix (Yang, Kwong [1996]; Yim, Chen, Bajscy [2002]).

Figure 6.4 compares the schematic representation of WHTS OCDMA along with multiple access schemes based on WDM and optical time domain multiplexing (OTDM). In WDM systems, different users are assigned to different wavelengths while in OTDM, the bit period is divided into timeslots and each user is assigned a different timeslot. In WHTS OCDMA, all active users share the same wavelength and time domain space, providing a fair division of the

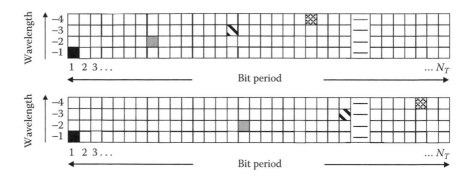

FIGURE 6.3 Two code sequences for WHTS OCDMA. The bit period is divided into N_T chips (N_T = code length) and code sequences are formed by placing w (w = code weight) pulses of different wavelengths in them.

bandwidth, as opposed to OTDM and WDM where only a small portion of the bandwidth is allocated to each user. While OTDM requires strict synchronization between users and network protocols with significant overhead, WHTS OCDMA provides truly asynchronous access, which in turn greatly simplifies network control and management.

Both WDM and TDM based schemes are limited in the scalability of the number of users by the number of available wavelengths and the number of time slots, respectively. On the other hand, the number of users in a WHTS OCDMA network has a soft limit with a graceful degradation of performance with increasing number of users. In addition, WHTS can offer differentiated service at the physical layer (Baby, Brès, Xu, Glesk, Prucnal [2004]; Stok, Sargent [2000]) by varying the code weight that is assigned to each code. Thus, performance can be easily changed to accommodate different classes of service. Code properties also allow the coexistence of users with multiple rates (Kwong, Yang [2004])—of significance to next generation networks that support a wide range of traffic types. Thus, WHTS provides tremendous flexibility in adapting to application requirements.

The rest of this chapter will focus on the implementation of WHTS OCDMA systems and is organized as follows. In Section 6.2, a generic system design is

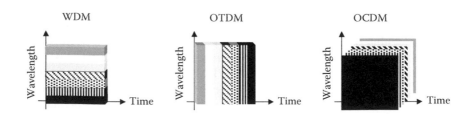

FIGURE 6.4 Schematic representation of WDM, OTDM, and WHTS OCDMA schemes.

presented, with an overview of transmitter and receiver design for WHTS and presents the physical constraints on the design of code matrices for WHTS. Section 6.3 focuses on the technologies, and covers the various technological options for source and en/decoders. In Section 6.4, experimental and simulation results from a WHTS testbed is presented. Finally, this chapter is concluded in Section 6.5.

6.2 WAVELENGTH-HOPPING TIME-SPREADING: SYSTEM ARCHITECTURE

6.2.1 TRANSMITTER AND RECEIVER DESIGN

Figure 6.5 shows the general schematic of a WHTS OCDMA network. The two main elements of the transmitter are the source and the encoder while the receiver consists of the decoder and the receiver electronics. Subsystems for reducing interference from other users can be optionally built into the receiver. The next section of the chapter will cover a more detailed description of the various technologies that can be used for the en/decoder. Here, we give a brief overview of the system components.

As mentioned previously, in WHTS OCDMA, the pulses are encoded in both time and wavelength. WHTS can therefore be implemented using laser sources that can be rapidly hopped from one wavelength to another. However, this method is limited to slow bit rates and results in signal degradation due to laser transients during wavelength hopping. Therefore, all recent implementations of WHTS have used multiwavelength lasers. The multiwavelength laser source can be created either by using lasers that lase at multiple wavelengths, an array of lasers each operating at a fixed wavelength, or by spectral slicing of a broadband source (Okuno, Onishi, Nishimura [1998]) such as erbium doped fiber lasers (Yegnanarayanan, Bhushan, Jalali, [2000]) or semiconductor laser diodes (Adam, Simova, Kavehrad, [1995]). Data modulation is preferably done before the encoder as it avoids any degradation to the codeword due to the rising and falling edges of the modulation window. However, in the cases where the source

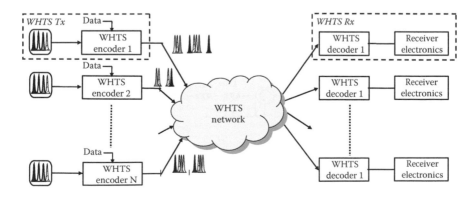

FIGURE 6.5 Schematic of a WHTS OCDMA network.

outputs different wavelengths on different fibers (e.g., a laser array source), it is more power efficient to do the modulation after the encoding. The multiwavelength output of the laser is fed into the encoder, which forms the WHTS code sequence, and the encoded data is sent onto the network.

The WHTS encoder selects the specific w wavelengths to be used in the code sequence, and positions them within N_T chips of the bit interval. Thus, the encoder essentially creates a combination of two patterns: a wavelength-hopping pattern and a time-spreading pattern. Encoder implementations using different technologies could differ in the coupling of the formation of these two patterns. As an example, WHTS encoders based on arrayed waveguide gratings (AWGs) and thin film filters (TFFs; Figure 6.6[a] and [b]) first create a wavelength-hopping pattern followed by a time-spreading pattern. The individual wavelengths of the code sequence are independently delayed, allowing the use of rapidly tunable timeslot tuners for dynamic variation of codes. However, encoders based on chirped Moiré gratings (CMGs) and holographic Bragg reflectors (HBRs) combine the formation of the wavelength-hopping and time-spreading patterns, thus forbidding independent control and delay of each wavelength. Encoders based on a linear array of fiber Bragg gratings (FBGs; Figure 6.6[c]) allow independent control of the wavelengths but require complex schemes for independent delay of each wavelength.

The WHTS receiver receives signals from all users on the network. The decoder discriminates between the intended and interfering data streams by correlating the received signal with the appropriate code matrix. The decoder thus undoes the time spreading for the appropriate code sequence and aligns its wavelength pulses back in time (see Figure 6.7). When the matched code sequence is received, the output of the decoder consists of an autocorrelation peak of height w (w is the code weight of the code). For the other received code sequences, since the decoder is not matched

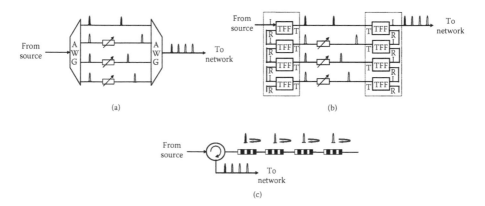

FIGURE 6.6 Schematic of a WHTS encoder with four wavelengths in the code and using (a) AWG, (b) TFF (I: Input port, T: Transmitted Port, R: Reflected port), and (c) FBG technologies.

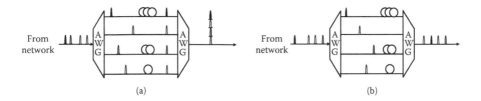

(a)   (b)

FIGURE 6.7 (a) Decoding of code sequence (1, 2, 3, 4) using a matched AWG-based decoder resulting in an autocorrelation peak. The numbers in parentheses show the chip positions of the four wavelengths, (b) Decoding of code sequence (1, 5, 4, 3) using the same decoder resulting in MAI.

to the code sequence, the code sequence is further spread over the bit period resulting in lower intensity multiple access interference (MAI). Since the decoder undoes the WHTS code sequence formed by the encoder, its structure is architecturally the same as the encoder. MAI due to the other network users can be reduced before the signal is passed onto the receiver electronics using different methods, which are covered in Section 6.3.4.

6.2.2 DERIVATION OF CODE MATRIX DIMENSIONS

As described in the introduction, a wavelength-hopping time-spreading (WHTS) OCDMA code sequence can be a viewed as a 2D matrix with time and wavelength on the horizontal and vertical axes, respectively. The performance of a WHTS system depends critically on both the code family used and the dimensions of the 2D code matrix. Defining the matrix size S of the code sequence to be

$$S = N_T N_\lambda,$$

where N_T is the number of chips used and N_l is the number of wavelengths used. All system performance metrics, including the maximum number of simultaneous users to satisfy a specific bit error rate (BER), cardinality and spectral efficiency, depend on S. In general, the performance is improved for higher values of S. For a given bit rate, the maximum value of S is limited by constraints related to the laser source and by requirements for acceptable crosstalk in both wavelength and time domains. Note that both WDM and OTDM systems are also limited by crosstalk in the wavelength and time domains respectively (Zhang, Yao, Chen, Xu, Chen, Gao [2000]; Zhou, Caddedu, Casaccia, Cavazzoni, O'Mahony [1996]; Uchiyama, Morioka, Kawanishi, Takara, Saruwatari [1997]). The influence of crosstalk in both time and wavelength domains must be taken into account for WHTS OCDMA.

Figure 6.8 shows the parameters of a pulse present in a WHTS code sequence, in the wavelength-time space. On the time axis, the pulse occupies a width t_p and

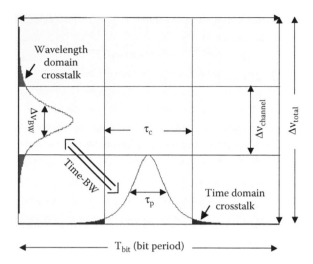

FIGURE 6.8 Parameters of a pulse in a WHTS code sequence in the wavelength-time space. Time domain and wavelength domain crosstalks are shown along with the time-bandwidth product relationship. (Adapted from C.-S. Brès, V. Baby, I. Glesk, L. Xu, D. Rand, P. R. Prucnal. [2004]. Proc. CLEO 2004, CWH7, with permission from OSA and IEEE [© 2004 IEEE].)

is placed in a chip of duration t_c. The bit period, T_{bit}, of the WHTS system is divided into N_T chips where $N_T = T_{bit}/\tau_c$. In the wavelength domain, the pulse occupies a bandwidth of Dn_{BW}. The pulses of other wavelengths of the codeword have the same bandwidth and are separated from each other by a channel separation of $Dn_{channel}$. The total bandwidth occupied by the code word is $\Delta v_{total} = N_\lambda \Delta v_{channel}$. Note that t_p and Dn_{BW} are related by the time-bandwidth product (obtained from Fourier transform) as,

$$A_{T\times BW} = \tau_p \times \Delta v_{BW}, \; \Delta v_{BW} \approx \Im(pulse).$$

For systems using optical gating after the decoder to reduce MAI, time domain crosstalk from MAI pulses can result from two factors: the finite extinction ratio of the gate, and the crosstalk from pulses in adjacent chips (Brès, Baby, Glesk, Xu, Rand, Prucnal [2004]). Both these factors are represented in Figure 6.9(a). In the wavelength domain, crosstalk from adjacent wavelength channels leads to a spreading of the signal energy over the bit period (see Figure 6.9[b]). This follows from the nature of the WHTS en/decoding: each wavelength is intended for a specific delay through the en/decoder pair. When wavelength crosstalk occurs, a fraction of the pulse energy at a specific wavelength undergoes a different delay and is therefore positioned in a nonintended chip. This event occurs at both the en/decoder and further accentuates the spreading of the MAI noise and the reduction in the autocorrelation peak.

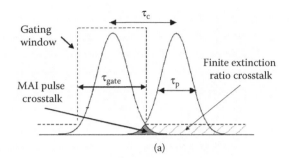

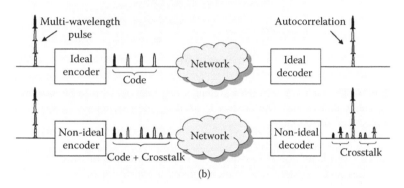

FIGURE 6.9 (a) Two contributors for time domain crosstalk—the finite gate extinction ratio and the MAI pulses in adjacent chips. (b) Effect of wavelength domain crosstalk in spreading of signal energy across the bit resulting in a reduction in the autocorrelation peak. MAI pulses undergo the same effect. (Adapted from C.-S. Brès, V. Baby, I. Glesk, L. Xu, D. Rand, P. R. Prucnal. [2004]. Proc. CLEO 2004, CWH7, with permission from OSA and IEEE [© 2004 IEEE].)

Defining the normalized chip width $B_{TD} \equiv \tau_c/\tau_p$ and the normalized channel spacing $C_{\lambda D} \equiv \Delta\nu_{channel}/\Delta\nu_{BW}$, the code matrix size S can be written as,

$$S = N_T N_\lambda = \left(\frac{T_{bit}}{\tau_c}\right)\left(\frac{\Delta\nu_{total}}{\Delta\nu_{channel}}\right) = \left(\frac{T_{bit}}{B_{TD}\tau_p}\right)\left(\frac{\Delta\nu_{total}}{C_{\lambda D}\Delta\nu_{BW}}\right)$$

$$\Rightarrow S = \frac{T_{bit}\Delta\nu_{total}}{A_{T\times BW} B_{TD} C_{\lambda D}} \tag{6.1}$$

The size of the code matrix depends on the total available bandwidth, the bit rate and the constants $A_{T\times BW}$, the time bandwidth product specific to the source, B_{TD}, the normalized chip width limited by time domain crosstalk, and $C_{\lambda D}$, the normalized channel spacing limited by wavelength domain crosstalk. Both B_{TD} and $C_{\lambda D}$ can be evaluated by a crosstalk analysis of the system (Brès, Baby, Glesk, Xu, Rand, Prucnal [2004]) and depends on the system requirements.

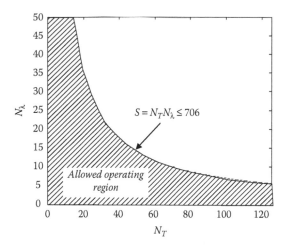

FIGURE 6.10 The trade-off between wavelength and time domains in the determination of the code matrix dimensions. The product is fixed by the requirements on crosstalk, laser characteristics and the bit rate.

As seen from Equation 6.1, system constraints in terms of bit rate, laser source characteristics and crosstalk requirements gives a constraint only on the size of the code matrix and therefore provide a trade-off between the wavelength and time dimensions of the code matrix. Figure 6.10 shows this tradeoff for system parameters of $A_{T \times BW} = 0.42$; $B_{TD} = 2$; $C_{ID} = 1.67$; $T_b = 400$ ps; $Dn_{total} = 2500$ GHz. The allowed code matrix dimensions for this system are indicated by the shaded region.

6.3 SUBSYSTEM TECHNOLOGIES

6.3.1 MULTIWAVELENGTH SOURCE

As previously mentioned, all recent implementations of WHTS have focused on using multiwavelength lasers for providing short pulses at multiple wavelengths. Multiwavelength sources using an array of lasers are limited by the higher complexity in controlling large numbers of lasers. System costs rise dramatically if a multiwavelength source is required at each node. Spectral slicing of a broadband source is a promising approach to achieve a multiwavelength source. Various broadband sources have been used: LED, amplified spontaneous emission (ASE) noise from an EDFA (Ben Jaafar, LaRochelle, Cortes, Fathallah [2001], and supercontinuum (SC) generation (Wada, Sotobayashi, Kitayama [2000]). Supercontinuum generation, in particular, is very promising for the generation of coherent broadband light.

Supercontinuum generation through the propagation of an intense optical pulse has been observed in various kinds of media (Alfano [1989]). Supercontinuum generation in optical fibers primarily involves the injection of high-powered short optical pulses into a span of fiber with carefully chosen dispersion and nonlinearity profiles. The high peak power of the optical pulses results in additional spectral components through nonlinear mechanisms, while the fiber dispersion profile

TABLE 6.1
Summary of Results of SC Generation in Fiber

Type of Fiber	Dl (nm)	Repetition Rate	Reference
DDF	150	—	Mori, Takara, Kawanishi, Saruwatari, Morioka [1997]
DFDF	280	—	Okuno, Onishi, Nishimura [1998]
DSF	420	15 MHz	Kim, Nowak, Boyraz, Islam [1999]
PCF	800	—	Hermann, Griebner, Zhavoronkov, Husakou, Nickel, Kom, Knight, Wadsworth, Russel [2002]
DFF	325	10 GHz	Sotobayashi, Kitayama [1998]
CW BRFL	100	—	Kim, Prabhu, Li, Song, Ueda, Allen [2000]

Note: DDF: dispersion decreasing fiber, DSF: dispersion shifted fiber, PCF: photonic crystal fiber, DFF: dispersion flattened fiber, DFDF: dispersion flattened and decreasing fiber, BRFL: Brillouin/Raman fiber laser.

maintains the temporal pulse width through the length of the fiber. The development of photonic crystal fibers (Knight [2003]) with greater control of the nonlinear and dispersive characteristics has brought a fresh impetus to this area. Various demonstrations have been made primarily differing in the type of nonlinear fiber and the pump. Table 6.1 summarizes these results.

A common and efficient method for SC generation uses dispersion decreasing fiber (DDF) (Mori, Takara, Kawanishi, Saruwatari, Morioka [1997]). The propagation of ultra-short pulses in optical fibers near the zero-dispersion wavelength implies short nonlinear interaction lengths and high peak powers, resulting in very large spectral broadening. Self phase modulation following self-focusing and optical breakdown as well as four wave mixing was thought to be the main mechanism (Bloembergen [1983]) for SC generation. For the anomalous dispersion regime, SC generation is now believed to be a result of the complex interplay of diverse phenomena due to the third-order nonlinear susceptibility c^3, as well as ionization and plasma generation (Schumacher [2002]). Over 200 nm of optical bandwidth was generated in DDF without serious degradation to the output optical pulse width (Mori, Takara, Kawanishi, Saruwatari, Morioka [1997]; Morioka, Kawanishi, Mori, Saruwatari [1994]). Figure 6.11 shows experimental results of SC generation obtained at 1.55 mm using 1 km of DDF fiber. The pump frequency at 1.55 mm appears as a peak in the spectrum.

The generation of SC in optical fibers with spectral filtering has been used for multiwavelength sources for applications in WDM/OTDM networks (Morioka, Takara, Kawanishi, Kamatani, Takiguchi, Uchiyama, Saruwatari, Takahashi, Yamada, Kanamori, Ono [1996]; Kawanishi, Takara, Uchiyama, Shake, Mori [1999]) and in WHTS OCDMA networks (Yegnanarayanan, Bhushan, Jalali [2000]; Glesk, Baby, Brès, Xu, Rand, Prucnal [2004]). The development of very broad SC entails

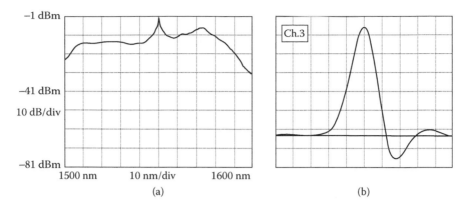

FIGURE 6.11 (a) Optical spectrum of SC generated using 1km of DDF. The pump for SC generation appears as the peak in the center. (b) Time domain shot of a pulse spectrally sliced from the SC.

that the scalability of the system is limited only by the number of channels in the wavelength MUX/DEMUX of the en/decoders. Currently, AWG technologies have achieved port counts of up to 80. Also, the spectral slicing approach implies that the system parameters such as the wavelength channel spacing and the number of wavelength channels can be changed by replacing the wavelength MUX/DEMUX at the en/decoders without requiring the entire source to be rebuilt.

6.3.2 FILTER TECHNOLOGIES

6.3.2.1 Fiber Bragg Gratings

Fiber Bragg gratings (FBGs) have been used for various applications in optical networks in recent years (Hill, Meltz [1997]; Kashyap [1999]). The applications include add/drop multiplexing and wavelength routing in WDM networks, dispersion compensation and wavelength stabilization techniques in fiber lasers. Fiber Bragg gratings have been used to implement various OCDMA encoding schemes such as spectral amplitude coding (Magne, Wei, Ayotte, Rusch, LaRochelle [2003]), spectral phase coding (Grunnet-Jepsen, Johnson, Maniloff, Mossberg, Munroe, Sweetser [1999a, b]), temporal phase coding (Lee, Teh, Petropoulos, Ibsen, Richardson [2002]; Teh, Petropoulos, Ibsen, Richardson [2001]) and WHTS (Chen [2001]; Kutsuzawa, Minato, Oshiba, Nishiki, Kitayama [2003]). The use of FBG-based approaches to OCDMA provides the advantages of all-fiber en/decoding for increased efficiency and the potential to support reconfigurable systems using tunable devices.

Implementations of WHTS with FBGs use two structures—linear array of FBGs or chirped Moire gratings. Figure 6.6(c) shows the implementation of a WHTS code using a linear array of FBGs (Jaafar, LaRochelle, Cortes, Fathallah [2001]; Wada, Sotobayashi, Kitayama [2000]; Kutsuzawa, Minato, Oshiba, Nishiki, Kitayama [2003]; Tamai, Iwamura, Minato, Oshiba [2004]). As a generalization of this structure, parallel linear arrays can be used with a power divider (see Figure 6.12; Chen [2001]) to implement multiple pulses per column (MPPC) codes (i.e., multiple

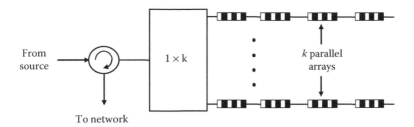

From source

$1 \times k$

k parallel arrays

To network

FIGURE 6.12 Parallel linear array of FBGs allows implementation of MPPC codes. k is the maximum number of wavelengths in the same column of the code matrix. (Adapted from L. R. Chen. [2001]. *IEEE Photon. Technol. Lett.* 13 (11): 1233–1235, with permission from IEEE [© 2004 IEEE].)

wavelengths occupying the same chip). The tunability of the FBG center wavelength using strain allows the variation of the code weights. Chirped Moire gratings (CMGs), composed of superposed linearly chirped FBGs, enable the use of a single passive grating structure to implement WHTS en/decoding (Chen, Smith [2000]; Chen, Smith, Sterke [1999]). However, the linear relationship between wavelengths and time and the associated inherent coupling of the wavelength-hopping and time-spreading patterns in CMGs implies that the number of codes that can be implemented is limited. Detailed descriptions of the different implementations of WHTS using FBGs and the different network demonstrations are covered in Chapter 5.

6.3.2.2 Arrayed Waveguide Gratings

Arrayed waveguide gratings (AWGs) have attracted much attention since the early 1990s due to the potential for low insertion loss and fabrication using the silicon, InP, or LiNbO$_3$ technologies (Agrawal [2002]; Smit, van Dam [1996]; Okayama, Kawahara, Kamijoh [1996]; Dragone [1998]). It consists of a phased array of optical waveguides that acts as a grating (see Figure 6.13). The input signal is coupled into an array of

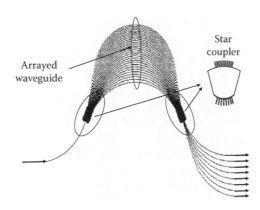

Arrayed waveguide

Star coupler

FIGURE 6.13 Schematic of an arrayed waveguide grating (AWG).

planar waveguides after passing through a free-propagation region. The waveguides are designed to be of different lengths to provide different phase shifts to the signal in each waveguide. Also, these phase shifts are wavelength dependent because of the frequency dependence of the mode-propagation constant. As a result, different wavelength channels focus to different output waveguides when the light exiting from the array diffracts in another free-propagation region. The input and output waveguides, the multiport couplers, and the arrayed waveguides are all fabricated on a single substrate.

The AWG has been used as an N ¥ 1 wavelength multiplexer and also as a 1 ¥ N wavelength demultiplexer in many applications. These include N ¥ N wavelength routers, (AWGs modified to provide multiple input ports and called waveguide-grating routers (WGRs) are used for this application) (Ramaswami, Sivarajan [2002]), optical crossconnects (Koga, Hamazumi, Watanabe, Okamoto, Obara, Sato, Okuno, Suzuki [1996]) and multiwavelength WDM sources and receivers. For WDM sources, AWGs were used either as spectral slicers of broadband light (Nuss, Knox, Koren [1996]) or within the laser cavity in both fiber and integrated lasers (Monnard, Srivastava, Doerr, Essiambre, Joyner, Stulz, Zirnbigl, Sun, Sulhoff, Zyskind, Wolf [1998]; Bellemare, Karasek, Rochette, LaRochelle, Tetu, [2000]). As WDM receivers, AWG-based demultiplexers have been integrated with photodiode array on the same chip (Chandrasekhar, Zirnbigl, Dentai, Joyner, Storz, Burrus, Lunardi, [1995]). AWGs with more than 100 ports have been reported (Sun, NcGreer, Broughton [1998]) while devices with more than 50 ports are commercially available.

Demonstrations of WHTS using AWGs include en/decoding using a single AWG each in a feedback configuration with fiber loops for delays (Yegnanarayanan, Bhushan, Jalali [2000]), both encoding and decoding using a single AWG with mirrored fiber delay lines in a foldback configuration (Yu, Shin, Park [2000]) and a bidirectional AWG-based en/decoding scheme utilizing the bidirectional nature of the AWGs. Relatively, the AWG is a more mature technology than the FBG and has the advantage that the entire WHTS en/decoder is waveguide integratable, resulting in a smaller footprint.

6.3.2.3 Integrated Holographic En/Decoder

A new integration technology platform has been recently developed based on the principle of volume holography. It utilizes the capability of standard tools such as deep ultraviolet (DUV) photolithography and laser written reticles to produce almost arbitrarily structured 2D spatial patterns of nanometer-scale feature size with essentially perfect spatial coherence over centimeter scales (Greiner, Iazikov, Mossberg [2004a]). Thus, essentially perfect 2D volume holograms can be constructed to interact with specific signal beams and provide signal routing and wave-front transformations with powerful spectral control. Photonic crystals (Prather [2002], Scherer, Painter, Vuckovic, Loncar, Yoshie [2002]) form a subset of such volume holograms (Mossberg, Greiner, Iazikov [2004]). The technology has been demonstrated for applications in coarse and dense WDM systems (Iazikov, Greiner, Mossberg [2004a]; Greiner, Iazikov, Mossberg [2004b]) and for spectral comparison (Mossberg, Iazikov, Greiner [2004]), using silica as the substrate.

The basic building block of this platform is the holographic Bragg reflector (HBR) (Greiner, Iazikov, Mossberg [2004a]; Mossberg [2001]). An HBR images the signal

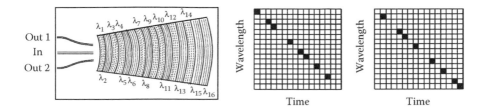

FIGURE 6.14 HBR-based WHTS encoder. (a) Top view of input, output waveguides and HBRs, dashed (dotted) segments with upper (lower) resonance wavelengths correspond to HBRs 1 (HBR 2) and outputs 1 (2), (b), (c) 16 ¥ 16 WHTS code matrix representation of code sequences 1 and 2 respectively. (Adapted from Y. -K. Huang, C. M. Greiner, D. Iazikov, V. Baby, I. Glesk, L. Xu, P. R. Prucnal, T. W. Mossberg, 2004, Proc. LEOS 2004, ThCC5, with permission from IEEE [© 2004 IEEE].)

from an input port to an output port while applying a specific spectral filtering function. It provides excellent channel-specific pass-band control like thin film filters and FBGs and its spatially distinct input and output ports obviate the need for bulky circulators or lossy power splitters to separate the input and output signals. The ability of HBRs to provide both spectral slicing and temporal delay simultaneously implies that the footprint of HBR-based WHTS en/decoders can be made very small. Multiple HBRs can be superimposed, interleaved, or overlaid to design multiport devices (Mossberg, Greiner, Iazikov [2004]).

Figure 6.14 shows the details of a HBR-based device used for WHTS en/decoding (Huang, Greiner, Iazikov, Baby, Mossberg, Prucnal [2004]). It consists of two HBRs—HBRs 1 and 2, heavily interleaved along the input beam direction and consisting of eight interleaved grating segments indicated by dashed and dotted lines respectively in Figure 6.14(a)—connecting the input port (IN) to the output ports OUT 1 and 2 with a unique spectral transfer function corresponding to a WHTS code. The HBRs were designed to encode 16 ¥ 16 WHTS matrices (Figure 6.14[b]) for the two ports, with an adjacent wavelength separation of 1 nm and time chips separated by 11 psec and approximately 6 psec wide. Tailored amplitude and phase apodization (Iazikov, Greiner, Mossberg [2004]; Greiner, Mossberg, Iazikov [2004]) was imposed on the grating segments to isolate adjacent spectral chips and reduce coherent beating.

Figure 6.15 compares the simulated and measured performance of this device. The simulations were done by applying Fresnel-Huygens diffraction theory (Born, Wolf [1999]) to an exact numerical model of the HBR structure. Measurements were done using a ~500-fsec-long (FWHM) pulse with 80-nm spectral width, generated through fiber-based supercontinuum generation, input into the device and a 30-GHz sampling oscilloscope for detecting the output. The large response time of the oscilloscope causes pulses in adjacent time chips to appear as single peaks of higher intensity. The dashed lines in the lower trace of Figure 6.15(b_2) shows the simulated HBR temporal response when the finite response time of the oscilloscope is taken into account, showing excellent agreement with the measured trace.

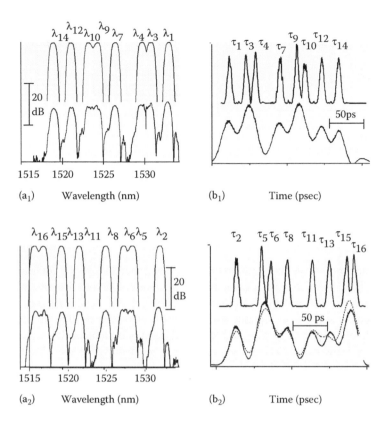

FIGURE 6.15 Designed and measured encoder performance. (a_1, a_2) Spectral transfer functions, (b_1, b_2) Time-domain responses, (a_1, b_1) HBR 1, (a_2, b_2) HBR 2. In all graphs, upper and lower lines show simulated and measured (using TM-polarized light) performance. (Adapted from Y. -K. Huang, C. M.Greiner, D. Iazikov, V. Baby, P. R. Prucnal, T. W. Mossberg. *IEEE Photon. Technol. Lett.,* accepted for publication in April 2005, with permission from IEEE [© 2004 IEEE].)

6.3.2.4 Thin-Film Filters

Thin-film filters (TFFs), in the most common use of the term, are devices that can be considered as Fabry-Perot (FP) etalons, with multiple reflective dielectric thin-film layers used to form the mirrors surrounding the FP cavity. The principles of thin-film optics (Knittl [1976]) which govern the performance of TFFs have been studied for a variety of other applications, including edge filters, anti-reflection coatings and high-reflectance coatings (MacLeod [1986]). Similar to a FP etalon, a TFF acts as a bandpass filter. In an extension of this idea, thin-film resonant multicavity filters have been built by having two or more cavities separated by reflective dielectric thin-film layers. By varying the number of cavities, different features of the transmission spectral profile such as the shape of the pass band and the steepness of its falling and rising edges can be varied.

Multiplexers and demultiplexers can be formed by cascading a number of these filters. For example, when used as a demultiplexer, the first filter in the cascade passes one wavelength and reflects all the others onto the second filter. The second filter passes another wavelength and reflects the remaining ones, and so on. Figure 6.6(b) shows the use of a TFF-based multiplexer and demultiplexer to form a WHTS encoder.

Tunability of the center wavelengths of TFFs has been demonstrated using different mechanisms (Lequime, Parmentier, Lemarchand, Amra [2002]). These include the change of angular position of the filter with respect to the incident beam (Dicon Fiber Optics), current injection techniques in InGaAsP/InP waveguides (Tsang, Mak, Chan, Soole, Youtsey, Adesida [1999]), change of optical thickness using the piezo-electric effect (e.g., Ta_2O_5 on fused-silica substrate; Parmentier, Lemarchand, Cathelinaud, Lequime, Amra, Labat, Bozzo, Bocquet, Charai, Thomas, Dominici [2002]), and the thermo-optic effect in silicon (Iodice, Cocorullo, Della Corte, Rendina [2000]) and amorphous silicon (Domash [2004]).

Thin-film filters have emerged as a dominant filter technology in commercial markets due to their flexibility, low loss, and passive temperature compensation. They provide the lowest-loss solution at low channel count, and similar loss to flat-top arrayed-waveguide gratings at high channel counts. In addition, the device is insensitive to the polarization of the signal and extremely stable to temperature variations. TFFs have been used for various optical networking applications such as optical switches (Sumriddetchkajorn, Chaitavon [2003]) and switchable add-drop filters (Domash [2004]).

The use of TFFs in WHTS systems imply that the chip size is limited only by the optical pulse width, unlike fiber Bragg grating (FBG)-based en/decoders in which the spacing between adjacent gratings imposes an additional constraint (Kutsuzawa, Minato, Oshiba, Nishiki, Kitayama [2003]). The resulting increase in code matrix size can thus be used for better system performance. TFFs have been used recently for WHTS en/decoders in various experiments (Baby, Brès, Xu, Glesk, Prucnal, [2004]; Baby, Glesk, Runser, Fischer, Huang, Brès, Kwong, Curtis, Prucnal [2004]).

6.3.3 TUNABLE TIME-DELAY ELEMENTS

The en/decoder architectures shown in Figure 6.6(a) and (b) use the wavelength MUX/DEMUX to separate and combine the different wavelengths, with separate time delay elements used for each coded wavelength. While this structure is not as compact as the architectures based on FBGs or HBRs mentioned in the previous section, it provides the possibility of utilizing delay line technologies that have been developed with the capability to tune the pulse time delay within the order of nanoseconds. This ultra-fast tunability has many implications. It enables fast address selection and, therefore, the dynamic nature of the multiaccess network. Also, it allows dynamic variation of code sequences, which leads to enhanced data confidentiality (Shake [2005]). This section provides an overview of the various kinds of delay lines that have been developed.

6.3.3.1 Switched Serial Delay Line

A switched serial delay line was demonstrated (Prucnal, Krol, Stacy [1991]; Mac-Donald [1987]) that provides a tunable delay to the pulse by switching the pulse between paths of fixed different delays (see Figure 6.16). The feed-forward structure (Jackson, Newton, Moslehi, Tur, Cutler, Goodman, Shaw [1985]) of this delay

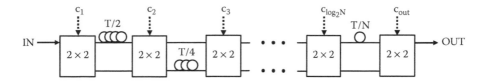

FIGURE 6.16 Schematic of variable integer delay line T: bit period, N: number of chips, $\{c_1, c_2, \overset{\circ}{\underset{}{}} c_k, \overset{\circ}{\underset{}{}}, c_{\log_2 N}\}$ defines the chip position of the pulse, c_{out} is the output control bit. (Reprinted from V. Baby, C.-S. Brès, L. Xu, I. Glesk, P. R. Prucnal. [2004]. *Electron. Lett.* 40 (12), 755–756, with permission from IEE, and from V. Baby, P. R. Prucnal, C.-S. Brès, L. Xu, I. Glesk, 423-064, Proc. OCSN (WOC) 2004, with permission from IASTED.)

line implies that there are $\log_2 N$ delay stages, $k = 1, 2, \overset{\circ}{\underset{}{}}, \log_2 N$, and an output stage, where N is the total number of accessible timeslots. Each stage has a 2 ¥ 2 optical switch that can rapidly be set in either the bar or cross state, as decided by the control bit at the electrical control port, to choose paths of different lengths.

The state of the coder is set by the control sequence $\{c_1, c_2, \overset{\circ}{\underset{}{}} c_k, \overset{\circ}{\underset{}{}}, c_{\log_2 N}\}$, where control bit c_k sets the state of the kth stage. The control sequence for the jth slot is generated from the binary representation of the integer j, $(b_1, b_2, \overset{\circ}{\underset{}{}}, b_{\log_2 N})$ where b_1 is the most significant bit, according to the rule, $c_1 = ` b_1$ and for $i = 2, \overset{\circ}{\underset{}{}}$, $\log_2 N$, $c_i = 0$, if $b_i = b_{i-1}$, otherwise $c_i = 1$. The control bit for the output stage, c_{out}, is set equal to "0" if the parity of the control sequence is odd, or to "1" if the parity of the control sequence is even. This ensures that the delayed pulse always exits at the chosen output of the 2 ¥ 2 switch.

The tuning latency depends on the switching technology used, e.g., $LiNbO_3$ switches can attain tuning speeds of at least a few GHz (Prucnal, Krol, Stacy [1991]). Commercial devices are available that can be tuned on the order of a few 100 kHz using thermo-optic polymers in a planar lightwave circuit design (Little Optics).

6.3.3.2 Gated Serial Delay Line

The gated serial delay line (Deng, Glesk, Kang, Prucnal [1997a]) consists of a k-stage feed-forward delay lattice structure and two modulators at the input and output of the delay lattice (see Figure 6.17). Each delay stage is a Mach-Zehnder

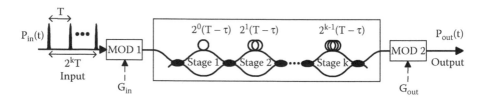

FIGURE 6.17 Schematic of gated serial delay line *T*: laser source period, *k*: number of stages in the lattice structure. The number of chips is equal to 2^{2k} and the control patterns $\{G_{in}, G_{out}\}$ to the modulators defines the chip position of the pulse. (Adapted from K.-L. Deng, K. I. Kang, I. Glesk, P. R. Prucnal, *IEEE Photon. Technol. Lett.* 9 (11), with permission from IEEE [© 2004 IEEE].)

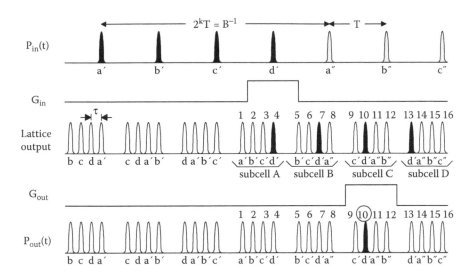

FIGURE 6.18 The timing diagram for a two-stage gated serial delay line for accessing 16 chips. The modulation gating functions, G_{in} and G_{out}, are set to tune the output channel to chip 10. By sliding the gating functions in time, any one of 16 chips are rapidly accessible.

lattice with a differential delay of $T - t$, $2(T - t)$, $4(T - t)^2$, $2^{k-1}(T - t)$ between the upper and lower paths. Here T is the input optical clock period, t is the chip width and the output bit period is $2^k T$. The two modulators at the ends of the delay lattice are controlled at the input optical clock rate to set the state of the delay line. The total number of accessible chips, N, is related to the number of stages by $N = 2^k \times 2^k$.

The operation of the delay line is summarized in Figure 6.18. As the input clock period is T, a total of $K = 2^k$ pulses at $t = iT$, $i = 0, 1, 2,$, $K - 1$, are available at the front-end modulator. Only one of these is selected by the input modulator and fed into the delay lattice. An incoming pulse is split into K daughter pulses as a result of the $k + 1$ cascaded 2×2 couplers. After the delay lattice, these pulses obtain different time delays of $T - t$, $2(T - t)$, $3(T - t)^2$, $(2^k - 1)(T - t)$, giving a constant delay of $(T - t)$ between adjacent daughter pulses. By combining the daughter pulses from all possible input clock pulses, the delay line is able to generate pulses with period t. By controlling the modulation pattern to the input and output modulators, pulses at different delay positions can be selected.

The gated serial delay line has $\frac{1}{2}\log_2 N$ stages for N chips, and the number of active components is independent of N. These two factors result in a reduction in loss and complexity. The main constraint on N is due to coherent crosstalk caused by the finite extinction ratios of the modulators. Calculations have shown scalability of N to hundred, with modulation extinction ratios of 30 dB (Deng, Glesk, Kang, Prucnal [1998]). However, the trade off for this reduced complexity and loss is the higher bandwidth requirement for both the source laser and the control electronics. The average tuning latency of this delay line is the output bit rate (Deng, Glesk, Kang, Prucnal [1997a]).

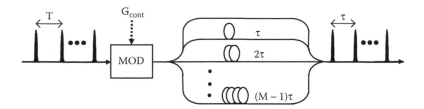

FIGURE 6.19 Schematic of the compression-based gated delay line T: laser source period, t: output bit period, M: compression factor. The control pattern $\{G_{cont}\}$ to the modulator defines the chip position of the pulse. (Adapted from V. Baby, B. C. Wang, L. Xu, I. Glesk, P. R. Prucnal. [2003]. *Opt. Commun.* 218 (4-6): 235–242, with permission from Elsevier.)

6.3.3.3 Compression-Based Gated Delay Lines

Compression-based gated delay lines (Wang, Glesk, Runser, Prucnal [2001]; Baby, Wang, Xu, Glesk, Prucnal [2003]) consist of a modulator and a parallel fiber delay lattice (Figure 6.19). For N chips, the parallel delay lattice has M arms with the constraint $NT = Mt$, where T is the input clock period to the delay line and t is the output bit period. Since the delay lattice has a uniform differential delay t between its arms, the information input to the lattice is compressed in the temporal domain by the factor of M. Initial demonstration of the delay line used $M = N + 1$ (Wang, Glesk, Runser, Prucnal [2001]).

The operation of the compression-based gated delay line is summarized in Figure 6.20 for $N = 4$ and $M = 5$. The parallel lattice combines shifted replicas of the pulse train coming out of the modulator. The gating pattern, G_{cont}, was $\{0001\}$ in this case. Since adjacent arms of the lattice have the same differential delay t, the output pulse train has a period of t. By varying the modulation pattern, pulses

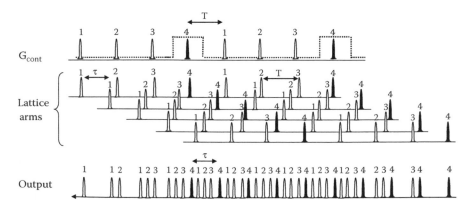

FIGURE 6.20 The timing diagram for with $N = 4$, $M = 5$. The gating modulation function, G_{cont}, is $\{0001\}$ and set to select the pulse in chip 4.

in any of the chips can be selected. For example, for $G_{cont} = \{0001\}$, pulses in chip 4 are chosen.

For proper operation of this delay line, it was shown that the number of chips N, and the compression factor M should be co-prime (Baby, Wang, Xu, Glesk, Prucnal [2003]). Note that $M = N + 1$, satisfies this constraint as $N + 1$ is always co-prime with N. By choosing M to be a composite number, the compression factor of M can be obtained by a cascade of smaller compression factors M_i, resulting in a serial-parallel structure. This serial-parallel structure can give better scalability for higher number of chips. By choosing M to be greater than N, the control electronic bandwidth can be less than the output bit rate.

Other delay lines have also been developed (Chan, Chen, Cheung [1996]). Various technology platforms have shown promise for the integration of these different delay lines. These include polymer-based materials (Little Optics), ring resonator–based technologies (Lenz, Eggleton, Madsen, Slusher [2001]), and planar lightwave circuit technology (Ohmori [1991]). Advanced networking features such as multicasting can also be implemented to a limited extent using these delay lines (Deng, Runser, Glesk, Prucnal [2000]).

6.3.4 TECHNOLOGIES FOR REDUCTION OF MULTIPLE ACCESS INTERFERENCE

All OCDMA systems are primarily limited in their performance by multiple access interference (MAI; Lam [2000]). This is also true in WHTS systems and is a direct result of the nonzero cross-correlation between different codes. MAI increases with the number of simultaneous users, resulting in a deterioration in the bit error rate (BER) performance and eventually limits the maximum number of simultaneous users. Analysis has shown that MAI is the dominant effect in limiting the number of users even when optical beat noise between pulses of the same wavelength from different users is taken into account (Tancevski, Rusch [2000]). Research efforts to reduce MAI in these systems have taken many directions including development of various coding techniques (see Chapter 3). Here, we will focus on hardware techniques that either reduce MAI or provide improved performance even in the presence of MAI.

Complementary encoding with balanced detection (see Figure 6.21[a]) has been proposed to overcome the code orthogonality problem of standard incoherent systems. By transmitting the unipolar versions (i.e., "+1" elements) of bipolar code sequences (Khaleghi, Kavehrad [1996]; Nguyen, Dennis, Aazhang, Young [1997]) and subtracting signals using balanced detectors, code orthogonality can be achieved. An example using m-sequences of length 7 and seven wavelengths is illustrated in Figure 6.21(b) and (c). This technique has been adopted primarily for spectral amplitude coding (Kavehrad, Zaccarin [1995]; Lam, Tong, Wu, Yablonovitch [1998]; Nguyen, Aazhang, Young [1995]), but can also be extended to WHTS systems (Kwong, Yang, Chang [2004]).

Two other techniques that have been used to reduce MAI are optical gating and optical thresholding. In comparison to the technique of complementary encoding and balanced detection, these techniques have the advantage of being

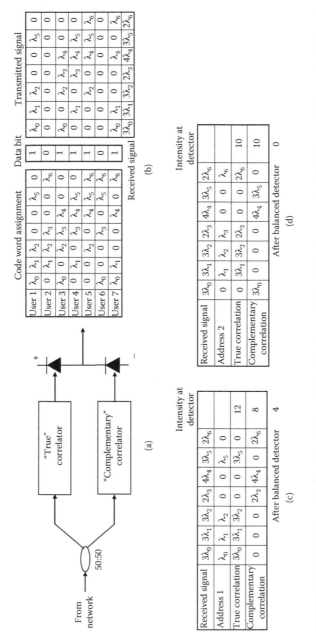

FIGURE 6.21 (a) Schematic for using balanced detection for reducing MAI. An example of MAI cancellation using balanced detection and complementary coding with m-sequence coding of length 7 and seven wavelengths: (b) code word assignment and transmitted signal for seven network users; (c), (d) operation of true and complementary correlators and balanced detection at two different nodes.

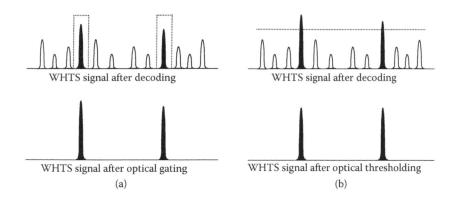

WHTS signal after decoding	WHTS signal after decoding
WHTS signal after optical gating	WHTS signal after optical thresholding
(a)	(b)

FIGURE 6.22 Reduction of MAI using (a) optical gating and (b) optical thresholding. The dotted lines show the switching window and threshold in (a) and (b), respectively. The dark pulse represents the autocorrelation peak and the light pulses show MAI from other users.

compatible to more types of code families and are usually suited for higher speeds. The use of optical techniques for reducing MAI provides ease in post-processing of the optical signal, such as OCDMA code conversion or an OCDMA-WDM gateway. In optical gating (Figure 6.22[a]), an optical clock train is used to open the switching window of an optical gate, thus choosing the desired autocorrelation peak while blocking the MAI noise that falls outside the switching window. Note that the use of the optical clock train does not require global network synchronization but requires optical clock recovery at the receiver for synchronization with the autocorrelation peaks. In contrast, an optical thresholder works by allowing the higher-intensity autocorrelation peak to pass while rejecting the lower intensity MAI. It does not require either global synchronization or optical clock recovery.

Nonlinear interaction between optical signals in a variety of materials e.g., fiber, semiconductor optical amplifiers (SOAs) and nonlinear waveguides such as periodically poled $LiNbO_3$ (PPLN), have been used to develop all-optical gates and thresholders (Glesk, Wang, Xu, Baby, Prucnal [2003]). A fiber-based solution—the nonlinear optical loop mirror (NOLM) (Doran, Wood [1988])—was used as an optical gate to reduce MAI in temporal-phase encoded systems (Lee, Teh, Petropoulos, Ibsen, Richardson [2002]). However, the NOLM suffers from the requirement of high optical powers and long interaction lengths due to the low nonlinear coefficient in fiber. In addition, as WHTS signals usually have very broad optical spectra, nonuniformity of fiber dispersion poses a challenge for operation of these signals in the long fiber lengths of the NOLM. Nonlinear wavelength shifts of optical signals in highly nonlinear fibers with filters have been used for thresholding spectral-phase encoded signals (Sardesai, Chang, Weiner [1998]) but are not as effective for WHTS signals due to their broad optical spectrum. Optical gating and thresholding using nonlinear waveguides have caused a lot of research interests recently, but still suffers from large waveguide losses and low nonlinear coefficients (Tajima, Nakamura, Sugimoto [1995]).

TABLE 6.2
Comparison of Various SOA-based Optical Switches

| | All-Optical Switches Based on SOA as Nonlinear Medium | | | |
Type of SOA-Based Switch	FWM in SOA	TOAD	Mach-Zehnder Interferometer	UNI
Demonstrated working speed	200 Gb/sec	250 Gb/sec	168 Gb/sec	100 Gb/sec
Integratability	Yes	Yes	Yes	Yes
Polarization sensitivity	Yes	No	No	Yes
Advantages	—	Low switching energy	Low switching energy	Low switching energy

Source: Adapted from I. Glesk, B. C. Wang, L. Xu, V. Baby, P. R. Prucnal, *Progress in Optics*, 45, 53–117, 2003. With permission from Elsevier.

SOA-based devices are very compact and offer the possibility for monolithic integration together with other photonic devices. The vast majority of these switches utilize the nonlinear gain and phase changes in a SOA between two interfering signal components to achieve signal processing. Various architectures have been implemented for all-optical gating (Glesk, Wang, Xu, Baby, Prucnal [2003]), e.g., semiconductor laser amplifier in a loop mirror (SLALOM), terahertz optical asymmetric demultiplexer (TOAD), symmetric Mach-Zehnder (SMZ), counter-propagating Mach-Zehnder (CPMZ), and the ultrafast nonlinear interferometer (UNI). In all these devices, the switching window duration for the gating operation is determined by the relative time delay with which the two interfering signal components pass through a SOA saturated by the control pulse, and is not limited by the recovery time of the SOA. An overview of SOA-based switches and their features is given in Table 6.2. Techniques have been proposed to improve the extinction ratio and reduce the noise (Diez, Ludwig, Weber [1999]), and to increase the repetition rate of switching (Manning, Davies, Cotter, Lucek [1994]) of these SOA-based switches. The high nonlinear coefficient of the SOA as compared to fiber, gain potential of the SOA, and the integratability along with its broad gain spectrum makes the SOA-based devices very attractive for handling OCDMA signals.

In this section, we will briefly describe the working principle of a SOA-based device, the TOAD (Sokoloff, Prucnal, Glesk, Kane [1993]), which has been used for both all-optical gating and thresholding.

6.3.4.1 Terahertz Optical Asymmetric Demultiplexer (TOAD)

As mentioned earlier, the operation of the majority of SOA-based devices, including the TOAD, depends on the variation in the gain and phase of the SOA due to optical signals. The gain in a SOA is caused by the creation of a carrier population inversion that ensures that the stimulated emission is more prevalent than absorption. The population inversion is usually achieved by electrical current injection in the p-n junction where the generated electron-hole pairs recombine by means of stimulated emission. The SOA gain in the presence of an optical signal can be described by the following differential equation (Agrawal, Olsson [1989]):

$$\frac{dh}{d\tau} = \frac{h_0 - h}{\tau_c} - \frac{P_{in}(\tau)}{E_{sat}}[\exp(h) - 1].$$

(6.2)

Here h is the integrated gain over the transverse direction of the SOA, h_0 is the small signal gain, τ_c is the SOA carrier lifetime, E_{sat} is the saturation energy, P_{in} is the input optical signal power and time t is measured in a reference frame moving with the input optical signal.

The calculated gain change using a short optical pulse at the input is shown in Figure 6.23. The parameter E_{sat} was chosen to be 1 pJ and h_0 was 4.5. The pulse width used in the calculations was 10 psec. When a short optical pulse is injected into a SOA, the carrier density decreases due to enhanced stimulated emission. For high energy pulses, the decrease in carrier density and hence the decrease in SOA gain can be very large. This process is known as gain saturation. After the short pulse has passed, the gain of the SOA recovers due to the injection of the carriers by the electrical current. The recovery time is decided by the carrier lifetime, which is typically several hundred picoseconds. However, the turn-on time for the saturation process can be as short as a few picoseconds.

The TOAD architecture consists of a SOA placed at an asymmetric position inside a Sagnac loop, as shown in Figure 6.24. The device has three ports: an input port, an output port, and a control port. When a data pulse enters the loop, it is split into clockwise (CW) and counterclockwise (CCW) traveling components by a 50:50 coupler. The two components pass through the SOA at different times as they counterpropagate around the loop and recombine interferometrically at the 50:50 coupler.

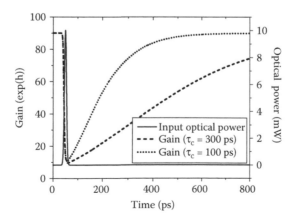

FIGURE 6.23 (**Color Figure 6.23 follows page 240.**) Dynamics of SOA gain change upon passing of a short optical pulse, for carrier recovery times of 100 ps and 300 ps. (Reprinted from I. Glesk, B. C. Wang, L. Xu, V. Baby, P. R. Prucnal. [2003]. *Progress in Optics*, 45, 53–117, with permission from Elsevier.)

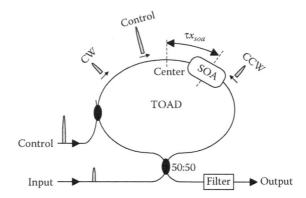

FIGURE 6.24 Structure of the TOAD as an optical gate. (Reprinted from I. Glesk, B. C. Wang, L. Xu, V. Baby, P. R. Prucnal. [2003]. *Progress in Optics,* 45, 53–117, with permission from Elsevier.)

For the TOAD geometry shown in Figure 6.24, the CCW data pulse reaches the SOA earlier than the CW data pulse. In the absence of a control pulse, both pulse components experience a similar effective medium as they propagate around the loop, and the data is reflected back toward the input port as in a loop mirror (Mortimore [1988]). If a high energy control pulse is injected into the loop, it saturates the SOA and changes the index of refraction. If the control pulse is timed such that it arrives at the SOA after the CCW data pulse and before the CW pulse, a differential phase shift can be achieved between the two counterpropagating data pulses. This differential phase shift can be used to switch the data pulses to the output port. Subsequent data pulses experience a gain and a refractive index that are slowly recovering; hence CW and CCW pulses have approximately zero phase difference and so are reflected on recombination at the coupler. A polarization or wavelength filter can be used at the output to discriminate the switched data signal from the control pulse.

Figure 6.25 shows the simulated phase change seen by the CW and CCW data pulses upon recombination at the coupler. For the simulations, the same SOA parameters as in Figure 6.23 were used. Due to the SOA offset, the two phase profiles are delayed with respect to each other and a time varying phase difference is obtained with two distinct values of phase differences. Figure 6.25(b) shows the resultant TOAD transmission. The duration of the switching window is determined by the offset of the SOA, Δx_{SOA} , from the midpoint of the loop. As this offset is reduced, the switching window duration decreases. The size of the nominal switching window duration, τ_{win}, is related to the offset position by $\tau_{win} = 2\Delta x_{SOA}/c_{fiber}$. By precisely controlling the offset position of the SOA, short switching windows can be achieved which are much shorter than the recovery time of the SOA. A 3.8 psec switching window was achieved (Glesk, Wang, Xu, Baby, Prucnal [2003]).

The TOAD structure can also be used for thresholding operation without the need for optical control pulses (Deng, Glesk, Kang, Prucnal [1997b]; Xu, Glesk, Baby, Prucnal [2004]). The difference in structure between the TOAD as a gate and

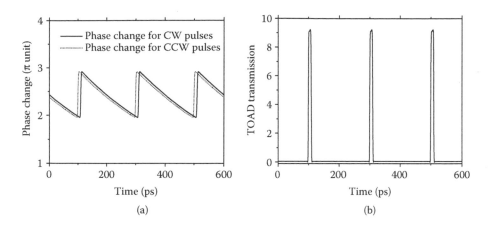

FIGURE 6.25 Calculated phase evolution for CCW and CW pulses, (b) the corresponding TOAD transmission. (Adapted from I. Glesk, B. C. Wang, L. Xu, V. Baby, P. R. Prucnal. [2003]. *Progress in Optics*, 45, 53–117, with permission from Elsevier.)

as a thresholder lies in the number of ports and at the input coupler to the device—the thresholding device has two ports and uses an asymmetric power splitting ratio for the input coupler (see Figure 6.26). Due to the asymmetry of the input coupler, the device is able to obtain different gain and phase changes to the CW and CCW signals even in the absence of a control pulse, resulting in a switching action. The degree of gain saturation and, therefore, transmission through the device depends on the intensity of the input pulse. The optimal splitting ratio for the thresholding operation input coupler varies based on the parameters of the SOA. Calculations were done previously for a set of SOA parameters showing an optimal splitting ratio of 80:20 (Deng, Glesk, Kang, Prucnal [1997b]). The threshold can be adjusted by adjusting the electrical current injected into the SOA, polarization within the loop, etc. A filter is used at the output of the device to reduce amplified spontaneous emission (ASE) from the SOA.

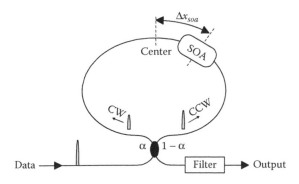

FIGURE 6.26 Structure of the TOAD as an optical thresholder.

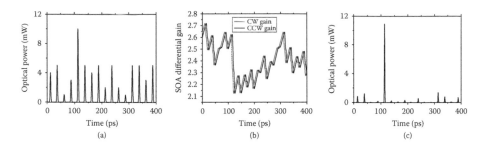

FIGURE 6.27 Simulated results for the TOAD in the thresholding operation. (a) Input waveform to the TOAD device, (b) SOA dynamic gain change, and (c) output signal of the TOAD device. (Adapted from L. Xu, I. Glesk, V. Baby, P. R. Prucnal, Proc. LEOS 2004, WP4, with permission from IEEE [© 2004 IEEE].)

Figure 6.27 shows simulation results for the thresholding operation of the TOAD, assuming the SOA parameters of Figure 6.22 and a splitting ratio of 80:20 at the input coupler. The SOA offset from the center of the loop was set at 1 mm resulting in a nominal switching window width of 5 psec. The input signal to the TOAD and the corresponding SOA gain change are shown in Figure 6.27(a) and (b), respectively. The output signal is shown in Figure 6.27(c), demonstrating significant suppression of the lower intensity signals in comparison to the higher-intensity signal. The ratio of high-intensity pulses to low-intensity pulses is increased from 3dB to 10dB.

A wavelength aware receiver was proposed (Baby, Brès, Glesk, Xu, Prucnal [2004]) which reduces false alarm probability by verifying the presence of all coded wavelengths in a peak before declaring it to be a valid autocorrelation peak. The principle of operation of this receiver is shown in Figure 6.28. The use of the wavelength aware receiver results in a doubling of the maximum spectral efficiency for carrier-hopping prime codes (Baby, Brès, Glesk, Xu, Prucnal [2004]).

Two different implementations of this receiver (see Figure 6.29) have been proposed which differ in their assignment of the additional receiver complexity to the electronic or optical domain, respectively. In the former implementation, the electronic AND gate ensures that a "1" is declared only if all wavelengths are present. Equivalently, an optical thresholder with prethresholding (aligning the wavelengths uniformly across the bit period) can be used to count the number of wavelengths in which the pulses are present. Note that this implementation scheme requires near uniform performance with wavelength of the thresholder. The TOAD-based thresholder can satisfy this due to the broad gain spectrum of the SOA.

6.4 EXPERIMENTAL DEMONSTRATION OF WAVELENGTH-HOPPING TIME-SPREADING

Experimental demonstrations of WHTS OCDMA have shown the capability for en/decoding using different technologies. Full network studies of WHTS include (1) a 16 user, 1.25Gb/sec per user FBG-based WHTS demonstration (Jaafar, LaRochelle, Cortes, Fathallah [2001]); (2) a network simulation study at 2.5Gb/sec (Mendez, Gagliardi, Feng,

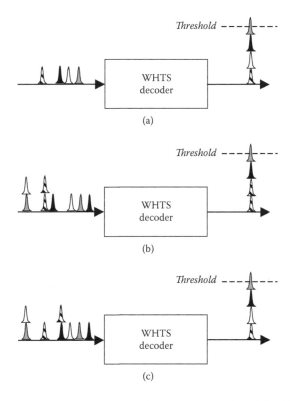

FIGURE 6.28 Principle of operation of wavelength aware receiver: (a) valid autocorrelation peak comprising of all coded wavelengths, MAI peaks that are possible error events for WHTS systems, (b) without wavelength awareness, (c) with and without wavelength awareness. (Adapted from V. Baby, C.-S. Brès, I. Glesk, L. Xu, P. R. Prucnal. [2004]. *Electron. Lett.* 40 (6), 385–387, with permission from IEE, and from V. Baby, C.-S. Brès, I. Glesk, L. Xu, P. R. Prucnal, Proc. CLEO 2004, CWH6, with permission from IEEE and OSA [© 2004 IEEE].)

Heritage, Morookian [2000]); and (3) a 4 user, 2.5Gb/sec per user demonstration at 253 Gchip/sec using TFFs (Glesk, Baby, Brès, Xu, Rand, Prucnal [2004]). The FBG-based demonstration has been described in detail in Chapter 5. In this section, we will describe results from the WHTS network demonstration based on TFFs.

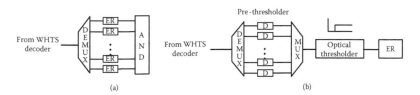

FIGURE 6.29 Two possible implementations of a wavelength aware receiver. The additional complexity is added in the (a) electronic and (b) optical domains. ER: Electronic Receiver, D: Optical Delay. (Adapted from V. Baby, C.-S. Brès, I. Glesk, L. Xu, P. R. Prucnal. [2004]. *Electron. Lett.* 40 (6), 385–387, with permission from IEE, and from V. Baby, C.-S. Brès, I. Glesk, L. Xu, P. R. Prucnal, Proc. CLEO 2004, CWH6, with permission from IEEE and OSA [© 2004 IEEE].)

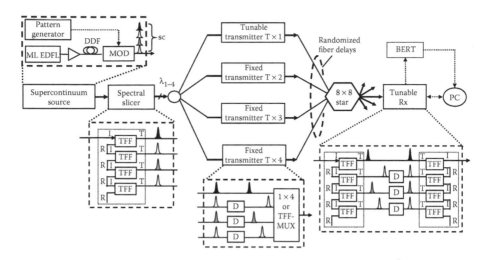

FIGURE 6.30 Schematic of the 4-node, 2.5Gb/sec testbed. ML-EDFL: Mode-locked Erbium doped fiber laser, Tx/Rx: transmitter/receiver, TFF: thin film filter, MOD: modulator, BERT: bit error rate tester.

The schematic of the experimental setup is shown in Figure 6.30. The multiwavelength pulse source was obtained by spectral slicing 20 nm of SC using a four-port TFF-based WDM DEMUX. The SC was generated by passing the output of a mode-locked erbium-doped fiber laser (ML-EDFL) through 1 km of dispersion decreasing fiber (DDF). A 2^{31}-1 pseudo-random bit sequence (PRBS) was modulated on the supercontinuum. Four pulse trains were obtained from the wavelength demultiplexer at center wavelengths of 1546 (l_1), 1550 (l_2), 1554 (l_3) and 1558 nm (l_4), each with a pulse width of 1.6 ps. Four copies of each pulse train were made using four 1 ¥ 4 power splitters and one copy of each wavelength was fed into each of the four transmitter (Tx1 – 4) modules for encoding.

For the WHTS encoding, the chip duration was set to be ~3.9 psec, corresponding to a code length of 101 and a chip rate of 253 Gchip/sec 4-wavelength, 101-time slot (chip) carrier-hopping prime codes were used in the testbed. Carrier-hopping prime codes provide zero autocorrelation sidelobes and a maximum cross correlation bound of one. The independence of the cross correlation bound to the code weight and code length enables both multirate and multiQoS communications (Yang, Kwong [2002]). For each transmitter, the wavelengths were multiplexed either using a TFF-based MUX (Txs 1 and 2) or a 1 ¥ 4 power coupler (Txs 3 and 4) after appropriate coding delays.

For Txs 2 - 4, the codes were generated using appropriate delays (D), fusion spliced to an accuracy of 2 ps and set to produce the sequences (1, 16, 31, 46), (1, 24, 47, 70) and (1, 31, 61, 91), respectively. In this notation, the numbers in parentheses (c_1, c_2, c_3, c_4) are the chip positions of the four wavelengths l_1 - l_4, respectively. For Tx1, the delays were made tunable using manually controlled opto-mechanical delay lines. The experimental results reported here correspond to Tx1 having the code (1, 9, 17, 25). Variable in-line attenuators were used to ensure power uniformity

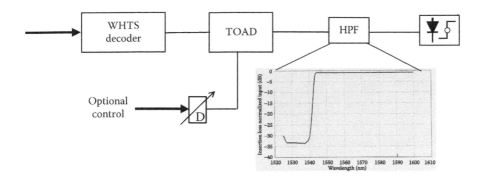

FIGURE 6.31 Schematic of the WHTS receiver with optional TOAD gating. D: Tunable Delay, HPF: High Pass Filter. The inset is the spectral profile of the HPF used for control/data separation after TOAD gating. (Adapted from I. Glesk, V. Baby, C. -S. Brès, L. Xu, D. Rand, P. R. Prucnal. *Acta Physica Slovaka* 53 (3), 245–250, with permission from Slovak Academy of Sciences (SAS), and from I. Glesk, V. Baby, C. -S.Brès, L. Xu, D. Rand, P. R. Prucnal, Proc. CLEO 2004, CWH5, with permission from OSA and IEEE [© 2004 IEEE].)

across all wavelengths and transmitters. The outputs of the transmitters were multiplexed and distributed using an 8 ¥ 8 star coupler to the WHTS network. The independence of the PRBS patterns for the four transmitters was ensured by delaying the output of the transmitters with respect to each other before the star coupler.

At the receiver, shown in Figure 6.31, TFF-based MUX and DEMUX were used for splitting and combining wavelengths in the decoder. The delays for three of the wavelengths were made tunable using computer controlled opto-mechanical delay lines and enabled the decoder to access any codes over the entire code space. The TOAD-based techniques for all-optical gating and thresholding previously mentioned were used optionally before detection to reduce the MAI. Part of the primary output of the ML EDFL (at 1539 nm) used to generate the SC was used as the control pulse train for the TOAD when used as an all-optical gate. The gating position was controlled using a manually controlled tunable delay line in the path of the control pulse. The output of the OCDMA receiver was split using a 90:10 power splitter (not shown): 90% was sent to a bit-error-tester through an attenuator (HP8157A) followed by a DC-15 GHz detector (HP11982A); 10% was sent to the digital sampling oscilloscope (Tektronix CSA8000) with a 30 GHz front-end detector.

Figure 6.32(a) shows the oscilloscope measurements for simultaneous operation of all four transmitters at constant weight without all-optical gating, with the decoder tuned to each of the four transmitters. The peak indicates the autocorrelation; the remaining pulses are due to MAI from the other users. Adjacent peaks are separated by ~400 ps, corresponding to the bit period. The bit error rate (BER) measurement with the WHTS decoder tuned to Tx1 is shown in Figure 6.32(b). Error-free operation is achieved for one, two, three, and four simultaneous users. The abscissa represents the total power received from all four simultaneous users and is not the power received from Tx1 alone. Therefore, the power penalty of ~6.5 dB is mainly an

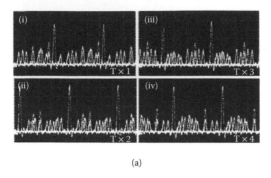

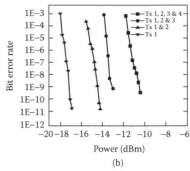

(a)

(b)

FIGURE 6.32 (a) The decoded signal for the operation of four nodes with the decoder tuned to (i) Tx1 (ii) Tx2 (iii) Tx3 and (iv) Tx4, respectively. (b) BER performance of Tx1 for the operation of multiple transmitters. (Adapted from V. Baby, C. -S. Brès, L. Xu, I. Glesk, P. R. Prucnal. [2004]. *Electron. Lett.* 40 (12), 755–756, with permission from IEE, and from V. Baby, P. R. Prucnal, C. -S.Brès, L. Xu, I. Glesk, [2004]. Proc. OCSN (WOC) 2004, 423-064, with permission from IASTED.)

artifact of the power of the MAI pulses (equivalent to 6 dB penalty for four users), and implies an effective penalty only of ~0.5 dB.

Figure 6.33 shows the use of the TOAD as an all-optical gate for MAI suppression. Figure 6.33(a) shows the signal at the output of the WHTS decoder (tuned to Tx3) before input to the TOAD and Figure 6.33(b) shows the signal at the output of the TOAD. For this operation, the switching window of the TOAD was set to 4 psec, and the control signal timed to ensure that only the autocorrelation pulse is gated out by the TOAD. The control pulse train at 1539 nm was eliminated at the output of the TOAD using the TFF-based high-pass filter (HPF), which had 35 dB suppression for all wavelengths less than 1542 nm.

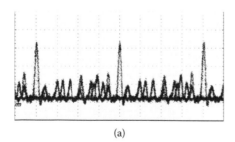
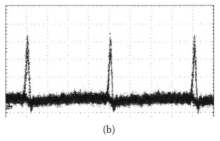

(a)   (b)

FIGURE 6.33 The use of the TOAD as a gate to eliminate MAI. (a) Received signal before TOAD with decoder tuned to transmitter Tx3, (b) Signal after TOAD. (Reprinted from I. Glesk, V. Baby, C.-S. Brès, L. Xu, D. Rand, P. R. Prucnal, *Acta Physica Slovaca* 53 (3), 245–250, with permission from Slovak Academy of Sciences (SAS), and from I. Glesk, V. Baby, C.-S. Brès, L. Xu, D. Rand, P. R. Prucnal, Proc. CLEO 2004, CWH5, with permission from OSA and IEEE. [© 2004 IEEE].)

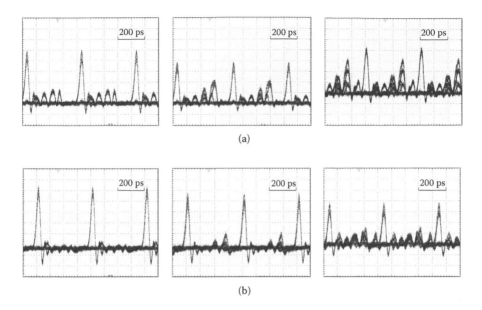

(a)

(b)

FIGURE 6.34 WHTS signal (a) before and (b) after TOAD based thresholding with two, three and four nodes in the network respectively. (Adapted from L. Xu, I. Glesk, V. Baby, P. R. Prucnal [2004]. Proc. LEOS, WP4, with permission from IEEE [© 2004 IEEE].)

Figure 6.34 shows the suppression of the MAI by using the TOAD as an all-optical thresholder (Xu, Glesk, Baby, Prucnal [2004]) for two, three, and four nodes transmitting simultaneously on the network. The signal to MAI ratio shows an increase from 5.7 dB to 8.8 dB for two users, 2.2 dB to 5.7 dB for three users, and 1.0 dB to 3.4 dB for four users.

As mentioned previously, carrier-hopping prime codes have the property of maintaining cross-correlation bounds even for codes of different weights. This opens up the possibility of using code weight variation (obtained by turning wavelengths on/off at the Tx) to satisfy varying quality of service (QoS) requirements. Ability to provide differentiated service within the same fiber channel will be suitable for next-generation networks that should satisfy a wide range of services with differing QoS requirements, such as voice-over-IP or video-on-demand. For example, signals with higher priority or QoS requirement can be assigned a higher code weight, while signals with lower priority or QoS requirement can be assigned a lower code weight.

In order to investigate the ability of the testbed to provide differentiated service, the BER was monitored as the code weight was changed. Figure 6.35 summarizes the operation of the network without all-optical gating, using different code weights for each node. Figure 6.35(a) shows the output of the decoder tuned to Tx2 with the code weights of the transmitters (Tx1, Tx2, Tx3, Tx4) as (1) (4, 3, 3, 3), (2) (4, 4, 4, 4), (3) (3, 4, 4, 4), and (4) (3, 3, 3, 3), respectively. For each transmitter with code weight of 3, λ_2 was not used in the code; the results are

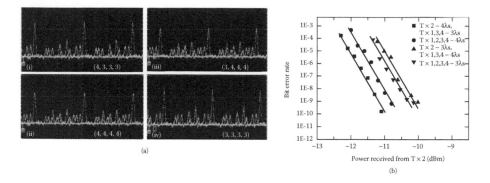

(a)

(b)

FIGURE 6.35 (a) Output signal from the WHTS decoder tuned to Tx2, code weights of Tx1, 2, 3, 4 being (i) (4, 3, 3, 3), (ii) (4, 4, 4, 4), (iii) (3, 4, 4, 4), (iv) (3, 3, 3, 3), respectively. (b) BER performance of Tx2. (Adapted from V. Baby, C. -S.Brès, L. Xu, I. Glesk, P. R. Prucnal. [2004]. *Electron. Lett.* 40 (12), 755–756, with permission from IEE and from V. Baby, P. R. Prucnal, C. -S.Brès, L. Xu, I. Glesk, Proc. OCSN (WOC) 2004, 423–064, with permission from IASTED.)

independent of the removed wavelength. Figure 6.35(b) shows the BER performance of Tx2 for these four cases. All wavelengths were monitored to ensure power uniformity among all users and wavelengths. As shown in Figure 6.35(a), reducing the number of wavelengths of a transmitter results in two effects: (1) it reduces the magnitude of the autocorrelation peak for that transmitter, resulting in lower performance — compare (4, 4, 4, 4) and (3, 4, 4, 4) — and (2) it reduces the MAI for decoders tuned to other transmitters, resulting in improved performance for those receivers — compare (4, 3, 3, 3) and (4, 4, 4, 4) —. BER variation of up to five orders of magnitude is obtained by code weight variation, as seen in Figure 6.35(b), thus showing the capability of this system to provide differentiated service.

It is important to note that in this demonstration, differentiated service is provided in the physical layer itself, unlike other access schemes in which it is usually handled in the higher network layers. Implementing differentiated service in the lower network layers guarantees, to some degree, a reduction in processing overhead and a simplification in the network design (Stok, Sargent [2000]).

6.5 CONCLUSIONS

While WDM, OTDM and WDM/OTDM have emerged as the choice for current long haul and wide-area networks, WHTS provides various promising features for next generation access networks. These include capability to provide asynchronous access and support for multiple traffic types that either require different bit rates or quality of service guarantees. The soft limit on the number of network users with graceful degradation of performance with the number of users allows for both flexibility in network management and expandability in the network size. It is arguable that WHTS systems can provide a higher utilization than WDM in access networks (Stok, Sargent [2002b]). The use of M-ary schemes in WHTS offers the ability to trade off excess cardinality for higher spectral efficiency. The code properties, along with the ability to dynamically shuffle the assigned codes, may not provide a cryptographic level of security, but does provide an added degree of signal obfuscation that can be of advantage to specific platforms such as avionic networks (Uhlhorn [2004]).

While we have generally limited our discussion to WHTS in star networks, experiments have shown the promising developments of subsystem technologies such as code drop filters (Brès, Glesk, Runser, Prucnal [2004]), which can extend the use of WHTS to other network architectures such as ring and bus networks. The development of WHTS code converters enable multihop networks (Brès, Glesk, Prucnal [2005]) while format conversion subsystems can be used as OCDMA-to-WDM gateways (Baby, Glesk, Huang, Xu, Prucnal [2005]) for coupling WHTS-based access networks to WDM-based metro and long haul networks.

In conclusion, we have summarized the various incoherent schemes for OCDMA, with particular focus on WHTS. Different technologies for implementing WHTS were described, along with experimental results from a 4-node WHTS testbed showing the feasibility of WHTS technology. The application of this technology to real networks will require integration solutions. Various integration technologies used for WHTS en/decoders are summarized in Chapter 8. The advancements made in this area promise great potential for introduction of WHTS systems in commercial and military network environments in the near future.

ACKNOWLEDGMENTS

The authors would like to acknowledge many discussions with Wing C. Kwong, Robert J. Runser, Russell Fischer, Evgenii Narimanov, and interactions with Antonio Mendez, Lawrence Chen, and Jagdeep Shah.

REFERENCES

Adam, L., Simova, E., Kavehrad, M. (1995). Experimental optical CDMA system based on spectral amplitude encoding of non-coherent broadband sources. *Photonics East, Proc. SPIE* 2614:122–132.

Agrawal, G. P., Olsson, N. A. (1989). Self-phase modulation and spectral broadening of optical pulses in semiconductor laser amplifiers. *IEEE Journal of Quantum Electronics* 25(11):2297–2306.

Agrawal, G. P. (2001). *Nonlinear Fiber Optics*, 3rd ed. San Diego: Academic Press.

Agrawal, G. P. (2002). *Fiber-Optic Communication Systems,* 3rd ed. New York: Wiley Inter-Science.

Alfano, R. R. (1989). *The supercontinuum laser source.* New York: Springer-Verlag.

Andonovic, I., Tancevski, L. (1996). Incoherent optical code division multiple access systems. *IEEE 4th Int. Symposium on Spread Spectrum Techniques and Application Proc.* 1:422–425.

Baby, V., Brès, C.-S., Xu, L., Glesk, I., Prucnal, P. R. (2004). Demonstration of differentiated service provisioning with 4-node 253 Gchip/s fast frequency-hopping time-spreading OCDMA. *Electron. Lett.* 40(12):755–756.

Baby, V., Brès, C.-S., Glesk, I., Xu, L., Prucnal, P. R. (2004). Wavelength aware receiver for enhanced 2D OCDMA system performance. *Electron. Lett.* 40(6):385–387.

Baby, V., Wang, B. C., Xu, L., Glesk, I., Prucnal, P. R. (2003). Highly scalable serial-parallel optical delay line. *Optics Communications* 218(4–6):235–242.

Baby, V., Glesk, I., Huang, Y.-K., Xu, L., Prucnal, P. R. (2005). Demonstration of an All-Optical Interface between Wavelength-Hopping Time-Spreading Optical CDMA and WDM Networks. *Proceedings of IEEE LEOS Summer Topicals*, TuC4.2.

Baby, V., Glesk, I., Runser, R. J., Fischer, R., Huang, Y.-K., Brès, C.-S., Kwong, W. C., Curtis, T. H., Prucnal, P. R. (2005). Experimental demonstration and scalability analysis of a 4-node 102Gchip/s fast frequency-hopping time-spreading optical CDMA network. *IEEE Photon. Technol. Lett.* 17(1):253–255.

Bellemare, A., Karasek, M., Rochette, M., LaRochelle, S., Tetu, M. (2000). *J. Lightwav. Technol.* 18(6): 825–831.

Bloembergen, N. (1983). The influence of electron plasma formation on superbroadening in light filaments. *Optics Communications* 8(4):285–288.

Blow, K. J., Doran, N. J., Nayar, B. K., Nelson, B. P. (1990). Two-wavelength operation of the non-linear fiber loop mirror. *Optics. Lett.* 15(4):248–250.

Born, M., Wolf, E. (1999). *Principles of optics: Electromagnetic theory of propagation, interference and diffraction of light.* Cambridge: Cambridge University Press.

Brès, C.-S., Baby, V., Glesk, I., Xu, L., Rand, D., Prucnal, P. R. (2004). Scalability of Frequency-Hopping Time-Spreading OCDMA Code Matrix. Conference on Lasers and Electro-Optics, CLEO 2004.

Brès, C.-S., Glesk, I., Prucnal, P. R. (2005). Demonstration of a transparent router for wavelength-hopping time-spreading optical CDMA. Accepted for publication in *Optics Communications*.

Brès, C.-S., Glesk, I., Runser, R. J., Prucnal, P. R. (2004). All-optical OCDMA drop unit for transparent ring networks. Lasers and Electro-optic Society Annual Meeting, LEOS 2004.

Chan, C.-K., Chen, L.-K., Cheung, K.-W. (1996). A fast channel-tunable optical transmitter for ultrahigh-speed all-optical time-division multiaccess networks. *IEEE Journal on Selected Areas in Communication* 14(5):1052–1056.

Chandrasekhar, S., Zirnbigl, M., Dentai, A. G., Joyner, C. H., Storz, F., Burrus, C. A., Lunardi, L. M. (1995). *IEEE Photon. Technol. Lett.* 7(11): 1342–1344.

Chen, L. R., Benjamin, S. D., Smith, P. W. E., Sipe, J. E. (1998). Applications of ultrashort pulse propagation in Bragg gratings for wavelength-division-multiplexing and code-division multiple access. *J. Quantum Electron.* 34(11):2117–2129.

Chen, L. R. (2001). Flexible fiber Bragg grating encoder/decoder for hybrid wavelength-time optical cdma. *IEEE Photon. Technol. Lett.* 13(11):1233–1235.

Chen, L. R., Smith, P. W. E., de Sterke, C. M. (1999). Wavelength-encoding/temporal-spreading optical code-division multiple access system with in-fiber chirped Moire gratings. *Appl. Opt.* 38(21):4500–4508.

Chen, L. R., Smith, P. W. E. (2000). Demonstration of incoherent wavelength-encoding/time-spreading optical CDMA using chirped Moire gratings. *IEEE Photon. Technol. Lett.* 12(9):1281–1283.

Datta, S., Agashe, S., Forrest, S. R. (2004). A high bandwidth analog heterodyne RF optical link with high dynamic range and low noise figure. *IEEE Photon. Technol. Lett.* 16(7):1733–1735.

Deng, K.-L., Glesk, I., Kang, K. I., Prucnal, P. R. (1997a). A 1024-channel fast tunable delay line for ultrafast all-optical TDM networks. *IEEE Photon. Technol. Lett.* 9(11): 1496–1498.

Deng, K.-L., Glesk, I., Kang, K. I., Prucnal, P. R. (1997b). Unbalanced TOAD for optical data and clock separation in self-clocked transparent OTDM networks. *IEEE Photonics Technol. Lett.* 9(6):830–832.

Deng, K.-L., Glesk, I., Kang, K. I., Prucnal, P. R. (1998). Influence of crosstalk on the scalability of large OTDM interconnects using a novel rapidly reconfigurable highly-scalable optical time-slot tuner. *IEEE Photon. Technol. Lett.* 10(7): 1039–1041.

Deng, K.-L., Runser, R. J., Glesk, I., Prucnal, P. R. (2000). Demonstration of multicasting in a 100-Gb/s OTDM switched interconnect. *IEEE Photonics Technol. Lett.* 12(5): 558–560.

Diez, S., Ludwig, R., Weber, H. G. (1999). Gain-transparent SOA-switch for high-bitrate OTDM add/drop multiplexing. *IEEE Photonics Technol. Lett.* 11(1):60–62.

Dicon Fiber Optics. www.diconfiberoptics.com/products/scd0090/index.html, Richmond, CA (accessed July 2003).

Domash, L. H. (2004). Thin-films sing a new tune. *Photon. Spectra* (Nov. 2004):70–74

Doran, N. J., Wood, D. (1988). Nonlinear-optical loop mirror. *Optics Letters* 13(1):56–58.

Dragone, C. (1998). *J. Lightwav. Technol.* 16(10):1895–1906.

Fathallah, H., Rusch, L. A., LaRochelle, S. (1999). Passive optical fast frequency-hop CDMA communications system. *IEEE J. Lightwave Technol.* 17(3):397–405.

Glesk, I., Wang, B. C., Xu, L., Baby, V., Prucnal, P. R. (2003). Ultrafast all-optical switching in optical networks. In: Wolf, E., ed. *Progress in Optics*, v. 45, 53–117.

Glesk, I., Baby, V., Brès, C.-S., Xu, L., Rand, D., Prucnal, P. R. (2004). Experimental demonstration of 2.5 Gbit/s incoherent two-dimensional optical code division multiple access system. *Acta Physica Slovaca* 54(3):245–250.

Golmie, N., Ndousse, T. D., Su, D. H. (2000). A differentiated optical services model for WDM networks. *IEEE Commun. Mag.* 38(2):68–73.

Greiner, C. M., Iazikov, D., Mossberg, T. W. (2004a) Lithographically fabricated planar holographic Bragg reflectors. *J. Lightwav. Technol.* 22(1):136–145.

Greiner, C. M., Iazikov, D., Mossberg, T. W. (2004b). Wavelength-division multiplexing based on apodized planar holographic Bragg reflectors. *Appl. Opt.*, forthcoming.

Greiner, C. M., Mossberg, T. W., Iazikov, D. (2004). Bandpass engineering of Lithographically scribed channel-waveguide. *Opt. Lett.* 29(8):806–808.

Griffin, R. A., Simpson, D. D., Jackson, D. A. (1995). Coherence coding for photonic code-division multiple access networks. *J. Lightwav. Technol.* 13(9):1826–808.

Grunnet-Jepsen, A., Johnson, A. E., Maniloff, E. S., Mossberg, T. W., Munroe, M. J., Sweetser, J. N. (1999a). Demonstration of all-fiber sparse lightwave CDMA based on temporal phase encoding. *IEEE Photon. Technol. Lett.* 11(10):1283–1285.

Grunnet-Jepsen, A., Johnson, A. E., Maniloff, E. S., Mossberg, T. W., Munroe, M. J., Sweetser, J. N. (1999b). Fiber Bragg grating based spectral encoder/decoder for lightwave CDMA. *Electron. Lett.* 35(13):1096–1097.

Hassan, A. A., Hershey, J. E., Riza, N. A. (1995). Spatial Optical CDMA. *IEEE JSAC.* 13(3):609–613.

Herrmann, J., Griebner, D., Zhavoronkov, N., Husakou, A., Nickel, D., Korn, G., Knight, J. C., Wadsworth, W. J., Russel, P. S. J. (2002). Experimental evidence for supercontinuum generation by fission of higher-order solitons in photonic crystal fibers. *Proc. Quantum Electron. and Laser Science (QELS '02)*:165–166.

Hibino, Y. (2002). Recent advances in high density and large scale AWG multi/demultiplexers with higher index-contrast silica-based PLCs. *IEEE J. Select. Topics Quantum Electron.* 8(6):1090–1101.

Hill, K. O., Meltz, G. (1997). Fiber Bragg grating technology fundamentals and overview. *IEEE J. Lightwav. Technol.* 15(8):1263–1276.

Huang, W., Nizam, M. H. M., Andonovic, I., Tur, M. (2000). Coherent optical CDMA (OCDMA) systems used for high-capacity optical fiber networks—System description, OTDM comparison and OCDMA/WDMA networking. *IEEE J. Lightwav. Technol.* 18(6):765–778.

Huang, Y.-K., Greiner, C. M., Iazikov, D., Baby, V., Mossberg, T. W., Prucnal, P. R. (2004). Integrated holographic encoder/decoder for 2D optical CDMA. LEOS 2004.

Hui, J. Y. (1985). Pattern code modulation and optical decoding—A novel code-division multiplexing technique for multifiber networks. *IEEE JSAC*. SAC-3(6):916–927.

Iazikov, D., Greiner, C. M., Mossberg, T. W. (2004a). Apodizable integrated filters for coarse WDM and FTTH-type applications. *IEEE J. Lightwav. Technol.* 22(5):1 402–1407.

Iazikov, D., Greiner, C. M., Mossberg, T. W. (2004b). Effective gray scale in lithographically scribed planar holographic Bragg reflectors. *Appl. Opt.* 43(5):1149–1155.

Iodice, M., Cocorullo, G., Della Corte, F. G., Rendina, I. (2000). Silicon Fabry-Perot filter for WDM systems channels monitoring. *Opt. Commun.* 183(5–6): 415.

Jaafar, H. B., LaRochelle, S., Cortes, P.-Y., Fathallah, H. (2001). 1.25Gbit/s transmission of optical FFH-OCDMA signals over 80km with 16 users. *Proc. Opt. Fiber Commun. Conf. (OFC '01)* 2:TuV3-1–TuV3-2.

Jackson, K. P., Newton, S. A., Moslehi, B., Tur, M., Cutler, C. C., Goodman, J. W., Shaw, H. J. (1985). Optical fiber delay-line signal processing. *IEEE Trans. Microwave Theory Tech.* MTT-3(3):193–209.

Jung, P., Baier, P. W., Steil, A. (1993). Advantages of CDMA and spread-spectrum techniques over FDMA and TDMA in cellular mobile radio applications. *IEEE Transactions on Communications* 41(3):231–242.

Kashyap, R. (1999). *Fiber Bragg gratings*. San Diego:Academic Press.

Kavehrad, M., Zaccarin, D. (1995). Optical code-division-multiplexed systems based on spectral encoding of noncoherent systems. *J. Lightwave Technol.* 13(3):534–545.

Kawanishi, S., Takara, H., Uchiyama, K., Shake, I., Mori, K. (1999). 3 Tbit/s (160 Gbit/s ¥19 channel) optical TDM and WDM transmission experiment. *Electron. Lett.* 35(10): 826–827.

Khaleghi, F., Kavehrad, M. (1996). A new correlator receiver for noncoherent optical CDMA networks with bipolar capacity. *IEEE Trans. Commun.* 44(10):1335–1339.

Kim, J., Nowak, G. A., Boyraz, O., Islam, M. N. (1999). Low energy, enhanced supercontinuum generation in high nonlinearity dispersion-shifted fibers. *Proc. Conf. Lasers and Electro-Optics (CLEO '99)*:224–225.

Kim, S., Yu, K., Park, N. (2000). A new family of space/wavelength/time spread three-dimensional optical code for OCDMA networks. *J. Lightwave Technol.* 18(4):502– 511.

Kim, S., Park, S., Choi, Y., Han, S. (2000). Incoherent bidirectional fiber-optic code division multiple access networks. *IEEE Photon. Technol. Lett.* 12(7):921–923.

Kim, N. S., Prabhu, M., Li, C., Song, J., Ueda, K.-I., Allen, C. (2000). 100nm supercontinuum generation centered at 1483.4nm from Brillouin/Raman fiber laser. *Conf. Lasers and Electro-Optics, CLEO 2000*:263–264.

Kitayama, K.-I. (1994). Novel spatial spread spectrum based fiber optic CDMA networks for image transmission. *IEEE J. Sel. Areas Commun.* 12(4):762–772.

Knight, J. C. (2003). Photonic crystal fibers. *Nature* 424(6950):847–851.

Knittl, Z. (1976). *Optics of thin films*. New York: John Wiley.

Koga, M., Hamazumi, Y., Watanabe, A., Okamoto, S., Obara, H., Sato, K. I., Okuno, M., Suzuki, S. (1996). *J. Lightwav. Technol.* 14(6): 1106–1119.

Kutsuzawa, S., Minato, N., Oshiba, S., Nishiki, A., Kitayama, K. (2003). 10Gb/s ¥ 2ch signal unrepeated transmission over 100km of data rate enhanced time-spread/wavelength-hopping OCDMA using 2.5Gb/s-FBG en/decoder. *IEEE Photon. Technol. Lett.* 15(2):317–319.

Kwong, W. C., Yang, G.-C. (1998). Image transmission in multicore-fiber code-divison multiple access networks. *IEEE Commun. Lett.* 2(10):285–287.

Kwong, W. C., Yang, G.-C. (2004). Multiple-length multiple-wavelength optical orthogonal codes for optical CDMA systems supporting multirate multimedia services. *IEEE J. Sel. Areas in Commun.* 22(9):1640–1647.

Kwong, W. C., Yang, G.–C., Chang, C.-Y. (In press). Wavelength-hopping time-spreading optical CDMA with bipolar codes. *J. Lightwav. Technol.*

Lam, C. F., Tong, D. T. K., Wu, M. C., Yablonovitch, E. (1998). Experimental demonstration of bipolar optical CDMA system using a balanced transmitter and complementary spectral encoding. *IEEE Photon. Technol. Lett.* 10(10):1504–1506.

Lam, C. F. (2000). To spread or not to spread—the myths of optical CDMA, *IEEE LEOS Annual Mtg.* 2:13–16.

Lee, J. H., Teh, P. C., Petropoulos, P., Ibsen, M., Richardson, D. J. (2002). A grating-based OCDMA coding-decoding system incorporating a nonlinear optical loop mirror for improved code recognition and noise reduction. *IEEE J. Lightwav. Technol.* 20(1):36–46.

Lenz, G., Eggleton, B. J., Madsen, C. K., Slusher, R.E. (2001). Optical delay lines based on optical filters. *IEEE J. Quantum Electron.* 37(4):525–532.

Lequime M., Parmentier R., Lemarchand F., Amra C. (2002). Toward tunable thin-film filters for wavelength division multiplexing applications. *Appl. Opt.* 41(16): 3277–3284.

Little Optics, www.littleoptics.com/delay, Annapolis Junction, MD (accessed July 2003).

MacDonald, R. I. (1987). Switched optical delay-line signal processors. *IEEE J. Lightwav. Technol.* 5(6):856–861.

MacLeod, H. A. (1986). *Thin-film optical filters,* 2nd ed. Bristol: A Hilger Ltd.

Magne, J., Wei, D.-P., Ayotte, S., Rusch, L. A., LaRochelle, S. (2003). Experimental demonstration of frequency-encoded optical CDMA using superimposed fiber Bragg gratings. *Proc. OSA Top. Mtg. on Bragg gratings, photosensitivity, and poling.*

Manning, R. J., Davies, D. A. O., Cotter, D., Lucek, J. K. (1994). Enhanced recovery rates in semiconductor-laser amplifiers using optical-pumping. *Electron. Lett.* 30(10):787–788.

Marhic, M. E. (1993). Coherent Optical CDMA Networks. *IEEE J. Lightwav. Technol.* 11(5/6):854–863.

Mendez, A. J., Gagliardi, R. M., Feng, H. X. C., Heritage, J. P., Morookian, J.-M. (2000). Strategies for realizing optical CDMA for dense, high-speed, long span, optical network applications. *IEEE J. Lightwav. Technol.* 18(12):1685–1696.

Mendez, A. J., Lambert, J. L., Morookian, J.-M., Gagliardi, R. M. (1994). Synthesis and demonstration of high speed, bandwidth efficient optical code division multiple access (CDMA) tested at 1 Gb/s throughput. *IEEE Photon. Technol. Lett.* 6(9):1146–1148.

Monnard, R., Srivastava, A. K., Doerr, C. R., Joyner, C. H., Stulz, L. W., Zirnbigl, M., Sun, Y., Sulhoff, J. W., Zyskind, J. L., Wolf, C. (1998). 16-channel 50GHz channel spacing long-haul transmitter for DWDM systems. *Electronics Letters.* 34(8):765–767.

Mori, K., Takara, H., Kawanishi, S., Saruwatari, M., Morioka, T. (1997). Flatly broadened supercontinuum spectrum generated in a dispersion decreasing fiber with convex dispersion profile. *Electron. Lett.* 33(21):1806–1808.

Morioka, T., Kawanishi, S., Mori, K., Saruwatari, M. (1994). Nearly penalty-free, < 4 ps supercontinuum Gbit/s pulse generation over 1535–1560 nm. *Electron. Lett.* 30(10): 790–791.

Morioka, T., Takara, H., Kawanishi, S., Kamatani, O., Takiguchi, K., Uchiyama, K., Saruwatari, M., Takahashi, H., Yamada, M., Kanamori, T., Ono, H. (1996). 1 Tbit/s (100 Gbit/s ¥10 channel) OTDM/WDM transmission using a single supercontinuum WDM source. *Electron. Lett.* 32(10):906–907.

Mossberg, T. W., Greiner, C. M., Iazikov, D. (2004). Holographic Bragg reflectors, photonic bandgaps and photonic integrated circuits. *Opt. Photon. News* 15(5):26–33.

Mossberg, T. W. (2001). Planar holographic optical processing devices. *Opt. Lett.* 26(7): 414–416.

Mossberg, T. W., Iazikov, D., Greiner, C. M. (2004). Planar-waveguide integrated spectral comparator. *J. Opt. Soc. America* 21(6):1088–1092.

Mortimore, D. B. (1988). Fiber loop reflectors. *IEEE J. Lightwav. Technol.* 6(7):1217–1224.

Narimanov, E., Kwong, W. C., Yang, G.-C., Prucnal, P. R. (2005). Shifted carrier-hopping prime codes for multicode keying in wavelength-time O-CDMA. *IEEE Transactions on Communication.* In press.

Nguyen, L., Aazhang, B., Young, J. F. (1995). All–optical CDMA with bipolar codes. *Electron. Lett.* 31(6):469–470.

Nguyen, B., Dennis, T., Aazhang, B., Young, J. F. (1997). Experimental demonstration of bipolar codes for optical spectral amplitude CDMA communication. *IEEE J. Lightwav. Technol.* 15(9):1647–1653.

Nuss, M. C., Knox, W. H., Koren, U. (1996). *Electron. Lett.* 32(14):1311–1312.

Ohmori, Y. (1991). Passive and active silica waveguides on silicon, Proc. ECOC '93. MoP1.1

Okayama, H., Kawahara, M., Kamijoh, T. (1996). *J. Lightwav. Technol.* 14(6): 985–990.

Okuno, T., Onishi, M., Nishimura, M. (1998). Generation of an ultra-broad-band supercontinuum by dispersion-flattened and decreasing fiber. *IEEE Photon. Technol. Lett.* 10(1):72–74.

Park, E., Mendez, A. J., Garmire, E. M. (1992). Temporal/spatial optical CDMA networks— Design, demonstration, and comparison with temporal networks. *IEEE Photon. Technol. Lett.* 4(10):1160–1162.

Parmentier, R., Lemarchand, F., Cathelinaud, M., Lequime, M., Amra, C., Labat, S., Bozzo, S., Bocquet, F., Charai, A., Thomas, O., Dominici, C. (2002). Piezoelectric tantalum pentoxide studied for optical tunable applications. *Appl. Opt.* 41(16):3270–3276.

Pfeiffer, T., Deppisch, B., Witte, M., Heidemann, R. (1999). Operational stability of a spectrally encoded optical CDMA system using inexpensive transmitters without spectral control. *IEEE Photon. Technol. Lett.* 11(7):916–918.

Prather, D. W. (2002). Photonic crystals—An engineering perspective. *Opt. Photon. News.* 13(6):16–19.

Prucnal, P. R., Santoro, M. A., Fan, T. R. (1986). Spread spectrum fiber-optic local area network using optical processing. *IEEE J. Lightwav. Technol.* 4(5):547–554.

Prucnal, P. R., Krol, M. F., Stacy, J. L. (1991). Demonstration of rapidly tunable optical time-division multiple-access coder. *IEEE Photon. Technol. Lett.* 3(2):170–172.

Salehi, J. A. (1989). Code divison multiple-access techniques in optical fiber networks: Part I: Fundamental principles. *IEEE Trans. Commun.* 37(8):824–833.

Salehi, J. A., Weiner, A. M., Heritage, J. P. (1990). Coherent ultrashort light pulse code-division multiple access communication systems. *IEEE J. Lightwav. Technol.* 8(3):478–491.

Sardesai, H. P., Chang, C.-C., Weiner, A. M. (1998). A femtoseconds code-division multiple-access communication system test bed. *IEEE J. Lightwav. Technol.* 16(11):1953–1964.

Scherer, A., Painter, O., Vuckovic, J., Loncar, M., Yoshie, T. (2002). Photonic crystals for confining, guiding, and emitting light. *IEEE Trans. Nanotechnol.* 1(1):4–11.

Schumacher, D. (2002). Controlling continuum generation. *Optics Lett.* 27(6):451–453.

Shake, T. H. (2005). Security performance of optical CDMA against eavesdropping. *Journal of Lightwave Technology,* 23(2): 655–670.

Smit, M. K., van Dam, C. (1996). *IEEE J. Sel. Topics Quantum Electron.* 2(2):236–250.

Sokoloff, J. P., Prucnal, P. R., Glesk, I., Kane, M. (1993). A terahertz optical asymmetric demultiplexer. *IEEE Photon. Technol. Lett.* 5(7):787–790.

Sotobayashi, H., Kitayama, K. (1998). 325nm bandwidth supercontinuum generation at 10Gb/s using dispersion-flattened and non-decreasing normal dispersion fibre with pulse compression technique. *Electron. Lett.* 34(13):1336–1337.

Stok, A., Sargent, E. H. (2002a). The Role of Optical CDMA in Access Networks. *IEEE Commun. Mag.,* 83–87.

Stok, A., Sargent, E. H. (2000). Lighting the local network: Optical code division multiple access and quality of service provisioning. *IEEE Networks* 14(6):42–46.

Stok, A., Sargent, E. H. (2002b). System performance comparison of optical CDMA and WDMA in a broadcast local area network. *IEEE Commun. Lett.* 6(9):409–411.

Stok, A., Sargent, E. H. (2003). Comparison of diverse optical CDMA codes using a normalized throughput metric. *IEEE Communication Lett.* 7(5):242–244.

Sumriddetchkajorn, S., Chaitavon, K. (2003). A reconfigurable thin-film filter-based 2 ¥ 2 add-drop fiber-optic switch structure. *IEEE Photon. Technol. Lett.* 15(7): 930–932.

Sun, Z. J., NcGreer, K. A., Broughton, J. N. (1998). Demultiplexer with 120 channels and 0.29nm channel spacing. *IEEE Photon. Technol. Lett.* 10(1):

Tamai, H., Iwamura, H., Minato, N., Oshiba, S. (2004). Experimental study on time-spread/wavelength-hop optical code division multiplexing with group delay compensating en/decoder. *IEEE Photonics Technology Letters.* 16(1):335–367.

Tsang, H. K., Mak, M. W. K., Chan, L. Y., Soole, J. B. D., Youtsey, C., Adesida, I., (1999). Etched cavity InGaAsP/InP waveguide Fabry-Perot filter tunable by current injection. *J. Lightwav. Technol.* 17 (10): 1890–1895.

Ramaswami, R., Sivarajan, K. N. (2002) *Optical networks—A practical perspective,* 2nd ed., San Diego: Morgan Kaufman

Tajima, K., Nakamura, S., Sugimoto, Y. (1995). Ultrafast polarization-discriminating Mach-Zehnder all-optical switch. *Applied Physics Letters* 67(25):3709–3711.

Takiguchi, K., Shibata, T., Itho, M. (2002). Encoder/decoder on planar lightwave circuit for time-spreading/wavelength-hopping optical CDMA. *Electron. Lett.* 38(10):469–470.

Tanceski, L., Andonovic, I. (1994). Wavelength hopping/time spreading code division multiple access systems. *Electron. Lett.* 30(9):721–723.

Tanceski, L., Rusch, L. A. (2000). Impact of the beat noise on the performance of 2-D optical CDMA systems. *IEEE Commun. Lett.* 4(8):264–266.

Teh, P. C., Petropoulos, P., Ibsen, M., Richardson, D. J. (2001). A comparative study of the performance of seven- and 63-chip optical code-division multiple-access encoders and decoders based on superstructured fiber Bragg gratings. *IEEE J. Lightwav. Technol.* 19(9):1352–1365.

Uchiyama, K., Morioka, T., Kawanishi, S., Takara, H., Saruwatari, M. (1997). Signal-to-noise ratio analysis of 100 Gb/s demultiplexing using nonlinear optical loop mirror. *IEEE J. Lightwav. Technol.* 15(2):194–201.

Uhlhorn, B. L. (2004). Optical code division multiple access. Avionics Fiber-Optics and Photonics Workshop AVFOP.

Wada, N., Sotobayashi, H., Kitayama, K. (2000). 2.5Gbit/s time-spread/wavelength-hop optical code division multiplexing using fiber Bragg grating with supercontinuum light source. *Electron. Lett.* 36(9):815–817.

Wang, B. C., Glesk, I., Runser, R. J., Prucnal, P. R. (2001). A fast tunable parallel optical delay line. *Optics Express* 8(11):599–604.

Xu, L., Glesk, I., Baby, V., Prucnal, P. R. (2004). Multiple access interference (MAI) noise reduction in a 2D optical CDMA system using ultrafast optical thresholding. LEOS 2004. WP4.

Yang, G., Shen, Y. R. (1984). Spectral broadening of ultrashort pulses in a non linear medium. *Optics Lett.* 9(11):510–512.

Yang, G. C., Kwong, W. C. (1996). Two-Dimensional Spatial Signature Patterns. *IEEE Trans. Commun.* 44(2):184–191.

Yang, G. C., Kwong, W. C. (2002). *Prime codes with applications to CDMA optical and wireless networks,* Norwood, MA: Artech House.

Yang, G. C. (1996). Variable-weight optical orthogonal codes for CDMA networks with multiple performance requirements. *IEEE Trans. Commun.* 44 (1):47–55.

Yegnanarayanan, S., Bushan, A. S., Jalali, B. (2000). Fast wavelength-hopping time-spreading encoding/decoding for optical CDMA. *IEEE Photon. Technol. Lett.* 12(5):573–575.

Yim, R. M. H., Chen, L. R., Bajcsy, J. (2002). Design and performance of 2-D codes for wavelength-time optical CDMA. *IEEE Photon. Technol. Lett.* 14(5):714–716.

Yu, K., Shin, J., Park, N. (2000). Wavelength-time spreading Optical CDMA System using wavelength multiplexers and mirrored fiber delay lines. *IEEE Photon. Technol. Lett.* 12(9):1278–1280.

Zaccarin, D., Kavehard, M. (1993). An optical CDMA system based on spectral encoding of a LED. *IEEE Photon. Technol. Lett.* 4(4):479–482.

Zhang, J., Yao, M., Chen, X., Xu, L., Chen, M., Gao, Y. (2000). Bit error rate analysis of OTDM system based on moment generation function. *IEEE J. Lightwav. Technol.* 18(11):1513–1518.

Zheng, Z., Weiner, A. M. (2000). Novel optical thresholder based on second harmonic generation in long periodically poled lithium niobate for ultrashort pulse optical code division multiple-access. *Optical Fiber Communication Conference* 3:317–319.

Zhou, J., Cadeddu, R., Casaccia, E., Cavazzoni, C., O'Mahony, M. J. (1996). Crosstalk in multiwavelength optical cross-connect networks. *IEEE J. Lightwav. Technol.* 14(6):1423–1435.

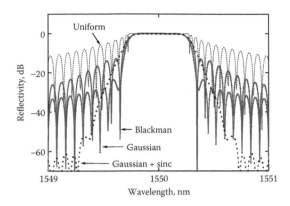

COLOR FIGURE 4.6 Comparison of the spectral response of a uniform FBG to an FBG with Gaussian, Blackman, or Gaussian-sinc apodization.

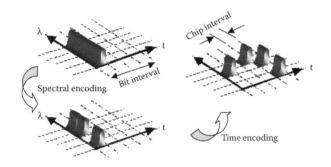

COLOR FIGURE 4.17 Schematic of encoding of 2D wavelength-time code.

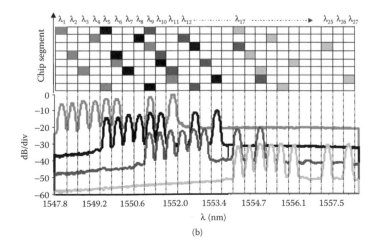

COLOR FIGURE 4.21 (b) Codes used in demonstrator system and spectral response of grating encoders.

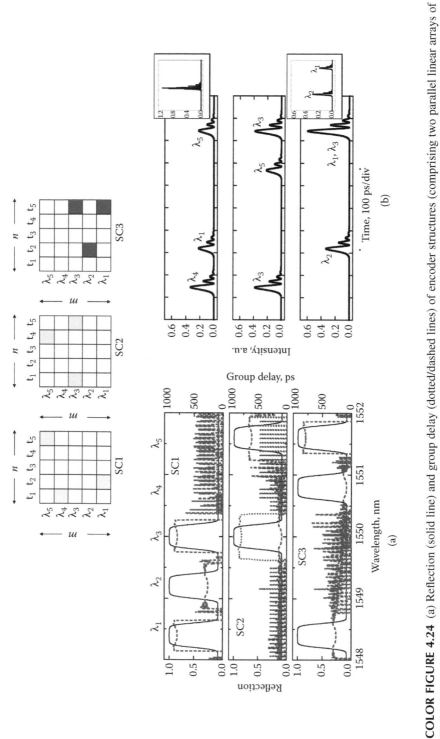

COLOR FIGURE 4.24 (a) Reflection (solid line) and group delay (dotted/dashed lines) of encoder structures (comprising two parallel linear arrays of FBGs) for the spatial codes shown in the inset. (b) Corresponding encoded signals obtained by reflecting a 1-psec Gaussian pulse from the grating encoders. The inset in the top figure shows the corresponding decoded signal of SC1 and the inset in the bottom figure shows the encoded signal corresponding to only one of the two arrays in the encoder structure. For the insets, the time scale is 100 psec/div. The slight variation in the waveform peaks is due to the nonuniform power distribution of the broadband input.

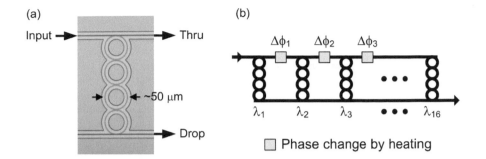

(a)

Input → Thru

←→ ~50 μm

→ Drop

(b)

$\Delta\phi_1$ $\Delta\phi_2$ $\Delta\phi_3$

λ_1 λ_2 λ_3 • • • λ_{16}

☐ Phase change by heating

COLOR FIGURE 5.12 (a) A three-ring MRR tunable filter, and (b) optical circuit for a MRR-based 16-channel coder.

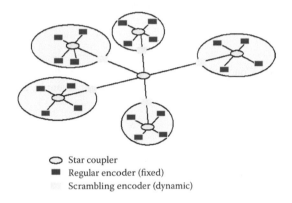

◯ Star coupler
■ Regular encoder (fixed)
▨ Scrambling encoder (dynamic)

COLOR FIGURE 5.23 Secure islands with conventional Hadamard coders interconnected with dynamic "scrambling" encoders for increased data obscurity between islands.

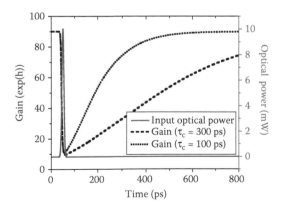

COLOR FIGURE 6.23 Dynamics of SOA gain change upon passing of a short optical pulse, for carrier recovery times of 100 ps and 300 ps. (Reprinted from I. Glesk, B. C. Wang, L. Xu, V. Baby, P. R. Prucnal. [2003]. *Progress in Optics,* 45, 53–117, with permission from Elsevier.)

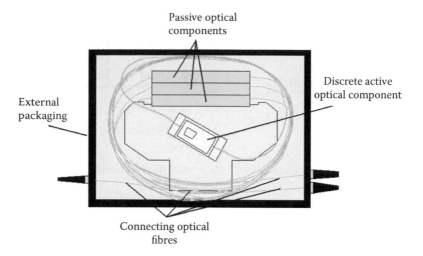

COLOR FIGURE 8.1 Conceptual schematic showing the module integration of a range of commercial-off-the-shelf (COTS) components to yield, for example, a line card that can be rack mounted.

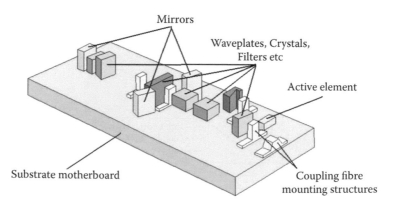

COLOR FIGURE 8.2 Conceptual schematic illustrating hybrid integration based on micro-optic platform.

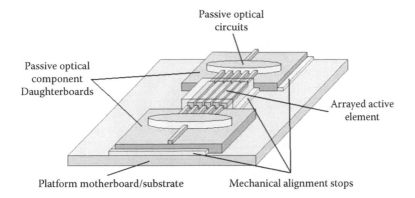

COLOR FIGURE 8.3 Conceptual schematic illustrating hybrid integration based on the assembly of active semiconductor devices with passive waveguides supported on a common micro-bench.

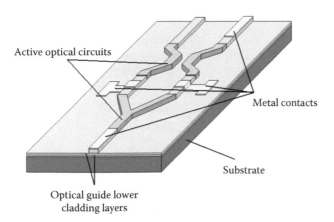

COLOR FIGURE 8.4 Conceptual schematic illustrating the monolithic integration of active and passive optical devices in a single material system.

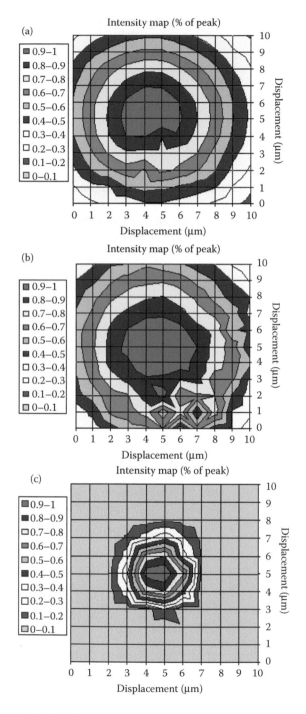

COLOR FIGURE 8.6 Coupling tolerance contour maps for (a) silica-on-silicon-SMF28 coupling, (b) mode-matched SOA-SMF28 coupling, and (c) nonmode matched SOA-SMF28 coupling.

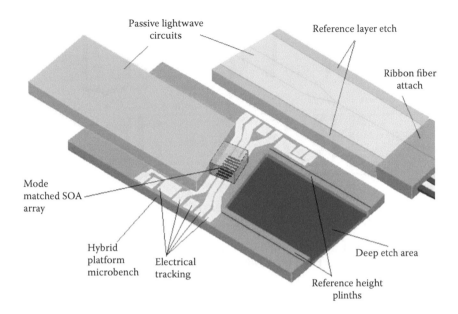

COLOR FIGURE 8.7 Hybrid assembly schematic [42].

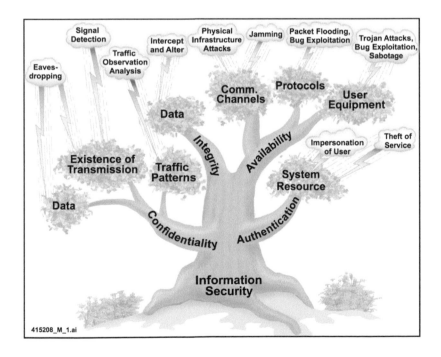

COLOR FIGURE 9.1 The security tree. Security is often divided into branches of confidentiality, integrity, availability, and authentication. Each of these branches has assets to be protected (in foliage). Some threats for the various assets are identified (in ominous clouds).

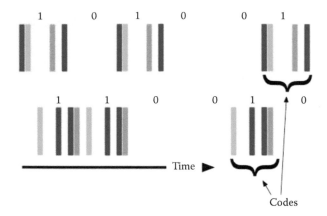

COLOR FIGURE 9.3 Schematic of two, asynchronous incoherent OCDMA channels. On–off-key modulation is being used. Note that no decoding is needed if only one channel is present.

7 Hybrid Multiplexing Techniques (OCDMA/TDM/WDM)

Hideyuki Sotobayashi

CONTENTS

7.1 INTRODUCTION

As global network infrastructures expand to support various type of traffic, photonic networks are expected to take an important role. The increasing demand for bandwidth forces network infrastructures to be of large capacity and reconfigurable. The efficient utilization of bandwidth is a major design issue for ultrahigh-speed photonic networks. The two main techniques for multiplexing data signals are currently time division multiplexing (TDM) and wavelength division multiplexing (WDM). Optical code

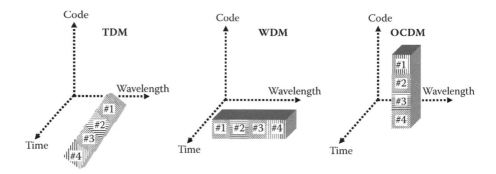

FIGURE 7.1 Various photonic multiplexing.

division multiple access (OCDMA) is an alternative method, which performs encoding and decoding through an optical signature code, in order to allow the selection of a desired signal so that different users can share the same bandwidth. In such a system, data signals overlap both in time and wavelength as shown in Figure 7.1.

The roots of OCDMA are found in spread spectrum communication techniques. Spread spectrum was developed in the mid-1950s mainly as a novel form of transmission, overcoming the grid restrictions in radio bandwidth allocation [1–3]. It is based on the idea of spreading the spectrum of the narrow band message over a much wider frequency spectrum by means of digital codes [4,5]. Due to the spreading, the transmitted signal arrives at the receiver as a noiselike signal. And message recovery is impossible unless the original code is known. The received signal is correlated by the authorized receiver with a local code, which is a replica of transmission code. Despreading and signal recovery in the presence of the interference from other users can be accomplished. As a result, spread spectrum found application in military communications, as a mechanism of transmitting signals in a very noisy environment with very high security. Furthermore, with the emergence of satellite and mobile communications, spread spectrum was considered as a basis of a new multiple access technique named CDMA [6]. The first work in OCDMA occurred in the late 1970s in the area of fiber delay lines for signal processing [7]. In OCDMA, each channel is optically encoded with the specific code. Only an intended user with the corrected code can recover the encoded information. A proper choice of optical codes allows signals from all connected network nodes to be carried without interference between signals. Therefore, simultaneous multiple access can be achieved without a complex network protocols to coordinate data transfer among the communicating nodes. As a result, multiusers can share the same transmission bandwidth. The selection of the desired signal is based on matched-filtering, followed by thresholding. OCDMA can provide advantages when applied to photonic networks, such as asynchronous transmission, a potential of communication security, soft capacity on demand, and high degree of scalability.

The performance of TDM systems is limited by the time-serial nature of the technology. Each receiver should operate at the total bit rate of the system. The allocation of dedicated time slots does not allow TDM to take advantage of statistical multiplexing gain, which is significant when the data traffic occurs in bursts. In spite of the fact that TDM technologies are well matured and developed at up to some tens of Gbits/s, another problem for the ultra–high speed TDM system is the upper limit of electronic circuit operation speed. Although the basic concept of ultrahigh speed TDM has been proposed, there is still need for further progress, including optical logic gates.

In WDM systems, the available optical bandwidth is divided into fixed wavelength channels that are used concurrently by different users. Thus, an issue with WDM is wavelength granularity — it is limited in that it can only handle traffic on an optical channel of the wavelength path. This may waste the wavelength resources. One of the main applications of a WDM system is its large capacity in long-haul transmission systems, due to its relatively simple transmission technology.

OCDMA offers an interesting alternative because neither time management nor frequency management at the transmitting nodes is necessary. OCDMA can operate asynchronously and does not suffer from packet collisions; therefore, very low latencies can be achieved. In contrast to TDM and WDM, in which the maximum transmission capacity is determined by the total number of time slots or wavelength channels, respectively, OCDMA allows flexible network design — the signal quality depends on the number of active channels.

Each multiplexing format has its own merits and application area. The motivation to research hybrid multiplexing techniques is to enhance network flexibility and scalability. Hence, hybrid multiplexing techniques would be utilized in various ways in which each multiplexing technique represents its advantage. In this chapter, applications of hybrid multiplexing techniques are reviewed. Section 7.2 describes hybrid multiplexing transmission systems such as TDM/WDM and OCDMA/WDM. Section 7.3 discusses the multiplexing format conversions between TDM and WDM, and also between OCDMA and WDM. Section 7.4 describes applications of OCDMA/WDM hybrid multiplexing for optical path technologies. A brief summary is given in section 7.5.

7.2 HYBRID MULTIPLEXING TRANSMISSION SYSTEM

7.2.1 Large Capacity TDM/WDM Transmission System

To meet the increasing demand for multiterabit/sec transmissions, 40 Gbit/sec TDM-based WDM systems are being researched as the next large capacity system. With the advent of optical amplifiers, the wavelength bands available for WDM transmission have expanded to the S-, C-, and L-bands. Expansion of the possible number of WDM channels naturally results in larger capacity, but also increases the complexity of the system. There are several modulation formats available for WDM transmission such as return-to-zero (RZ), non-return-to-zero (NRZ), and multilevel phase/amplitude modulations [8,9]. As one of the WDM transmissions ready for a practical use, this section describes a frequency standardized simultaneous wavelength-band generation methods in carrier-suppressed return-to-zero (CS-RZ) format [10]. Critical issues in WDM transmissions are the linear and nonlinear cross talk from adjacent channels. The CS-RZ format exhibits lower spectrum bandwidth compared to the conventional RZ format and good tolerance against nonlinear effects [11,12]. Another merit of CS-RZ format is a reduction of interchannel interference in the time domain. Because the relative carrier-phase of adjacent channel in CS-RZ format is shifted by π in the time domain, interchannel interference due to the pulse broadening is reduced.

7.2.1.1 Key Technologies

Unlike conventional 40 Gbit/sec CS-RZ generation using a complex ETDM setup, we generate 40 Gbit/sec CS-RZ using the simple configuration of an optical multiplexer integrated as a planar lightwave circuit (PLC), as shown in Figure 7.2(a). A 10 Gbit/sec

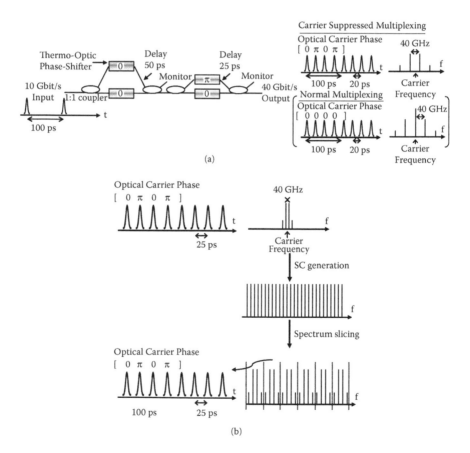

FIGURE 7.2 Operating principle of (a) carrier suppressed RZ generation in the optical domain using a time-delayed optical multiplexer with phase shifter, and (b) simultaneous wavelength-band generation of frequency standardized DWDM of CS-RZ format using SC generation and spectrum slicing.

RZ signal is time-delayed optically and multiplexed to form a 40 Gbit/sec signal. The phase of the optical carrier of each delayed adjacent pulse is changed by π using optical phase shifters. As a result, a multiplexed 40 Gbit/sec CS-RZ format is obtained. Simultaneous multiwavelength 40 Gbit/sec CS-RZ multiplications are obtained from a supercontinuum (SC) source, which is directly fed by the 40 Gbit/sec CS-RZ signal. Since the SC is generated by the accumulation of frequency chirp due to nonlinear propagation in a normal dispersion fiber [13–15], no coherent degradation occurs and the relative phase between the adjacent pulses is conserved [16,17]. By slicing the obtained spectrum using an arrayed waveguide grating (AWG), simultaneous multiwavelength CS-RZ can be generated as shown in Figure 7.2(b). In addition to its simplicity, another advantage of this method is the precise control of the WDM channel spacing, which is strictly locked by the microwave mode-locking frequency of the source laser. When the center wavelength of the original source pulse for SC generation is locked to an ITU grid frequency, the resulting multiwavelength channels can be simultaneously locked to ITU grids [18]. The complex wavelength control of each WDM channel used in conventional DWDM systems is not necessary in this multiwavelength generation method. Thus, the proposed technique would be favorable for cost-effective, multichannel frequency standardization.

7.2.1.2   Large Capacity TDM/WDM Transmission System

Figure 7.3 shows a system configuration for the generation and transmission of a 3.24 Tbit/sec (81 WDM × 40 Gbit/sec) wavelength-band in CS-RZ format. A 10 GHz, 1.5 psec pulse train at 1530.33 nm was modulated with a 10 Gbit/sec data and optically multiplexed into a 40 Gbit/sec CS-RZ data stream by using an optical multiplexer with phase shifter. After being amplified, the multiplexed signal was launched into the super-continuum fiber (SCF) [14]. The SC generated by the 40 Gbit/sec CS-RZ signal was sliced and recombined by AWGs with a 100 GHz channel spacing to generate multiwavelength 40 Gbit/sec CS-RZ signals. The transmission lines were

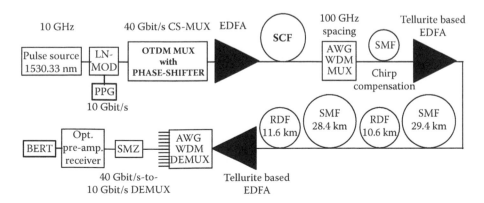

FIGURE 7.3 System configuration of simultaneous generation and transmission of 3.24 Tbit/s (81 WDM × 40 Gbit/sec) wavelength-band.

two pairs of a single mode dispersion fiber (SMF) and a reversed dispersion fiber (RDF). The total length was 80 km. Tellurite-based, erbium-doped fiber amplifiers (T-EDFAs) were used for signal amplification of 66 nm continuous band in the C– and L-bands [19,20]. Signals were spectrally demultiplexed by an 81 channel AWG with 100 GHz spacing (channel 1: 1535.04 nm–channel 81: 1600.60 nm). The resulting WDM DEMUX 40 Gbit/sec CS-RZ signal was temporally demultiplexed into the original 10 Gbit/sec stream by using a symmetric Mach-Zehnder all-optical switch [21].

Figure 7.4(a) shows the optical spectra of signals before transmission. Figures 7.4(b) through (d) show the measured optical spectra after 80-km transmission and spectral decomposition of channel 1, channel 41, and channel 81, respectively. It is clearly shown that optical carriers were suppressed even after 80-km transmission. For all measured channels, the bit error rate (BER) was less than 10^{-9}.

The scheme used a single SC source for WDM signal generation, which was directly pumped by an optically multiplexed CS-RZ signal. Transmission of simultaneously generated 3.24 Tbit/sec (81 WDM × 40 Gbit/sec) CS-RZ over 80-km dispersion compensated link was demonstrated. In this system, the spectral efficiency was 0.4 bit/sec/Hz.

7.2.2 Spectrally Efficient OCDMA/WDM Transmission System

OCDMA can be overlaid onto the existing WDM networks in order to expand the transmission capacity and add some additional functions in the networks. OCDMA/ WDM has proposed in the mid-1980s [22] and demonstrations of OCDMA/WDM transmission have been performed by several groups [23–28].

Spectral efficiency is considered to be a key factor in optical networks. One large motivating factor is the efficient expansion of use in the limited available spectrum. As one of the advantages of OCDMA, a highly spectrally efficient OCDMA/WDM hybrid system is proposed and demonstrated [27,28]. In order to achieve a highly spectral-efficient OCDMA system, the following three key technologies are applied: (1) quaternary phase shift keying (QPSK) optical encoding/decoding is used to increase the number of multiplexing channels in OCDM having a superior orthogonality between the codes, (2) optical time-gating is used to achieve a large processing gain in the optical decoding, and (3) optical hard thresholding is used to obtain further increase of the signal-to-noise ratio.

7.2.2.1 Key Technologies of the Spectrally Efficient OCDMA Transmission

Two schemes can encode the OCDMA signal: encoding in the frequency domain and in the time domain. In the time-domain encoding an optical short pulse, having much higher frequency spectrum than the data bandwidth, is spread over one bit timeframe of T by the optical encoding. The sequence of the phase of N chip pulses in one bit represents an optical code sequence. Matched filtering in the optical domain is a basis of the dispreading. The impulse response of matched filter of the

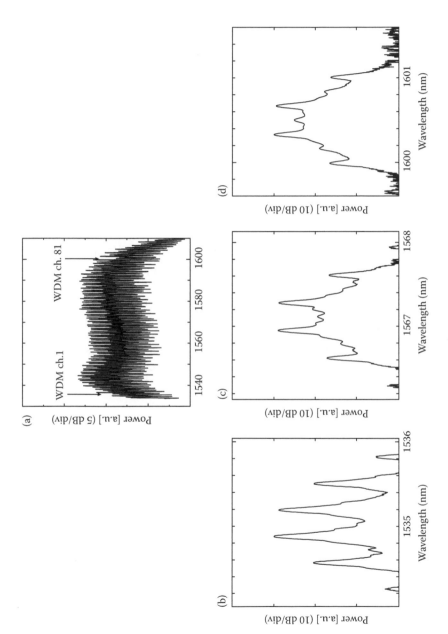

FIGURE 7.4 Measured optical spectra of signals (a) before transmission and after 80-km transmission and WDM DEMUX of (b) channel 1, (c) channel 41, and (d) channel 81.

decoder side $h_d(t)$ along with its Fourier spectrum $H_d(\omega)$ is the complex conjugate of optical encoder, and it is given by

$$H_d(\omega) = H_e(\omega)^* e^{-j\omega t_0} \tag{7.1}$$

$$h_d(t) = h_e(t_0 - t) \tag{7.2}$$

where the impulse response of the optical encoder is $h_e(t)$, and its Fourier spectrum is $H_e(\omega)$. The output of matched filter is expressed by the convolution of the impulse responses of the encoder and the decoder as

$$
\begin{aligned}
Output &= \int_{-\infty}^{\infty} H_e(\omega) H_d(\omega) e^{j\omega t} df \\
&= \int_{-\infty}^{\infty} |H_e(\omega)|^2 \, e^{j\omega(t-t_0)} df \\
&= \int_{-\infty}^{\infty} h_e(t') h_e(t' - t + t_0) dt \\
&= \Psi(t - t_0)
\end{aligned}
\tag{7.3}
$$

where $\Psi(t)$ represents the autocorrelation function of the input optical code $h_e(t)$ [29]. The matched filtering response can be realized by time reversing the in/output of the optical encoder. When the optical codes between the encoder and decoder match, the decoded time-despread signal reconstructs the original short pulse as an auto-correlation. While on the other hand, unmatched codes remain spread over one bit time flame of T after the decoding as a cross-correlation waveform.

Figure 7.5 shows the operating principle of the QPSK-code OCDMA system with optical time-gating and optical hard thresholding [27,28]. The time spread QPSK pulses

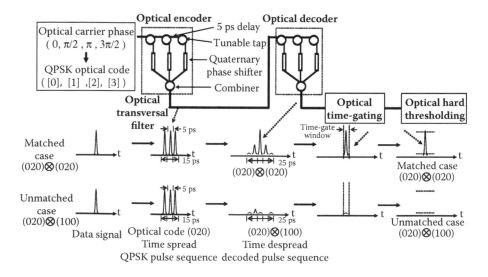

FIGURE 7.5 The operating principle of QPSK code-OCDM with optical time-gating and optical hard thresholding.

are used as the optical codes and optical transversal filters are used as optical encoders and decoders. The transversal filter consists of tunable taps, 5 psec delay lines, programmable quaternary optical phase shifters, and a combiner. An optical code sequence of 3-chip QPSK pulses with a chip interval of 5 psec is generated. The QPSK-encoded signal is time-despread by the decoder at the receiver side. The QPSK-decoded output shows the correlation waveform having a 5-pulse sequence in the time domain. When the receiver's optical code is matched with the transmitter's optical code, an auto-correlation waveform, characterized by a sharp central peak, is observed. On the other hand, in the case of unmatched codes, the cross-correlation waveform is formed. One bit of the despread signal occupies a 25 psec timeframe. Therefore, 40 Gbit/sec QPSK optical encoding/decoding is possible. The impulse response of the optical encoding has a frequency periodicity of 200 GHz, due to the 5 psec chip interval of the code [30]. As a result, simultaneous multiwavelength encoding with 200 GHz WDM channel spacing can be achieved using a single optical encoder when a broadband coherent optical pulse, such as a super-continuum pulse, is used as a chip pulse [24].

In coherent OCDMA systems, interference noise is a severe problem. Although optical matched filtering is the optimum beat-noise limiting receiver, it is only ideal in the absence of inter-symbol interference [31]. By introducing optical time gating, the interference noise outside the time-gate window can be rejected. Time gating can realize a strict sense of the processing gain in OCDMA systems [29]. As shown in Figure 7.5, the time-gate opens the time-window of $\Delta\tau$ at the repetition rate of $1/T$ to extract only the mainlobe of the auto-correlation. The time-gate samples only the data-bearing fraction of each bit, thereby rejecting the sidelobes of auto-correlation and the interference codes that fall outside the gate interval. This operation is equivalent to the narrow-bandpass filtering in the frequency domain CDMA. Placing the optical time gate before the photodetector has a substantial advantage because the requirement for the detector bandwidth is significantly relaxed to the bit rate of $1/T$. If we place an electrical time-gate after the photodetector, the detector bandwidth needs the chip rate of $1/\Delta\tau$. When the signal powers of the channels are equal, the signal-to-interference power at the receiver is given by

$$SNR = \frac{S}{(K-1)S+\sigma^2} = \frac{1}{(K-1)+\sigma^2/S} \tag{7.4}$$

where K is the number of channels, S is the signal power, and σ^2 is the variance of additive noise power including the ASE noise of the optical amplifier, shot noise, and thermal noise. The bit energy-to-interference density ratio E_b/N_0 [32] after time-gating is obtained by multiplying S in the numerator by T and by multiplying the noise by the time-window $\Delta\tau$ as

$$\frac{E_b}{N_0} = \frac{ST}{(K-1)S\Delta\tau+\sigma^2\Delta\tau} = \frac{T/\Delta\tau}{(K-1)+\sigma^2} \tag{7.5}$$

Compared with Equation 7.4, the bit energy-to-interference density ratio is enhanced by $T/\Delta\tau$. Therefore, the process gain of $T/\Delta\tau$ is can be achieved by introducing the optical time-gating. For ultrafast operation, a 100 m long nonlinear dispersion-shifted fiber (HNLF) based nonlinear optical loop mirror (NOLM) was used [33,34]. The control signal consisted of a pulse train with 1.5 psec pulses at a repetition rate of 10 GHz, and hence the optical time gating window ranged from 1.5 psec to 1.8 psec for all WDM channels.

Even after optical time-gating, the chances are that there still remains interference noise inside the time-gate window. As shown in Figure 7.5, we reduce this interference noise by introducing optical hard thresholding [27,28]. Optical hard thresholding is achieved by using the nonlinear transmission response of a second NOLM [35,36]. The second NOLM is an optical self-switching device, which acts as a pulse shaper by setting the proper threshold level. It reflects lower intensity signals and only transmits the higher intensity signal by limiting its intensity to an appropriate level. By adjusting the power of the input signal, interference noise inside the auto-correlation mainlobe after optical time gating is suppressed in both "0" and "1" bits. Therefore, the power variations in the eye diagrams of the received signal were reduced, which improved bit-error rate characteristics. For the ultra-high speed and wideband operation, the device length and hence the group delay must be reduced. For this purpose, we use a shorter length, 50 m, HNL-DSF as a hard thresholding NOLM. The group delay of the WDM wavelength range (1553–1564 nm) is less than 200 fsec, resulting in ultra-high speed operation as well as wideband operation.

7.2.2.2 Spectrally Efficient OCDMA/WDM Transmission System

Figure 7.6 shows the system configuration of a 6.4 Tbit/sec OCDMA/WDM (4 OCDM $\times$ 40 WDM $\times$ 40 Gbit/sec) system with 1.6 bit/sec/Hz spectral efficiency [27,28]. A 10 Gbit/sec signal at 1532 nm with 1.5 psec pulse width was four times multiplexed to create a 40 Gbit/sec data stream. The signal was launched into a SCF [14] to generate 40 Gbit/sec SC spectrum, which covers the entire C-band. The generated 40 Gbit/sec SC signal was linearly polarized by a polarizer and split into eight paths, each serving as the light source for simultaneous multiwavelength optical encoding in the QPSK optical encoders. Two groups of polarized signals, each having four different optical encoded signals (optical code #1–4 and #5–8), were generated. These two groups of encoded signals were polarization multiplexed. Because each group had a 160 Gbit/sec (4 OCDM $\times$ 40 Gbit/sec) signal with a WDM channel spacing of 200 GHz, the spectral efficiency of the multiplexed signal was 1.6 bit/sec/Hz in total. The transmission line was composed of two spans of 10.6 km RDF and 29 km SMF. After 80-km transmission, the signal was split into two paths and each path was WDM demultiplexed by a 20-channel AWG with a 200 GHz channel spacing. The pass band wavelengths of the two AWGs were separated by 100 GHz (odd WDM channels; channel 1: 1532.7 nm–channel 39: 1563.1 nm, even WDM channels; channel 2: 1533.5 nm–channel 40: 1563.9 nm). After WDM

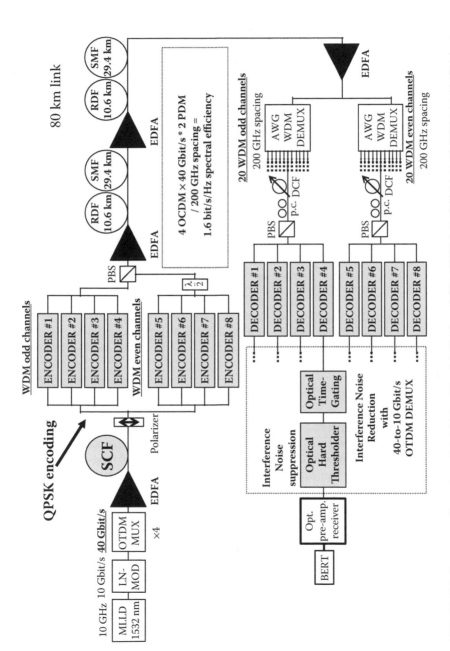

FIGURE 7.6 System configuration of 1.6 bit/sec/Hz, 6.4 Tbit/sec OCDM/WDM (4 OCDM × 40 WDM × 40 Gbit/sec) transmission system.

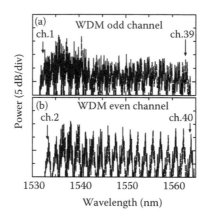

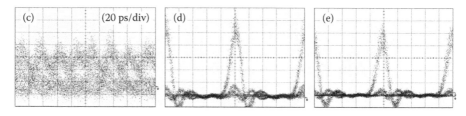

FIGURE 7.7 Optical spectra of (a) odd WDM channels, and (b) even WDM channels after WDM and polarization demultiplexing. Also shown are eye diagrams of optical code #5 in WDM channel 2 (c) after optical decoding, (d) after optical time-gating, and (e) after optical hard thresholding.

DEMUX, residual dispersion was compensated and the signal was polarization demultiplexed. The signal was then decoded by using the OCDMA decoder. Optical time-gating was performed at 10-GHz repetition rate. The optical time-gating rejects the interference noise outside the gating window while demultiplexing the 40 Gbit/sec OTDM signal back to the original 10 Gbit/sec data. Optical hard thresholding was achieved by using the second NOLM.

Figure 7.7(a) and (b) show odd WDM channels and even WDM channels, respectively, after WDM and polarization demultiplexing. Figure 7.7(c) shows the eye diagram of WDM channel 2 after optical decoding using optical code #5. The interference noise from the other three unmatched codes severely distorts the signal to noise ratio. Figure 7.7(d) shows the signal after optical time-gating. It can be seen that interference noise outside the time-gate window was rejected. In addition, as shown in Figure 7.7(e), by introducing optical hard thresholding, interference noise inside the time-gate window was greatly reduced both on bits "0" and "1," resulting in a clear eye opening. For all measured 160 channel × 40 Gbit/sec signals, the BERs were less than 10^{-9}. A 6.4 Tbit/sec OCDMA/WDM (4 OCDM × 40 WDM × 40 Gbit/sec) system with a spectral efficiency of 1.6 bit/sec/Hz was achieved.

7.3 PHOTONIC GATEWAY: MULTIPLEXING FORMAT CONVERSION

The increasing demand for bandwidth has led to a shift in the telecommunications industry toward IP-centric networks. The first generation of photonic networks dealt with bandwidth expansion issues. As the next generation of photonic networks is emerging, the main focus has turned toward dynamic allocation of bandwidth based on demand. In the scenario of IP growth, IP over photonic networks will perform routing and switching in the optical layer [37]. The objective of IP over photonic networks is to provide packet switching at rates much higher than electronic circuits can support. In the current IP networks, electronic routers and gateways play key roles for packet data transfer. On the one hand, routers forward packets between networks in which the forwarding decisions are based on network layer information and routing tables. Photonic routers have been proposed for ultrafast photonic networking [38,39]. On the other hand, electronic gateways interconnect heterogeneous networks that use different protocols. Unlike conventional electronic gateways, "photonic gateways" can provide multiplexing format transparency to photonic network nodes. As shown in Figure 7.8(a), photonic gateways perform data rate conversions in the optical layer [40]. The photonic gateways also convert multiplexing formats between TDM, WDM, and OCDMA as shown in Figure 7.8(b).

In this section, multiplexing format conversions of both TDM-to-WDM-to-TDM and OCDMA-to-WDM-to-OCDMA are explained. The demonstrations are based upon ultrafast photonic processing in both time and frequency domain. A potential for ultra-high-speed operation, as well as large scalability, distinguishes these conversion schemes from those in the electronic domain.

7.3.1 PHOTONIC GATEWAY: BI-DIRECTIONAL TDM-WDM CONVERSIONS

The multiplexing conversions from TDM to WDM have been demonstrated by using cross-phase modulation of SOAs in a Mach-Zehnder interferometer [41], a NOLM [42], four-wave mixing (FWM) in an optical fiber [43,44], FWM in an SOA [45]. On the other hand, the multiplexing conversions from WDM to TDM have been demonstrated by using a cross-gain compression in a SOA [46] and a NOLM [47]. The bidirectional conversion and reconversion have been demonstrated in [48–50].

Figure 7.9 illustrates the operating principle of multiplexing conversion between TDM and WDM [50]. The proposed photonic conversion scheme is based on ultrafast photonic processing both in the time and frequency domains; that is, optical time-gating and time-shifting in the time domain, and SC generation followed by spectrum-slicing in the frequency domain. Four 10 Gbit/sec OTDM signals are multiplexed to form a 40 Gbit/sec OTDM user i being positioned in time slot T_i ($i = 1,2,3,4$). After super-continuum generation, each multiplexed signal yields a multiwavelength signal in the same time slot, as shown in Figure 7.9(b). The generated SC is spectrally sliced to create WDM channels with center wavelengths of λ_i ($i = 1,2,3,4$) as shown in Figure 7.9(c). Each WDM signal is then time-shifted so that the time position T_i

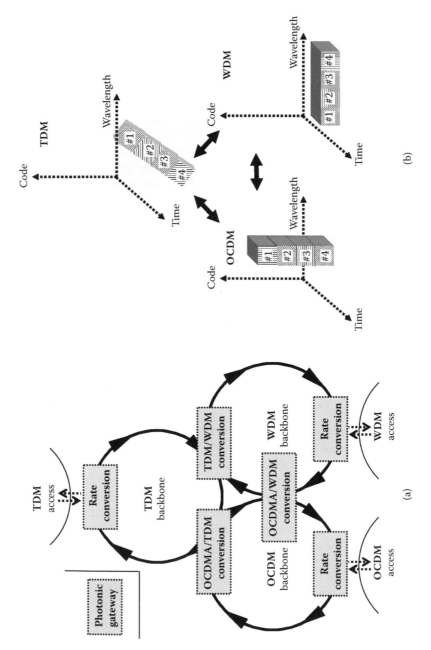

FIGURE 7.8 (a) Photonic gateways in photonic networks and (b) multiplexing format conversions.

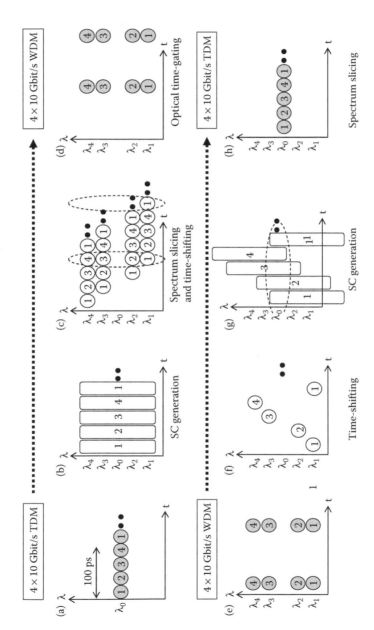

FIGURE 7.9 Operational principle of photonic conversion and reconversion of OTDM-to-WDM-to-OTDM.

at λ_i aligns within the same timeframe as shown in Figure 7.9(c). Finally, they are optically time-gated at 10 GHz repetition rate. Thus, each 40 Gbit/sec TDM data with time slot T_i is converted into 4×10 Gbit/sec WDM data of wavelength λ_t as shown in Figure 7.9(d). For the WDM-to-TDM conversion, the four WDM signals are time-shifted to position the λ_i WDM signal in the T_i time-frame, as shown in Figure 7.9(f). The multiplexed 4×10 Gbit/sec WDM signals are used to generate SCs as shown in Figure 7.9(g). By spectrum-slicing the obtained SC at the original wavelength of λ_0, the 4×10 Gbit/sec WDM signals are simultaneously converted into a 40 Gbit/sec TDM signal as shown in Figure 7.9(h).

Figure 7.10 shows the system configuration of a 40 Gbit/sec TDM-to-WDM-to-TDM conversion [50]. In the OTDM-to-WDM conversion, 40 Gbit/sec data at λ_0 (= 1553.9 nm) is generated by multiplexing four 10 Gbit/sec signals. The SC spectrum is generated in SCF#1 [14]. An AWG with channel spacing of 350 GHz is used for 4 WDM (λ_1: 1549.7 nm–λ_4: 1558.2 nm) demultiplexing and multiplexing. The time positions of the four WDM channels at λ_1–λ_4 are individually aligned using optical delay lines as shown in Figure 7.9(c). The time-gating at the repetition rate of 10 GHz was achieved using a semiconductor saturable absorber (SA), driven by a 10 GHz repetition rate, 2 psec width pulse train at 1532 nm [51,52]. The time-gated signal was WDM demultiplexed using the 4-channel, 350 GHz spacing AWG. Thus, 4×10 Gbit/sec OTDM-to-WDM was accomplished. The WDM-to-TDM reconversion then takes place. After time positioning of the four WDM channels at λ_1-λ_4 using optical delay lines, 4×10 Gbit/sec WDM signals were multiplexed using a star coupler. The SC spectrum is generated in SCF#2. The generated multiwavelength 40 Gbit/sec SC was spectrum-sliced at λ_0 (= 1553.9 nm) using a 3-nm optical filter to convert into a 40 Gbit/sec TDM signal [53]. Thus, 4×10 Gbit/sec WDM-to-TDM is completed.

The measured optical spectrum and the corresponding temporal waveform of the 40 Gbit/sec OTDM at the point indicated in Figure 7.10 are shown in Figure 7.11(a) and (b), respectively. Figure 7.11(c) and (d) show the converted 4×10 Gbit/sec WDM signals in the frequency domain and time domain, respectively. The experimental results of the WDM-to-TDM conversion, performed in series, are shown in Figure 7.11(e) and (f). The 40 Gbit/sec TDM at λ_0 is converted to 10 Gbit/sec WDM channels at λ_1–λ_4 and converted back to a 40 Gbit/sec OTDM at λ_0 with clear eye openings. All of the measured BERs of the four converted 10 Gbit/sec WDM channels, and four reconverted 10 Gbit/sec TDM channels were measured to be less than 10^{-9}.

7.3.2 PHOTONIC GATEWAY: BIDIRECTIONAL WDM-OCDMA CONVERSIONS

Figure 7.12 shows the system configuration for a 40 Gbit/sec (4×10 Gbit/sec) OCDMA-to-WDM conversion, where format conversion is performed after 80-km transmission of the OCDMA signals [54]. A 10 Gbit/sec optical pulse train at λ_0 (= 1542.8 nm) is split into four paths and each is optically encoded. The optical transversal filters described in Section 7.2 are used as optical encoders and decoders. An 8 chip-BPSK pulse code sequence with a chip interval of 5 psec is used as the optical code. The four different optical signals (OCDMA code #1–#4) are combined and the 4 OCDMA 10 Gbit/sec signals are transmitted over a dispersion-compensated link. After 80 km of

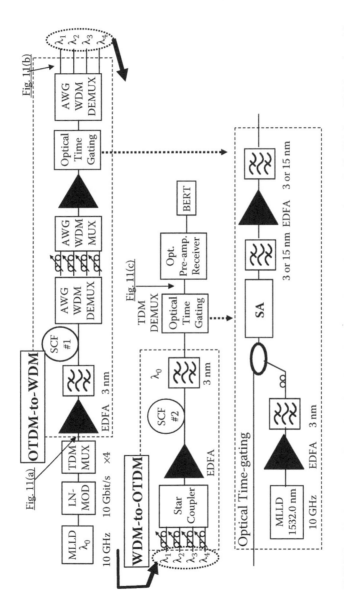

FIGURE 7.10 System configuration for photonic conversion and reconversion of 40 Gbit/sec TDM-to-4 × 10 Gbit/sec WDM-to-40 Gbit/sec TDM in series.

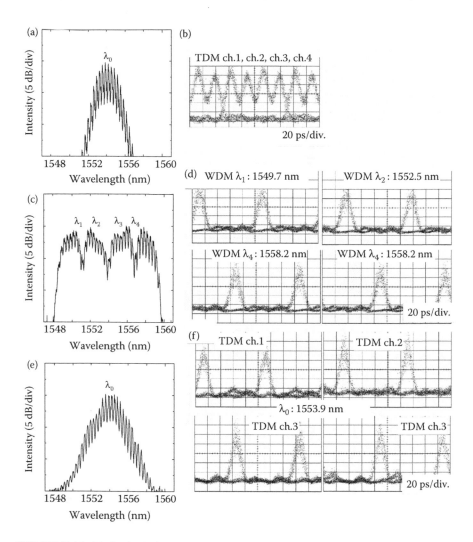

FIGURE 7.11 (a) Optical spectrum and (b) eye diagram of original 40 Gbit/sec TDM, (c) optical spectra and (d) eye diagram of converted 4 × 10 Gbit/sec WDM, and (e) optical spectrum and (f) eye diagram of reconverted 40 Gbit/sec TDM.

transmission, the traffic is split into four channels and each channel is converted from OCDMA-to-WDM. Conversion is realized by optical decoding, followed by wavelength conversion. The four optical signals are individually decoded by using optical decoders followed by optical time-gating. Optical time-gating at 10 GHz is performed by using a 100-m-long, HNL-DSF based NOLM switch [34]. Wavelength conversion is achieved by SC generation followed by spectrum-slicing [50]: the optically decoded 10 Gbit/sec signal is used as the seed source for the SCF for SC generation and the SC spectrum is spectrum-sliced by an AWG. A 4-channel AWG with 350 GHz channel spacing is used [(WDM channel 1: 1538.6 nm(λ_1)–channel

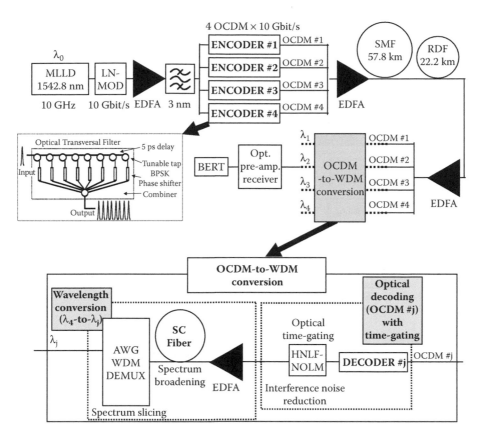

FIGURE 7.12 System configuration for 40 Gbit/sec (4 × 10 Gbit/sec) OCDMA-to-WDM conversion along with 80-km OCDM transmission.

4: 1546.9 nm(λ_4)]. As a result, OCDM #j signal is converted to WDM #j signal of λ_j ($j = 1$–4).

Figure 7.13(a) shows the optical spectra of the four 10 Gbit/sec OCDMA signals measured at the transmission side. Figure 7.13(b) shows the optical spectra of 4 WDM × 10 Gbit/sec after spectrum slicing at each WDM channel wavelength. These results show the four different OCDMA channels converted to four different WDM channels. The measured BER performances of OCDMA-to-WDM conversions are less than 10^{-9}.

Figure 7.14 shows the system configuration of 4 × 10 Gbit/sec WDM-to-OCDMA gateway, where format conversion is done after an 80-km transmission of the WDM signals. At the transmission side, 4 WDM signals are generated by spectral slicing of a SC spectrum. The SC spectrum was obtained by propagation of a 10 Gbit/sec signal at 1535.0 nm through a nonlinear fiber [14]. The demultiplexing and multiplexing of the WDM signals is done using 4-channel AWGs (λ_1:1538.6 nm–λ_4:1546.9 nm). After an 80-km transmission, the four WDM signals are wavelength demultiplexed using

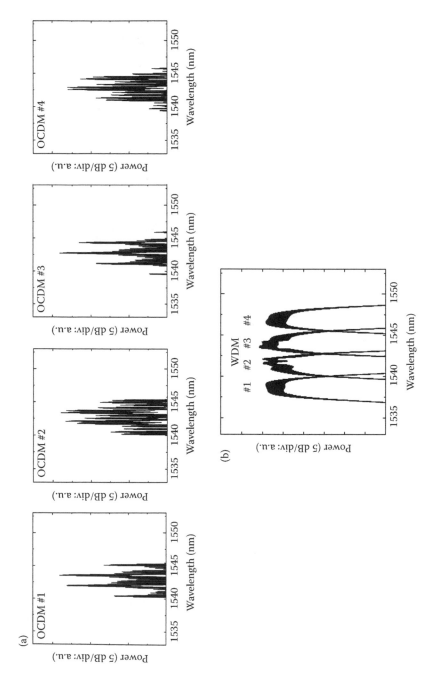

FIGURE 7.13 (a) Optical spectra of 4 OCDMA × 10 Gbit/sec signals at λ_0, (b) optical spectra of 4 WDM × 10 Gbit/sec after OCDMA-to-WDM conversion.

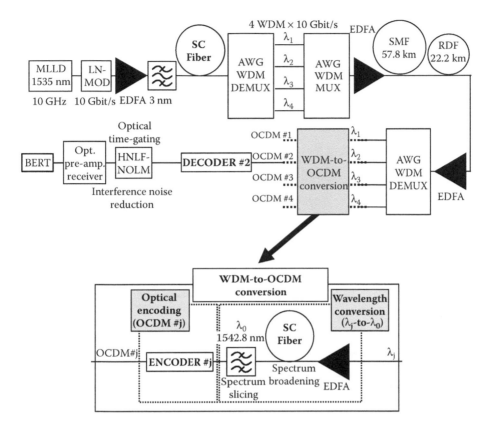

FIGURE 7.14 System configuration for 40 Gbit/sec (4 × 10 Gbit/sec) WDM-to-OCDMA conversion along with 80-km WDM transmission.

the same 4-channel AWG. Each WDM channel is WDM-to-OCDMA converted by wavelength conversion to λ_0 followed by optical encoding. Wavelength conversion is performed by the WDM #j (j = 1–4) signal induced SC along with spectrum slicing at λ_0(= 1542.8 nm) using a 3-nm optical filter. The j-th channel (OCDMA #j) was encoded by an optical encoder. As a result, WDM #j signal of λ_j is converted to OCDMA #j signal of λ_0 (j = 1–4).

Figure 7.15(a) shows the optical spectra of 4 WDM signals before transmission, while Figure 7.15(b) shows the optical spectra of four converted 10 Gbit/sec OCDMA signals. The BER performances of the converted OCDMA signals, measured after optical decoding followed by optical time-gating, were less than 10^{-9}.

7.4 OCDMA/WDM VIRTUAL OPTICAL PATH CROSS CONNECT

The OCDMA technique would be favorably applied not only to multiple access networks but also to path networks. The optical code path (OCP), defined as the logical path determined by the optical code (OC), has been proposed within a concept

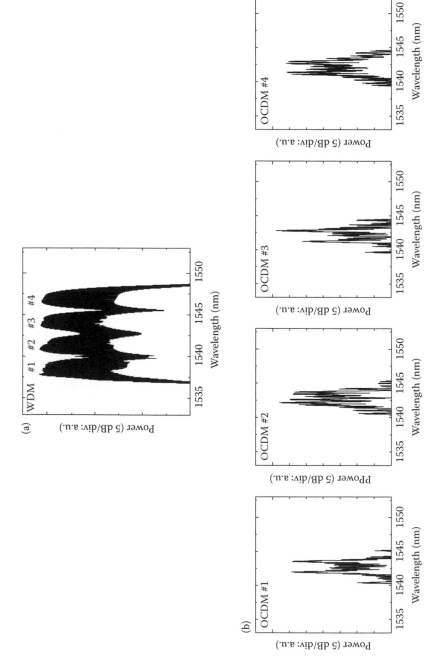

FIGURE 7.15 (a) Optical spectra of 4 WDM × 10 Gbit/sec signals and (b) four OCDMA signals at λ_0 after WDM-to-OCDMA conversion.

of OCDMA networks [55,56]. OCDMA can be overlaid onto the existing WDM path networks. The introduction of OCs provides soft capacity of networks and saves network resources. In future hybrid OCDMA/WDM networks, flexible OCs and wavelength conversion will be key technologies for establishing optical paths. In this section, we present the concept of the OCDMA/WDM virtual optical path network. A simultaneous OC and wavelength convertible node is obtained by SC generation in the operational bandwidth of 8.05 THz. An OC and wavelength convertible OCDMA/WDM virtual optical path network with a total link length of 180 km with four network nodes is also demonstrated [56].

7.4.1 OCDMA/WDM Virtual Optical Path Network

In OCDMA/WDM path networks, there are two approaches to OC and wavelength path assignment: with or without the OC and wavelength conversions. As shown in Figure 7.16(a), in a wavelength path (WP) network, a wavelength is assigned along the whole optical path, that is, the optical path is identified by a wavelength [57]. Similarly, in OCDMA, as shown in Figure 7.16(b), an OC is assigned along the entire optical path, that is the optical path is identified by an optical code. In both cases, to establish six optical paths requires six wavelengths or six optical codes, respectively. In a separate approach, referred to as a virtual optical code path/virtual wavelength path (VOCP/VWP), optical path provisioning is based upon OC and wavelength conversion. In this case, as shown in Figure 7.16(c), the OC and wavelength are allocated on a link by link basis. For example, to establish Optical path #3, OC2 at $\lambda 2$ is converted to OC1 at $\lambda 1$ at Node B. In the VOCP/VWP case, to establish the same six optical paths as in Figure 7.16(a) and (b), there need be only two wavelengths and two optical codes. As a result, the introduction of VOCP/VWP potentially solves the OC and wavelength path assignment problems, which may limit the network and optical path expansion. To maintain the scalability and reconfigurability of OCDMA/WDM path networks, simultaneous OC and wavelength conversions are key technologies, analogous to the roles of wavelength conversion for VWP networks and OC conversion for VOCP networks.

7.4.2 OCDMA/WDM Virtual Optical Path Network Cross Connect

Figure 7.17(a) shows the system configuration of an OCDMA/WDM path network. For simplicity, only the data transports from Node A to Node C by optical path 1 and from Node A to Node E by optical path 2 and by optical path 3 in Figure 16(c) are demonstrated. Three types of OC and wavelength conversions must be done at Node B. For the data transport from A to C by path 1, a $\lambda 1$-OC1 to $\lambda 1$-OC2 conversion (OC conversion only) is needed. For the data transport from A to E by path 2, a $\lambda 1$-OC2 to $\lambda 2$-OC2 conversion (wavelength conversion only) is needed, and finally from A to E by path 3 a $\lambda 2$-OC2 to $\lambda 1$-OC1 conversion (simultaneous OC and wavelength conversions) is needed. Each network node was linked by a nonzero dispersion-shifted fiber with dispersion compensation. The total length is 180 km: the distance is 80 km from Node A to Node B, 50 km from Node B to

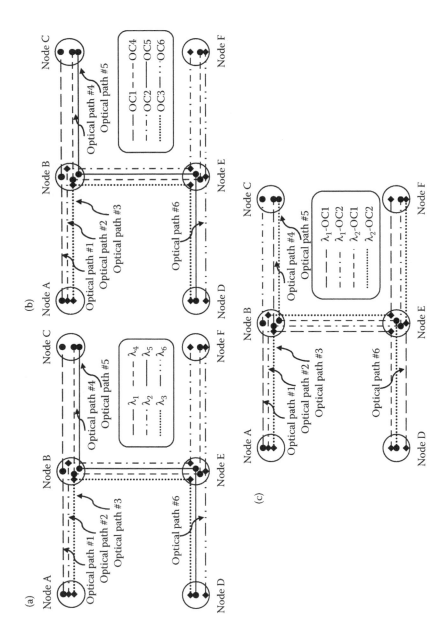

FIGURE 7.16 Optical paths in (a) WP network, (b) OCP network, and (c) VOCP/VWP network.

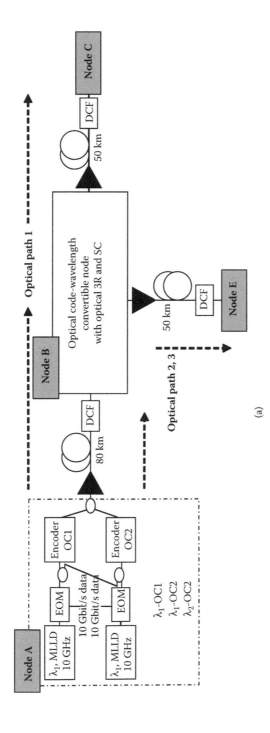

FIGURE 7.17 System configuration of (a) VOCP/VWP network by optical code and wavelength conversion with four network nodes, (b) OC and wavelength convertible Node B, and (c) receiving node of Node C and Node E.

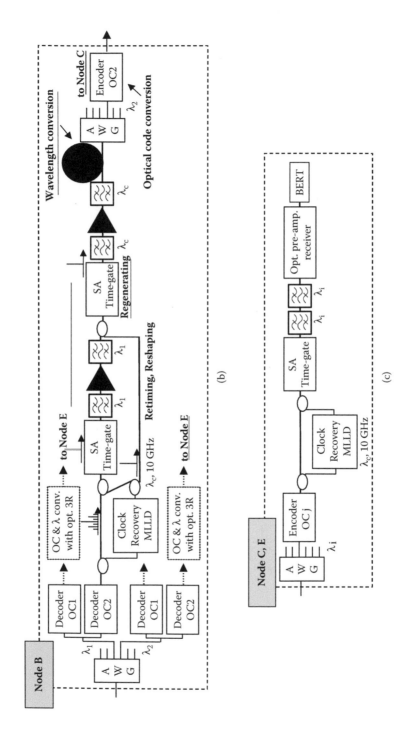

(b)

(c)

FIGURE 7.17 (Continued).

Node C, and 50 km from Node B to Node E. At Node A, $\lambda1$-OC1, $\lambda1$-OC2, and $\lambda2$–OC2 are generated and multiplexed at 10 Gbit/sec at 1549.7 nm ($\lambda1$) and 1552.5 nm ($\lambda2$). An 8-chip BPSK pulse sequence optical code with a chip interval of 5 psec is generated. OC1 is chosen to be [00000000] and OC2 to be [$0\pi0\pi\pi0\pi0$]. As shown in Figure 7.17(b), at Node B, received signals were decoded, 3R regenerated, wavelength converted by SC generation, and OC converted. First, the received signals are wavelength demultiplexed using an AWG with channel spacing of 350 GHz and decoded by optical decoders. Decoded signals are divided into two optical paths. One path leads to an injection locked mode-locked laser diode for 10-GHz clock recovery [51,52]. The clock recovery generates high SNR, negligible excess timing jitter, and coherent optical pulses at λc of 1555 nm. The recovered clock pulses are themselves divided into two paths. One path directs the clock pulses to be used as pump pulses for the SA time-gating device [51,52]. After time-gating, decoded signals are, in turn, used for pump pulses for the second SA time-gating to gate the clock pulse train. By controlling the optical time gate ON/OFF using decoded signals, the recovered clock pulses are modulated with the signal data. Thus, all-optical 3R is obtained. For wavelength conversion, the propagation of the 3R regenerated signals at λc in a nonlinear fiber generates a SC. After spectrum-slicing using AWG and selection of the targeted wavelength, the data is optically encoded with the targeted OC. Consequently, simultaneous OC and wavelength conversion is obtained. Figure 7.17(c) shows the schematic of the detection of the OC and wavelength converted signals at Node C and Node E after subsequent 50-km transmission. We confirmed that 24 error-free channels with total operational bandwidth of 8.05 THz, wavelength and OC conversions were demonstrated at Node B. Figure 7.18 shows the measured BERs of back-to-back, after 80-km transmission, after OC and wavelength conversion with optical 3R at Node B, and after subsequent 50-km

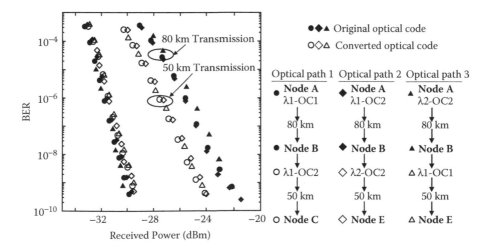

FIGURE 7.18 Measured BERs: back-to-back, after 80-km transmission, after OC and wavelength conversion with optical 3R, and after subsequent 50-km transmission.

transmission to Node C or Node E. The comparison of BERs back-to-back with the BERs OC and wavelength conversions at Node B shows an almost power penalty-free operation. An efficient OC, wavelength conversions, and 3R are demonstrated. In this experiment, an OCDMA/WDM virtual path network, along with OC and wavelength convertible signal transport on a 4-node network over a total link length of 180 km is demonstrated.

7.5 SUMMARY

In this chapter, system demonstrations of hybrid multiplexing techniques in photonic networks were reviewed. The demonstrated applications are the following: (1) hybrid multiplexing transmission systems, (2) bidirectional multiplexing format conversions, and (3) hybrid multiplexing techniques in optical path networks. Considering its unique attributes, OCDMA is a promising multiplexing technique available for hybrid multiplexing in order to enhance the flexibility of future photonic networks.

ACKNOWLEDGMENTS

The author wishes to thank professor Kenichi Kitayama of Osaka University for collaborating on OCDMA projects and for many helpful discussions, professor Takeshi Ozeki of Sophia University for the collaborating on TDM/WDM projects and discussions. The author also would like to thank professor Erich P. Ippen of Massachusetts Institute of Technology for his valuable suggestions and encouragements.

REFERENCES

[1]   Dixon, R. C. (1994). Spread spectrum systems with commercial application. New York: Wiley-Interscience.

[2]   Dixon, R. (1975).Why spread spectrum. *IEEE Commun. Soc. Mag.*13(4):21–25.

[3]   Scholtz, R. (1977). The spread spectrum concept. *IEEE Trans. Commun.* 25(8):748–755.

[4]   Pickholtz, R., Schilling, D., Milstein, L. (1982). Theory of spread spectrum communications-A tutorial. *IEEE Trans. Commun.* 30(5)855–884.

[5]   Utlaut, W. (1978). Spread spectrum: Principles and possible applications to spectrum utilization and allocation. *IEEE Commun. Soc. Mag.* 16(5):21–31.

[6]   Sust, M. Code division multiple access for commercial communication. In: Review of Radio Science 1992–1994, pp. 155–179, International Union of Radio Science (URSI).

[7]   Marom, E. (1978). Optical delay line matched filter. *IEEE Trans. Cir. Sys.* 25(6)360–364.

[8]   Desurvire, E., Bayart, D., Desthieux, B., Bigo, S. (2002). *Erbium-doped fiber amplifiers, device and system development.* New York: Wiley-Interscience, Chapter 7.

[9]   Winzer, P. J., Essiambre, R. (2003). Advanced optical molulation formats. In: Proc. the 29th European Conference on Optical Communication (ECOC 2003), Paper Th2.6.1.

[10] Sotobayashi, H., Konishi, A., Chujo, W., Ozeki, T. (2002). Wavelength-band generation and transmission of 3.24-Tbit/s (81-channel WDM40-Gbit/s) carrier-suppressed

return-to-zero format by use of a single supercontinuum source for frequency standardization. *OSA J. Opt. Soc. Am. B.* 19(11):2803–2809.

[11] Miyamoto, Y., Hirano, A., Sano, A., Toba, H., Murata, K., Mitomi, O. (1999). 320 bit/s
(8 × 40 Gbit/s) WDM transmission over 376-km zero-dispersion-flattened line with
120-km repeater spacing using carrier-suppressed return-to-zero pulse format. Tech.
Digest of Optical Amplifiers and their Applications (OAA '99), ODP4, PdP4-1–PdP4-4.

[12] Kobayashi, Y., Kinjo, K., Ishida, K., Sugihara, T., Kajiya, S., Suzuki, N., Shimizu,
K. (2000). A comparison among pure-RZ, CS-RZ and SSB-RZ format, in 1 Tbit/s
(50 × 20 Gbit/s, 0.4nm spacing) WDM transmission over 4,000 km. Tech. Digest of
26th European Conference on Optical Communication (ECOC 2000), PDP 1.7.

[13] Alfano, Robert R., ed., (1989). The Supercontinuum Laser Source. New York:
Springer Verlag.

[14] Sotobayashi, H., Kitayama, K. (1998). 325 nm bandwidth supercontinuum generation
at 10 Gbit/s using dispersion-flattened and non-decreasing normal dispersion fibre
with pulse compression technique. *IEEE Electron. Lett.* 34(13):1336–1337.

[15] Takushima Y., Kikuchi, K. (1999). 10-GHz, over 20-channel multiwavelength pulse
source by slicing supercontinuum spectrum generated in normal-dispersion fiber.
IEEE Photon. Technol. Lett. (11):322–324.

[16] Sotobayashi H., Kitayama, K. (1999). Broadcast-and-select OCDM/WDM network
by using 10Gbit/s spectrum-sliced supercontinuum BPSK pulse code sequences.
IEEE Electron. Lett. 35(22):1966–1967.

[17] Sotobayashi, H., Kitayama, K. (1999). Observation of phase conservation in multiwavelength BPSK pulse sequence generation at 10 Gbit/s using spectrum-sliced supercontinuum in an optical fiber. *OSA Optics Letters* 24(24):1820–1822.

[18] Cundiff, S. T., Ye, J. (2003). Colloquium: Femtosecond optical frequency combs.
Rev. Mod. Phys. 75(1):325–342.

[19] Yamada, M., Mori, A., Kobayashi, K., Ono, H., Kanamori, T., Oikawa, K., Nishida,
Y., Ohishi, Y. (1998). Gain-flattened tellurite-based EDFA with a flat amplification
bandwidth of 76 nm. *IEEE Photon. Technol. Lett.* 10(9):1244–1246.

[20] Makino, T., Sotobayashi, H., Chujo, W. (2002). 1.5 Tbit/s (75 × 20 Gbit/s) DWDM
transmission using Er3+-doped tellurite fiber amplifiers with 63 nm continuous signal
band. *IEEE Electron. Lett.* 27(17):1555–1557.

[21] Nakamura, S., Ueno, Y., Tajima, K., Sasaki, J., Sugimoto, T., Kato, T., Shimoda, T.,
Itoh, M., Hatakeyama, H., Tamanuki, T., Sasaki, T. (2000). Demultiplexing of 168-
Gb/s data pulses with a hybrid-integrated symmetric Mach-Zehnder all-optical switch.
IEEE Photon. Technol. Let. 12(4):425–427.

[22] Vannucci, G. (1989). Combining frequency division multiplexing and code division
multiplexing for high capacity optical network. *IEEE Network* 3(2):21–30.

[23] Teh, P. C., Ibsen, M., Lee, J. H., Petropoulos, P., Richardson, D. J. (2001). A 4-
channel WDM/OCDMA system incorporating 255-chip, 320 Gchip/s quaternary
phase coding and decoding gratings. In: Proc. Optical Fiber Communication Conference (OFC2001), PDP 37 PD37-1–PD37-3.

[24] Sotobayashi, H., Chujo, W., Kitayama, K. (2001). 1.52 Tbit/s OCDM/WDM (4
OCDM × 19 WDM × 20 Gbit/s) transmission experiment. *IEEE Electron. Lett.*
37(11):700– 701.

[25] Scott, R. P., Cong, W., Li, K., Hernandez, V. J., Kolner, B. H., Heritage, J. P., Yoo,
S. J. B. (2004). Demonstration of an error-free 4 × 10 Gb/s multiuser SPECTS O-
CDMA network testbed. *IEEE Photon. Technol. Lett.* 16(9):2186–2188.

[26] Jiang, Z., Seo, D. S., Yang, S. D., Leaird, D. E., Roussev, R. V., Langrock, C., Fejer,
M. M., Weiner, A. M. (2005). Four-User, 2.5-Gb/s, spectrally coded OCDMA system

demonstration using low-power nonlinear processing. *IEEE/OSA J. Lightwave Technol.* 23(1):143–158.

[27]   Sotobayashi, H., Chujo W., Kitayama, K. (2002). 1.6-b/s/Hz 6.4-Tb/s QPSK-OCDM/WDM (4 OCDM × 40 WDM × 40 Gb/s) transmission experiment using optical hard thresholding. *IEEE Photon. Technol. Lett.* 14(4):555–557.

[28]   Sotobayashi, H., Chujo, W., Kitayama, K. (2004). Highly spectral efficient optical code division multiplexing transmission system (invited). *IEEE J. Select. Top. Quant. Electron.* 10(2):250–258.

[29]   Kitayama, K., Sotobayashi H., Wada, N. (1999). Optical code division multiplexing (OCDM) and its application to photonic networks. *IEICE Transactions on Fundamentals of Electronics, Communications and Computer Sciences.* E82-A(12):2616–2626.

[30]   Sotobayashi, H., Kitayama, K. (1999). Transfer response measurements of a programmable bipolar optical transversal filter by using the ASE noise of an EDFA. *IEEE Photon. Technol. Lett.* 11(7):871–873.

[31]   Winzer P., Essiambre, R. (2003). Optical receiver design trade-offs. In: Proc. Optical Fiber Communications Conf. 2003 (OFC2003), Paper ThG1.

[32]   Gilhousen, K. S., Jacobs, I. M., Padovani, R., Viterbi, A. J., Weaver, L. A., Wheatley, C. E. (1991). On the capacity of a cellular CDMA system. *IEEE Trans. Veh. Tech.* 40(2):303–312.

[33]   Sakamoto, T., Futami, F., Kikuchi, K., Takeda, S., Sugaya, Y., Watanabe, S. (2001). All-optical wavelength conversion of 500-fs pulse trains by using a nonlinear-optical loop mirror composed of a highly nonlinear DSF. *IEEE Photon. Technol. Lett.* 13(5):502–504.

[34]   Sotobayashi, H., Sawaguchi, C., Koyamada, Y., Chujo, W. (2002). Ultrafast walk-off free nonlinear optical loop mirror by a simplified configuration for 320 Gbit/s TDM signal demultiplexing. *OSA Opt. Lett.* 27(17):1555–1557.

[35]   Mamyshev, P. V. (1998). All-optical data regeneration based on self-phase modulation effect. In: Proc. the 24th European Conference on Optical Communication (ECOC 1998) 1:475–476.

[36]   Lee, J. H., Teh, P. C., Yusoff, Z., Ibsen, M., Belardi, W., Monro, T. M., Richardson, D. J. (2002). A holey fiber-based nonlinear thresholding device for optical CDMA receiver performance enhancement. *IEEE Photon. Technol. Lett.* 14(6):876–878.

[37]   Murata M., Kitayama, K. (2001). A perspective on photonic multiprotocol label switching. *IEEE Network* 15(4):56–63.

[38]   Kitayama, K., Wada, N., Sotobayashi, H. (2000). Architectural considerations for photonic IP router based upon optical code correlation. *IEEE/OSA J. Lightwave Technol.* 18(12):1834–1844.

[39]   Sotobayashi, H., Kitayama, K. (2000). Optical code based label swapping for photonic routing. *IEICE Trans. Commun.* E83-B, 10:2341–2347.

[40]   Sotobayashi, H., Kitayama, K., Chujo, W. (2001). 40 Gbit/s photonic packet compression and decompression by supercontinuum generation. *IEEE Electron. Lett.* 37(2):110–111.

[41]   Duelk, St. Fischer, M., Puleo, M., Girardi, R., Gamper, E., Vogt, W., Hunziker, W., Gini, E., Melchior, H. (1999). 40-Gb/s OTDM to 4 × 10 Gb/s WDM Conversion in Monolithic InP Mach-Zehnder Interferometer Module. *IEEE Photonic Technol. Lett.* 11(10):1262–1264.

[42]   Uchiyama, K., Takara, H., Morioka, T., Kawanishi, S., Saruwatari, M. (1996). 100 Gbit/s multiple-channel output all-optical demultiplexing based on TDM-WDM conversion in a nonlinear optical loop mirror. *IEE Electron. Lett.* (32):1989–1991.

[43]   Morioka, T., Kawanishi, S., Takara, H., Saruwatari, M. (1994). Multiple-output, 100 Gbit/s all-optical demultiplexer based on multichannel four-wave mixing pumped by a linearly-chirped square pulse. *IEE Electron. Lett.* 30(23):1959–1960.

[44]   Sotobayashi, H., Chujo, W., Ozeki, T. (2001). 80 Gbit/s simultaneous photonic demultiplexing based on OTDM-to-WDM conversion by four-wave mixing with a supercontinuum light source. *IEE Electron. Lett.* 37(10):640–642.

[45]   Summerfield, M. A., Lacey, J. P. R., Lowery, A. J., Tucker, R. S. (1994). All-optical TDM to WDM conversion in a semiconductor optical amplifier. *IEE Electron. Lett.* 30(3):255–256.

[46]   Norte D., Willner, A. E. (1996). Demonstration of all-optical data format transparent WDM-to-TDM network node with extinction ratio enhancement for reconfigurable WDM networks. *IEEE Photon. Technol. Lett.* (14):1170–1182.

[47]   Daza, M. R. H., Liu, H. F., Tsuchiya, M., Ogawa, Y., Kamiya, T. (1997). All-optical WDM-to-TDM conversion with total capacity of 33 Gb/s for WDM network links. *IEEE Selected Topics in Quantum Electronics* (3):1287–1294.

[48]   Norte D., Willner, A. E. (1996). All-optical data format conversions and reconversions between the wavelength and time domains for dynamically reconfigurable WDM networks. *J. Lightwave Technol.* (14):1170–1182.

[49]   Sotobayashi, H., Chujo, W., Ozeki, T. (2001). Bi-directional photonic conversion between 4 × 10 Gbit/s OTDM and WDM by optical time-gating wavelength interchange. Optical Fiber Communication Conference (OFC 2001), WM5, pp. WM5-1–WM5-3, Anaheim.

[50]   Sotobayashi, H., Kitayama, K., Chujo, W. (2002). Photonic gateway: TDM-to-WDM-to-TDM conversion and reconversion at 40 Gbit/s (4 channels 10 Gbits/s). *OSA J. Opt. Soc. Am. B.* 19(11):2810–2816.

[51]   Kurita, H., Ogura I., Yokoyama, H. (1998). Ultrafast all-optical signal processing with mode-locked semiconductor lasers. *IEICE Trans. on Electron.* E81-C(2):129–139.

[52]   Kurita, H., Hashimoto, Y., Ogura, I., Yamada, H., Yokoyama, H. (1999). All-optical 3R regeneration based on optical clock recovery with mode-locked LDs. 25th European Conference on Optical Communication (ECOC 1999), PD3-6:6–57.

[53]   Sotobayashi, H., Chujo, W., Ozeki, T. (2001). Wideband tunable wavelength conversion of 10 Gbit/s RZ signals by optical time-gating of highly chirped rectangular shape supercontinuum light source. *OSA Opt. Lett.* 26(17):1314–1316.

[54]   Sotobayashi, H., Chujo, W., Kitayama, K. (2002). Photonic gateway: Multiplexing format conversions of OCDM-to-WDM and WDM-to-OCDM at 40 Gbit/s (4 × 10 Gbit/s). *IEEE/OSA J. Lightwave Technol.* 20(12)2002–2008.

[55]   Kitayama, K. (1998). Code division multiplexing lightwave networks based upon optical code conversion. *IEEE J. Select. Areas Commun.* 16(9):1309–1319.

[56]   Sotobayashi, H., Chujo, W., Kitayama, K. (2002). Transparent virtual optical code/wavelength path network. *IEEE J. Select. Top. Quant. Electron.* 8(3):699–704.

[57]   Sato, K. (1996). *Advances in transport network technologies*, Section IV. Boston: Artech House Publisher.

8 Integration Technologies

Ivan Andonovic

CONTENTS

8.1 INTRODUCTION

8.1.1 BACKGROUND

The introduction of wavelength division multiplexing (WDM) into the existing tele-communications infrastructure represents the first serious deployment of optical networking in the evolution of the modern-day network. There is a growing need for increased capacity, driven by the end user's desire to access a range of high-speed data and video services stimulated by the explosion in Internet usage. Optical networking has a combination of characteristics that make it a strong candidate for satisfying these dynamic and unpredictable demands by providing a transparent physical layer amenable to easy upgrade while supporting the legacy electronic infrastructure.

Recent history shows that the capacity exhaust problems brought about by the significant adoption of the Internet was the fundamental economic driver for the deployment of WDM point-to-point transmission, the most basic form of increasing the throughput of a single strand of optical fiber. The economics of this rudimentary upgrade stage was relatively easy to determine and proved compelling; however, subsequent stages of the roll-out of higher level all-optical networking concepts ultimately resulting in true, flexible end-to-end optical path provisioning are not as clear, with economics proving a fundamental barrier. These latter scenarios require additional functionalities to be executed directly in the optical domain through optical add/drop and optical cross-connects.

Although it is acknowledged that future network evolution will comprise a more extensive optical layer, all indications are that the deployment of the advanced infrastructure has stalled with system integrators, from necessity, keen on low-cost system solutions to provision enhanced functionality. This naturally leads to a requirement for low-cost componentry and higher levels of integration in all dimensions; the range of optical functionalities will need to be manufactured in a scaleable way to meet volume and cost demands. It is also fair to observe that, of the range of implementation options, a layer utilizing optical code division multiple access (OCDMA) techniques is very demanding and will only come to pass through increased complexity in the optical domain. Such increased complexity in the optical domain must be supported by increasingly sophisticated electronic control, and as the data rates increase, there will be a need to closely integrate optical and electronic components.

8.1.2 INTEGRATION

The benefits of integration to subsystems, systems, and network manufacturers are unambiguous. Replacing a large number of discrete components means less components to test, easier assembly, and fewer in-stock components, which facilitates inventory management. These factors result in the reduction of product cost that is then passed down the value chain through the network operators to the end user who ultimately pays for services. In addition to the reduced capital expenditure brought about by higher levels of integration, operational, and maintenance costs are reduced. The fewer the number of elements, the less potential for malfunction with smaller form factors, with corresponding benefits for power consumption and space. These cost benefits fundamentally change the economics of new deployments or upgrades to the telecommunications hierarchy and open up new application domains such as the metropolitan and access layers.

The aim of this chapter is to summarize the candidate approaches that will potentially yield viable strategies to integration and in doing so spawn a range of opto-electronic modules enabling the cost-effective implementation of a number of all-optical networking principles. To aid in clarity and to set the framework, a definition of the term *integration* is important at the outset; it is also necessary to note the crucial role of electronics in any strategy. The electronics supporting any network, subsystem, line card, or indeed chip enhances performance and manages the limitations of the optical physical layer. This observation is especially true in OCDMA, where the elegance of the technique is sustained through the appropriate use of electronics despite the complexity of the optical implementation.

The term *integration* can be interpreted in a number of ways; for the purposes of this chapter, integration strategies are those that go some way to lowering the cost of OCDMA implementations while satisfying the demanding performance requirements. Given that photonics lags behind electronics by some 10 to 20 years, the adoption of the following approaches reflects an evolution driven by the maturity of the technology:

- The system integration of existing off-the-shelf components and devices yields a conventional telecom wiring rack or closet producing technology demonstrators. Figure 8.1 is a conceptual schematic showing the integration of a range of commercial-of-the-shelf (COTS) components to yield for example, a line card that can be rack mounted.

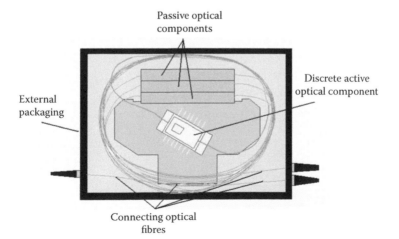

FIGURE 8.1 (Color Figure 8.1 follows page 240.) Conceptual schematic showing the module integration of a range of commercial-off-the-shelf (COTS) components to yield, for example, a line card that can be rack mounted.

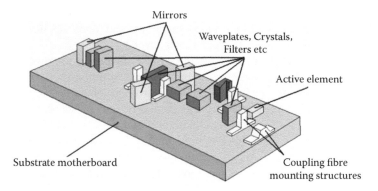

FIGURE 8.2 (Color Figure 8.2 follows page 240.) Conceptual schematic illustrating hybrid integration based on micro-optic platform.

- The utilization of hybrid integration platforms. Here there are two options worthy of consideration:
  - Micro-optic platforms based on a single planar mounting surface or "optical bench" akin to miniaturizing a complex opto-mechanical assembly comprising electronic and optical devices [1] (Figure 8.2).
  - Active semiconductor devices with passive waveguides that can support on a common micro-bench produced, for example, through micro-machining of silicon [2] (Figure 8.3).
- The monolithic integration of optical devices either in semiconductor (viz. InP) [3], silicon or glass, e.g., silica-on-silicon material systems [4] (Figure 8.4).
- The realization of opto-electronic integrated circuits that (OEICs) comprise optical and electronic functions on one material system [5].

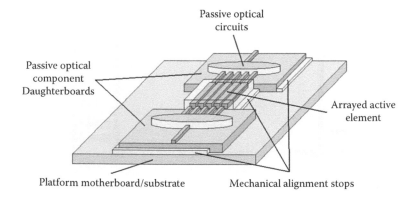

FIGURE 8.3 (Color Figure 8.3 follows page 240.) Conceptual schematic illustrating hybrid integration based on the assembly of active semiconductor devices with passive waveguides supported on a common micro-bench.

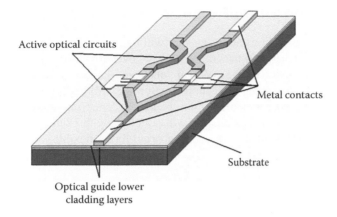

Active optical circuits

Metal contacts

Substrate

Optical guide lower
cladding layers

FIGURE 8.4 (Color Figure 8.4 follows page 240.) Conceptual schematic illustrating the monolithic integration of active and passive optical devices in a single material system.

The goal of any long-term realization strategy is the manufacture of a large number of elementary device functions in as few processes as possible and preferably in a single material system; the ideal solution produces highly functional chips (volume manufactured) in one process and on a single material. Optical integration is different, however, from electronics in two very important aspects. First, it will never achieve the density of VLSI semiconductor electronics since components are large compared with the wavelength (akin to microwave devices) routinely characterized by an "asymmetric" geometry, i.e., centimeters "long" and of the order of a wavelength wide and deep. Second, any ultimate integration strategy will incorporate the processing of not only optical but also electronic signals. Here compatibility of the processing steps in the production of the functions in the two domains poses real challenges.

It is worth noting that there are great hopes for a new generation of devices based on photonic crystals with circuits of similar density to VLSI claimed as potentially viable [6]. This technology is in its infancy and no real examples of extensive integration have occurred; therefore, it will not be covered in this chapter.

There is also a distinction to be made with respect to serial and parallel integration. In the former, several functions integrated into one module act on an optical signal as it propagates across the platform, for example, a CW laser in tandem with a modulator. In the latter, one function integrated into one module operates in parallel on multiple optical signals (e.g., wavelengths), leading to the integration of many photo-detectors in an array. Of course there is also the option of using both integration dimensions to yield, for example, an array of laser transmitters each able to be modulated independently by in-line modulators. The degree of integration has a bearing on the economic benefits that accrue. To date parallel integration has brought more significant cost savings than serial and the combination of both has, potentially, an even greater impact in the near term. This conclusion is intimately tied up with the maturity and complexity of the processes used in manufacture, the degree of automation that can be brought to bear, the yield, and the associated economies of scale.

The remaining sections of this chapter will be organized as follows. Each one of the candidate approaches to integration will be covered in more detail in Section 8.2, supported by examples of the types of functions realized with each approach in terms of optical networking in general. Section 8.3 will highlight the work done to date in identifying the geometries of the encoder/decoder function specific to a range of OCDMA approaches. The chapter will conclude with examples of the range of advanced functions needed in the goal of providing a flexible, high capacity cost-effective OCDMA network.

8.2 INTEGRATION STRATEGIES

8.2.1 TECHNOLOGY DEMONSTRATORS

Photonics has emerged as an important enabling technology primarily due to the giant strides that have been made in off-the-shelf components from advances in WDM optical networking. A set of predominately discrete photonic components like lasers, detectors, modulators, and filters capable of maintaining specifications under stringent operating environments are available on the marketplace. A number of networks exist that conform to well-defined standards for both systems and components; these standards will continue to evolve especially if more advanced network implementations are deployed efficiently. Standards are an important catalyst for any future rollout of optical networking, enabling equipment suppliers to develop products offering users interchangeable, cost-effective solutions. Qualification is also a barrier to implementation, placing additional demands on the development of any new technology that needs to exhibit long-term stability (> 25 years). It must also be fit for the deployed environment and be able to be maintained by the end user. Consideration must be given to the combination of size, modularity, and location of equipment such that it is readily accessible, achieving significant cost savings.

WDM point-to-point systems are relatively crude when compared to all-optical networking scenarios and to date the implementation costs associated with applications of advanced photonics are prohibitively high, thereby restricting deployment primarily to the transport layer of the network. Thus, at the outset, the natural approach to determining the potential benefits of this technology in new architectures and applications is through the utilization of the base technology derived from these initial commercial systems. The result is a technology demonstrator akin to a conventional telecom wiring rack or closet made with stock items allowing a diligent determination of the feasibility of any approach. This stage is best viewed as a means of simplifying the design by reducing the hardware to a small number of standard elements, the additional benefit accrued being the integration of a number of key functions such as management and control. In some applications it may prove advantageous to deploy these sorts of systems if significant cost/performance enhancements are proven. It is fair to conclude that OCDMA technology is at this demonstrator stage.

8.2.2 Hybrid Integration

Hybridization is a well-established route to capturing the benefits of integration through the combination of passive and active optical and electronic components that have been separately optimized using their respective processes, yielding multifunctional circuits otherwise difficult to manufacture in a single fabrication process. The price paid for the freedom to optimize the performance of each component independently through the tailoring of the properties of specific material systems is the need for high precision alignment processes in the assembly stages. From a commercial viewpoint, a hybridization platform allows the standardization of fabrication and assembly of multiple products using the same manufacturing line providing modules for a range of applications; it is with this standardization in mind that any integration platform must be designed.

The advantage this approach enjoys has its roots in the maturity of photonics components and the critical cost/performance balance that necessitates different base materials to implement different optical functions. While the adoption of monolithic integration is attractive, for the foreseeable future it can be argued that optimum subsystem functionalities are best realized through the utilization of separate chips.

It is easiest to classify the two main approaches to realizing hybrid integration platforms through the manner that the mounting of optical components is achieved and the method of interconnection between these components. These high level considerations define the enabling technologies required and the resultant restrictions on the techniques to attach the components, which in turn define the overall size and configuration of the module.

8.2.2.1 Micro-Optic Platforms

Micro-optic hybridization platforms have as their base a single planar mounting surface or optical bench, akin to a complex, compact optical-mechanical-electrical assembly.

The key enabling technology is a family of miniature mounting pads—both with and without the capability for active alignment—and electrical interconnects that allow the automated assembly of electronic, opto-electronic, and optical components. Flexures can be produced based on the LIGA process (Lithographie, Galvanofromung, Abformung [7]), which can then be soldered to the motherboard providing individual mounts for a set of elements in an optical train. In-module optical connectivity is achieved through free-space, which does not preclude the use of waveguide components in the train. As long as the resultant package conforms to industry-standard sizes and is within the mechanical stability limits placed on the height of the optical axis, the platform allows substantial flexibility in the component suite it can harness. The platforms are compatible to a wide range of chip sizes including MEMS, high power active devices, and passive devices of various form factors.

The platform assembly process must be largely automated—preferably software-driven—if products are to be assembled quickly while meeting performance and reliability requirements. Assembly begins with the mounting of the various elements using pick-and-place and vision recognition/alignment technologies, facilitating the attachment of individual elements to a micro-optical bench to relatively high accuracy. Achieving submicron tolerances requires an additional step in the process, active alignment. The use of robotic techniques plays an important role here, locating each (active) element and adjusting its micro-mechanical structure to execute postassembly "final" alignment using, for example, the coupling efficiency as feedback to the process. Using plastic deformation of the flexure, optimization is obtained by adjusting each element in turn.

The supporting optical bench must be mechanically rigid and exhibit high thermal conductivity. Thus although silicon has significant advantages with respect to fabrication technology, allowing the easy creation of additional features for alignment purposes (see Section 8.2.2.2), high conductivity ceramics such as aluminium nitride are preferred since they also provide a higher degree of flatness.

Module functionalities are implemented with the aid of a toolbox of optical elements, effectively miniaturized versions of many commonly available components. Precision micro-lenses and micro-optical components that direct and manipulate the optical signal are critical in order to match the optical modes from different components in the train, thereby achieving low-loss transfer of the optical signal between them. Low cost micro-lenses provide the capability to mode match multiple waveguide devices, all with different mode profiles. In addition to passive and active waveguide elements, the platform is compatible with agile MEMS devices able to perform filtering or routing of signals. The resultant module may comprise many discrete elements supported on a single thermo-electric cooler in a standard 14-pin butterfly package.

The attachment of the optical fiber to the package is also a challenging requirement not just for this approach but all others. Different thermal expansion coefficients for the fiber and the surrounding materials have to be considered in the design of the structure for fiber alignment and attachment. Otherwise, temperature-dependent forces will create unwanted movement of the guiding components, causing a lowering of the coupling efficiency.

One powerful example of a module realized through this hybridization approach is the optical channel monitor viz. monitors each individual wavelength channel of a multiplex for the purposes of management and control of the transmission. This is a fairly complex subsystem comprising around twenty elements including several lenses, fixed filters, beam combiners, steering mirrors, a broadband LED source, a photo-detector, and a MEMS-based tunable filter [7]. This component has undergone a rigorous qualification procedure and has found application in commercial WDM systems. Other modules such as a laser-external modulator tandem or preamplified photo-detectors can be realized in a similar manner. Another important device family is the relatively inexpensive tunable transmitter using vertical cavity semiconductor lasers (VCSELs) within an external cavity formed by a miniature movable mirror [8].

8.2.2.2 Hybrid Waveguide Platforms

8.2.2.2.1 General

An alternative approach to hybrid integration assumes similar principles but uses waveguide components solely. Although glass has been considered, the material used almost exclusively as the supporting bench is silicon for a number of well-documented reasons. It exhibits good mechanical properties, good thermal conductivity, is able to be patterned easily through mature processing, and is the base substrate for a family of passive and active waveguide components. The latter feature has relevance both for hybrid devices as well as monolithic devices, as will be covered in Section 8.2.3.

The utilization of silicon as an optical bench has its roots as far back as the late 1980s when it became the cornerstone for commercial laser and detector packaging. Thermal heat sinking, electrical tracks/pads for interconnection/bonding, etched features such as trenches to support ball lenses for efficient coupling, and support for electrical components for impedance matching represented a powerful set of advantages that lowered the cost of discrete optical components. The acceptance of silicon micro-machined piece parts in standard optical products has promoted its use in higher levels of integration [9].

The earliest examples of enhanced hybrid integration utilized silicon submounts equipped with mechanical alignment features to ensure precise relative positioning of optical fibers with the active elements. V-grooves provided passive alignment perpendicular to the optical axis while the position of the active component was locked by mechanical stand-offs and lateral stops. The coupling efficiency relies on the ability to precisely cleave the active component, implemented at the photolithographic stage of manufacture through the formation of channels at the edge of the device to define the scribe lanes [10]. This method allows the edges of the scribe lanes to be determined with an accuracy of +/− 0.2 μm. The dimensions of these features are defined by well-controlled etching processes along individual crystal planes in which the cleaved active chip alignment against the silica stops ensures the correct relative position to the V-groove. Solder bumps make contact between the active device and the electrical tracks on the submount. The relative position of the bumps and wetting pads on the active device together with the surface tension of the solder provides a small mechanical force that helps drive the chip into physical contact with the alignment features. This approach can be extended to the assembly of ribbon fiber and active arrays [11].

In its simplest form the extended platform comprises a silicon base unit, passive waveguide structures (in some cases the base unit is the substrate of the passive waveguide structure), and active optical devices. The complexity of the module, however, is far greater on consideration of thermal management issues, mechanical integrity, and electrical interconnection (especially at high data rates). Thus although highly functional silicon substrates have been qualified for use in certain applications, there are massive challenges to be mastered with respect to next generation hybrid modules if they are to enjoy widespread deployment.

A variety of approaches to hybridisation have been developed:

- Active alignment with automated pick and place to assemble elements on a common platform [12]
- Active and passive components attached to a featured silicon micro-bench base [13,14]; alignment can be achieved through fiducial or mechanical features and/or solder wettable pads

- Using registration features on both the active and passive to align on a separate motherboard [15,16]
- Precision holes have been etched in passive waveguide circuitry to house active devices. This can be extended to a two-stage process aligning the active on a separate silicon daughterboard and subsequently aligning the combination on the waveguide motherboard [17]

Of these approaches, active alignment and automatic pick-and-place techniques integrated with machine vision support have evolved with respect to accuracy and have yielded a number of noteworthy hybrid modules with significant complexity [18]. This approach differs from the micro-optic platform described in Section 8.2.2.1 since it only comprises waveguide elements.

8.2.2.2.2 Mode Expansion

The alignment tolerances associated with all the elements are demanding, especially so as they accumulate with increased module complexity. To decrease the sensitivity to alignment of standard buried heterostructure (BH) active semiconductor designs, it is desirable to expand the mode on-chip, adapting the strongly confined mode in the semiconductor waveguide (typical dimension ~1 μm) to the much wider mode of a fiber (typical dimension ~10 μm) [19,20]. Without the ability to manipulate the mode size, it is almost impossible to achieve coupling efficiencies below 1 dB, especially in a semi-automated assembly process. The problem proves largely insurmountable if an array of optical devices has to be coupled.

Mode expansion is achieved through tapered active regions (in the lateral and/or vertical dimensions) that alter the on-chip optical mode as it propagates through the device. In the case of a modulator or optical amplifier (two facet devices) the incoming optical mode—usually from a standard single mode optical fiber—can either be coupled directly to a tapered active region or through a loose confinement passive guide that passes the signal on to the active region. In the latter case a taper accepts this input optical mode and reduces its radius, confining it to the active region. It is then expanded by the output taper and passed through another passive section to match the output waveguide, in so doing optimizing the transfer of power to the output waveguide while maintaining an efficient interaction in the active region.

To illustrate the power of on-chip expansion, Figure 8.5 depicts the scans of the far-fields for the following optical modes: (1) a single mode passive silica-on-silicon waveguide, near optimized to a mode of a single mode optical fiber; (2) the optical field of SMF28 optical fiber; (3) a mode-expanded semiconductor optical amplifier (SOA), designed to match the optical field of a standard single mode optical fiber (SMF28); and (4) a typical BH device. "Standard" non–mode expanded devices diverge very quickly with a full-width-half-maximum (FWHM) of ~40°, due to their tightly confined optical mode. In comparison, the mode-expanded waveguiding sections exhibit mode fields which diverge with a FWHM of ~7°, closely matched to the fields of the SMF28 and single mode silica-on-silicon waveguides. In the latter instance, the modes are more compatible and coupling losses will be smaller as a result.

Characterization of the coupling tolerances between the passive waveguides and standard and mode expanded SOAs is summarized in Figure 8.6. These tolerance

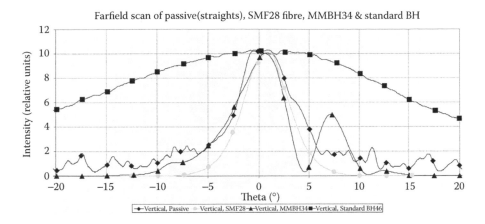

FIGURE 8.5 Far-field scans of the modes of (a) a SMF28 single mode fiber, (b) a single mode silica-on-silicon waveguide, (c) a mode-expanded SOA, and (d) a BH SOA.

maps of each facet show that both the mode matched SOA and silica-on-silicon passive waveguide exhibit similar tolerances with 1 dB loss occurring after a displacement of ±2 μm, confirming that the modes have comparable coupling properties. Without mode matching a much tighter tolerance to transverse displacements results, with 1 dB coupling loss after only ±0.5 μm misalignment because the two optical modes diverge at different rates. On-chip mode expansion can therefore be viewed as a powerful example of monolithic integration as it permits more tolerant interfacing in a number of integration techniques.

8.2.2.2.3 Platform Types

The Silica-on-Terraced-Silicon (STS) platform consisting of silica-on-silicon waveguides on a silicon substrate with terraced regions [21] acting as alignment planes for the active devices is one of the more mature approaches to achieving a hybrid waveguide platform.

Such a platform is manufactured by first creating an under-cladding layer by flame hydrolysis deposition (FHD) on a silicon substrate with terraced features realized through micro-machining. Silica-on-silicon passive waveguide circuits are then formed by a standard FHD/reactive ion-etching (RIE) step. The waveguide is partly etched by RIE of the waveguide layers, exposing the surface of the silicon terraces on which the active devices are to be attached. Differential etch rates between the silica and silicon facilitate the control of the etch depth such that alignment between the waveguide core and the mounted active element is within ~1 μm. In effect, the vertical reference plane is the surface of the silicon, which also provides the heat sinking requirements. An electrode pattern is then created on the bottom of the assembly region. Specific silicon motherboards are required for different module functionalities [22].

The assembly process relies on a number of alignment procedures. Making use of fiducial marks on the active and registration features on the motherboard provides alignment executed with the aid of an IR viewer [23]. The active device is flipped

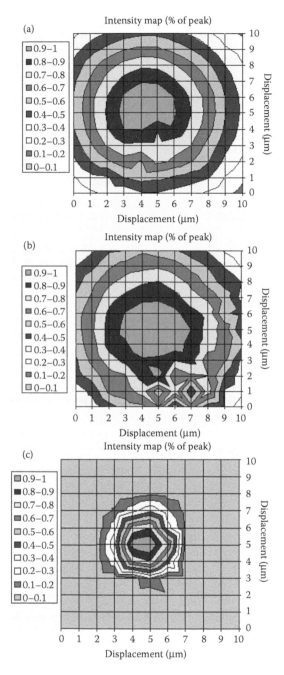

FIGURE 8.6 (Color Figure 8.6 follows page 240.) Coupling tolerance contour maps for (a) silica-on-silicon-SMF28 coupling, (b) mode-matched SOA-SMF28 coupling, and (c) non-mode matched SOA-SMF28 coupling.

onto the terrace and a derivative of the technique can incorporate active optimization by monitoring the power through the module. Alternatively, precision-etched active elements with wettable solder bumps in tandem with lateral and vertical mechanical stops can meet the alignment requirements for high-speed modules [24], which can also be extended to incorporate optical fiber attachment [25].

Use of registration features on all of the elements of the hybrid platform has resulted in a number of assembly strategies. It extends the early concepts of using silicon submounts for active assembly but now incorporates passive waveguide elements also. The platform is characterized by a fairly large silicon motherboard with appropriate registration and/or alignment features and registration features on the passive and/or active elements. Wafer bow is particularly problematic for relatively large devices.

A number of highly functional modules have been demonstrated with the above techniques. Transmitters [26–28], receivers [29], transceivers [30,31], external cavity lasers [32,33], differential receiver modules [34], optical switch matrices [25,35,36], wavelength selectors [37–39], all-optical wavelength converters [40], and multiwavelength sources [41] are a set of impressive proof-of-principle modules exhibiting competitive performance. This set of examples are varied and go some way to validating the versatility of the hybrid approach.

A recent example of this type of approach uses a structured silicon micro-bench as a base unit, formed by multistep etch procedures to define alignment features [42]. Reference height plinths were created by protecting sections of the wafer surface from all subsequent etching stages. All etch depths are related to this single surface, reference plane. The plinths also act as height-referenced mounts for attaching planar lightwave circuits (PLCs) during the final assembly stages. Solder well structures on a central cross spar and shallow sections connected to these wells, which define tracking, were etched. Gold was later deposited into these tracks in order to create electrical pathways connecting the active devices to the external packaging pins.

The PLCs were commercial silica-on-silicon waveguides fabricated using FHD and RIE techniques. Lower cladding and core layers were deposited in a controlled manner, producing layers varying in thickness by ±0.3 µm. The final thick upper cladding layer was then deposited to within ±2 µm to complete the guide and protect the waveguide core from the environment. Reference height rails were then etched on each PLC chip by removing all of the deposited layers (to within ±0.05 µm) along the length of the chip edge, leaving the substrate plane exposed. One waveguide end facet is polished to 8° in preparation for bonding to ribbon fiber. Processing of the other chip end requires a dicing process to create a facet that matches the angle at which the SOA is to be mounted onto the micro-bench. The angling of the PLC chip facet allows a butt joint with high coupling efficiencies to be made between all the active and passive components while limiting reflections that could be coupled back into the waveguides causing instabilities in the active region.

Assembly of the hybrid module requires the mode-expanded active device to be flip-chip mounted to the solder well structures. The active element is then powered temporarily to actively align the PLC components. Placement of the reference height rails on the passive waveguide chip onto the etched plinths of the micro-bench

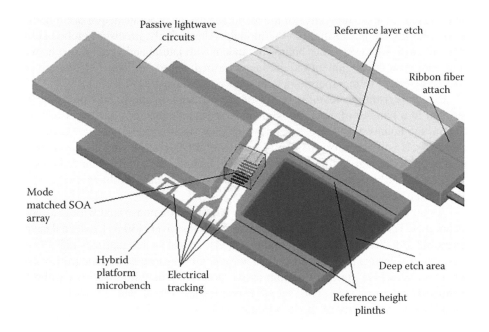

FIGURE 8.7 (Color Figure 8.7 follows page 240.) Hybrid assembly schematic [42].

eliminates the height, pitch, and roll degrees of freedom from the alignment process, thereby easing the coupling procedure. Yaw, longitudinal, and horizontal adjustments of the PLC chip are then used to optimize coupling to the active waveguide. Once optimally placed, the passive waveguide chip is bonded to the micro-bench and the process may be repeated to align and secure a second PLC chip. The hybrid module is subsequently secured into a 20-pin butterfly package, comprising a thermo-electric cooler (TEC). Permanent electrical connections to the external package pins can then be made from the electrical tracking on the micro-bench via gold wire bonds (Figure 8.7).

The assembly process introduces coupling misalignments due to the combined etch/growth height tolerances of the individual components. In the case above, the height of the guide core of the PLCs is known to within ± 0.5 μm from the reference height rails of the chip. Micro-bench solder-well etch depth can vary by ± 1.25 μm, and the center of the mode field of active component is known to within a tolerance of ± 0.5 μm. At worst, these tolerances add directly resulting in a tolerance-limited vertical offset of ± 2.25 μm between the peak of the mode fields of both the PLC and the active. Estimates made from coupling tolerance measurements indicate that this level of misalignment corresponds to an excess loss of ~1.5 dB. In most cases the misalignment will not be as severe; the RMS tolerance of ± 1.43 μm imposes an assembly associated, vertical coupling penalty of ~0.7 dB. The most significant use of this platform has been in the production of an all-optical wavelength converter through mode expanded semiconductor optical amplifiers in a passive Mach-Zehnder interferometer [42].

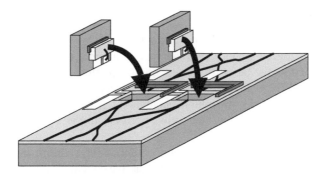

FIGURE 8.8 Precision-cleaved SOAs passively flip-chipped on to micro-machined silicon "daughterboards" that are subsequently passively assembled onto a machined silica on silicon waveguide motherboard. (Courtesy of G. Maxwell [44].)

Another approach adopts a slightly different strategy using the top (cladding) surface of the passive waveguide circuit as the vertical reference plane for alignment, simplifying the processing required for the waveguide motherboard since it is made via a standard manufacturing process [43]. Here the motherboard is the silica-on-silicon chip with precise machined holes to accommodate the active components — mode expanded and precision cleaved — supported on a silicon "daughterboard," a micro-machined silicon submount containing the features required to facilitate the passive assembly of the active devices with the lateral and vertical positioning assembly structures of the motherboard. Lateral positioning is achieved through mechanical end stops formed in photoresist onto the top of the cladding (Figure 8.8). This approach separates the alignment function between the waveguide motherboard and silicon daughterboard, which can therefore be optimized independently allowing higher yields. For high performance and yield, precision-cleaving active elements, control of cladding thickness, cladding uniformity, and wafer bow are crucial.

Electrical connections are made via transmission lines on the top surface of the motherboard and connecting pads on the daughterboard. The daughterboard also acts as an efficient heatsink for the active regions. Optical fiber pigtailing is achieved through a passive procedure facilitated through a V-grooved silicon submount capable of supporting single strands or arrays. The depth of the groove ensures that the mode height after assembly is equal to the distance between the center of the waveguide core and the top of the cladding on the motherboard, thereby providing the necessary vertical alignment. Etching "arrow head" vertical faces into this submount and subsequent alignment to features patterned on the cladding of the motherboard provides the lateral registration.

The most effective illustration of the capability of this hybrid assembly platform was the demonstration of penalty-free, 2R regeneration at 40 Gbit/sec using a nonlinear Mach Zehnder interferometer comprising mode expanded SOAs. Total fiber-waveguide-SOA coupling losses as low as 4.4 dB were recorded using this entirely passive alignment process [44].

8.2.3 MONOLITHIC INTEGRATION

Monolithic integration of a suite of optical functions in one substrate is a natural step of any integration technology roadmap. A number of material systems have been investigated as the basis for the integration platform; only Indium phosphide (InP) and silicon/silica-on-silicon (SOS) will be considered here. In addition, this summary will be confined to monolithic photonic integration with separate electronics; the final degree of integration-monolithic opto-electronic ICs (OEICs)—will be covered in Section 8.2.4.

Indium phosphide has a very strong potential for creating integrated optical devices, allowing the combination of different active optical elements that accomplish the generation, manipulation, and detection of the signal with passive waveguides that distribute the signal. In comparison, the other monolithic integration platform cited — silicon/SOS based — provides limited active functionality but nevertheless yields practical modules of note.

Indium phosphide monolithic integration potentially enables all of the possible functions required by future networks; together with its ternary (InGaAs) and qua-ternary (InGaAsP) derivatives the material yields devices that can operate across the extended low loss range of optical fibers (1.3 µm to 1.6 µm). The refractive indices of InP (and its derivatives) are relatively high compared to other optical materials, this translates into a size advantage since bends can be made much sharper. This in fact is one of the major advantages of this material system, yielding small devices with die sizes typically less than 5 mm. The energy band-gap is also closer to that of light energy and, as a consequence, electro-optical effects are stronger than in other materials, with corresponding benefits with respect to either shorter interaction regions or lower drive voltages. However, a disadvantage of the tightly confined on-chip modal fields is that it becomes more difficult to couple to optical fiber. As discussed in Section 8.2.2.2.2, this can be overcome by means of taper structures at the interfaces.

A number of complex circuits have been demonstrated with significant results. Integrated SOA Mach-Zehnder interferometer all-optical wavelength converter modules operating at data rates of 40 Gbit/sec [45] and compact wavelength selectors (AWGs and SOA arrays) [46] are two modules that illustrate the capability of the platform. The former device [47,48] can be realized with passive-active regions or fully active regions, the entire structure consisting of active (SOA) waveguides [49]. For the wavelength selector, active-passive integration is mandatory for low loss AWG operation. Two schemes are possible to realize this device: an active-passive butt-joining technique using self-aligned buried ridge stripes [50] or multistep epi-taxy processes [51].

Other significant examples are multiwavelength sources and substantial optical switch fabrics. The WDM drive to create an array of wavelength selectable lasers has resulted in the integration of DFBs with a multiplexer or power combiner. Complex tunable laser structures have also been produced based on classical mul-tisection DFB or DBR structures. Alternatively, integrating an array of SOAs with a multiplexer into a Fabry-Perot cavity provides simultaneously, a number of wave-lengths each able to be modulated independently. The wavelengths are automatically

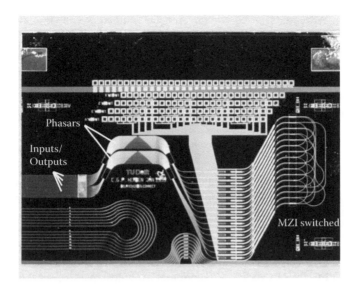

FIGURE 8.9 InP-based WDM cross-connect with dilated switches. (Courtesy of M. Smit [3].)

selected by the multiplexer passband [52]. Even more compact devices using a ring configuration have been developed [53]. Optical cross-connects for WDM-based networks are exemplars of the kind of complexity that can be achieved using InP. Utilizing electro-optical switches based on a pin-type waveguide operated in reverse bias combine very low power dissipation with a high switching speed [54]. Figure 8.9 depicts a dilated 2×2 cross-connect that has AWG mux/de-mux integrated on the same substrate [3].

These circuits can be created either by low pressure or selective regrowth technologies [55] or through postgrowth processing [56]. As has often been cited, the monolithic approach is relatively immature, suffers from poor yield, and often the integration of active elements results in reduced performance of at least one of the devices.

In addition to its more established role in electronics, silicon is also a material able to provide functionality in the optical domain [57]. For the wavelengths of interest, silicon is transparent and generally does not interact with light, making it a good candidate for interconnection between active components. The introduction of additional materials such as silicon dioxide, standard electronic dopants, and silicon-germanium (SiGe) alloys enhance electro-optic interactions, allowing the creation of active devices such as intensity modulators and photo-detectors. To date, however, efficient light sources have not been demonstrated in silicon. Again, a silicon waveguide ($n = 3.45$) system employs waveguide tapers to efficiently interface to optical fibers.

This versatile material system has yielded a number of useful optical functions [58]. A single mode, tunable external cavity laser has been realized by coupling a laser diode to a silicon-based waveguide Bragg grating. The lasing wavelength is selected by the grating and can be tuned through the thermo-optic effect (heating the grating). Waveguide-based silicon optical modulators can be realized through the thermo-optic effect, current injection (limited to ~20 MHz), or MOS capacitor-based geometries that potentially provide modulation bandwidths up to 10 GHz. Although silicon is an efficient photodetector < 1 μm, SiGe alloys are required to extend the responsivity out to wavelengths used for all-optical networking. There have been a range of passive components available commercially in silicon ranging from simple splitters through to large port count arrayed waveguide grating (AWG) filters with comparable performance to components produced in silica-on-silicon technologies. However, this family of components has not enjoyed significant traction in the marketplace.

Although covered in significant detail as an important element of hybrid waveguide integration, the silica-on-silicon waveguide system has a noted, albeit limited, capability in terms of monolithic integration. A number of commercial entities exist that offer a range of multifunctional modules based on this technology. Two well-understood deposition techniques are used in the manufacture of components ó flame hydrolysis deposition (FHD) and plasma enhanced chemical vapor deposition (PECVD). In addition to extremely low loss passive waveguide components, the material system exhibits photosensitivity with UV-induced refractive index changes permitting the production of gratings as well as limited activity through the thermo-optic effect. With suitable doping (erbium) it can potentially provide optical amplification although device efficiencies in high silica glasses still remain poor [59].

Modules produced in SOS waveguide systems include arrayed waveguide gratings (AWGs, sometimes defined as an example of highly parallel integration), thermo-optic (Mach Zehnder type) switches (TOSs), variable optical attenuators (VOAs), and other filtering functions such as interleavers. The platform can be extended to comprise any combination of these basic functions, e.g., AWG + TOS + VOA. Figure 8.10 demonstrates the extended monolithic capability of these systems

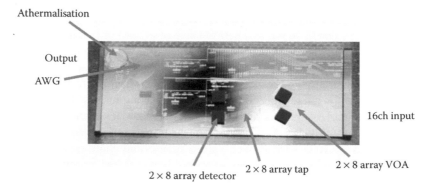

FIGURE 8.10 An example of a highly functional silica-on-silicon module exhibiting a number of basic functions integrated on one substrate (athermal operation). Also shown is the extended hybrid capability incorporating detectors developed by Alcatel Optronics in the U.K., which is now part of Gemfire Corp. (Photo courtesy of M. Hesketh of Gemfire Corp.)

but is further enhanced by the integration of monitoring photodiodes, which strictly falls under the hybrid classification as neither detectors nor indeed lasers can be created in this material. It is worth noting that the platform lends itself to hybridization of polymer waveguides as well as MEMS components.

8.2.4 OPTO-ELECTRONIC INTEGRATED CIRCUITS (OEICs)

Opto-electronic integrated circuits—the monolithic integration of electronics and optics—are the holy grail for all designers since they enable the range of functions as integrated building blocks facilitating complex network design, implementation, and maintenance. OEICs further reduce overall module size, power consumption, and unwanted electrical parasitics that compromise high speed operation.

Not only does InP hold the pole position as the material of choice for monolithic optical integration, but it is also beginning to become an important material for high-speed electronics. InP-based electronic devices such as heterojunction bipolar transistors (HBTs) and high electron mobility transistors (HEMTs) have surpassed the performance of devices created in competing material systems such as Gallium Arsenide (GaAs). InP devices have been shown to provide functions such as power amplifiers, low-noise amplifiers, mux/demux, clock data recovery, phase-locked loops, and laser drivers at bandwidths up to 100 GHz. Current development work on HBTs also indicates a strong long-term potential of combining optical elements with electrical drive circuitry [60].

The most likely applications for InP electronic devices as the technology matures are chip sets that support SONET-like 40 Gbit/sec transmission systems based on WDM, integrated (OEIC) receivers for standard 10 Gbit/sec optical communications systems promoting the expansion of the data rate in the metro layer, power amplifiers for mobile handsets, and MMICs for broadband wireless access. Combined, these application areas represent a substantial market and although a number of issues remain to be solved before large-scale manufacture can be carried out, InP devices will successfully penetrate a number of applications for which their performance is well matched.

Although integrated optical modules have not gained significant adoption in silicon, it should not be forgotten that specific applications may exist where the close integration of an optical component and an electronic circuit can provide a highly cost-effective system solution. It is a fact that in the electronic domain, silicon integrated circuits are the workhorse in all layers of the modern broadband infrastructure executing complex network intelligence functions. Although it is true that optical devices are more challenging to produce because material defects are more damaging to performance, it is still, nevertheless, a fact that with maturity this sector has acquired in high volume, high yield, and highly complex IC manufacture, it has the experience and know-how to apply the same rigorous methodology to the creation of a new breed of devices.

The severe disadvantages of elemental silicon for OEIC realization are its indirect band-gap and negligible electro-optic coefficient. However the introduction of strain through, for example, the growth of silicon germanium (SiGe) films on silicon improves the electronic properties and extends the capability of the material. Low-noise HBTs that are compatible with photo-detectors can be made, meaning that high-performance receivers become possible. Other band-gap engineering processes are also possible with the promise to deliver enhanced opportunities for OEIC realization [61,62].

8.3 ENCODING/DECODING FOR OCDMA SYSTEMS

8.3.1 GENERAL

It is not the intention in this section to detail the various approaches to implementing an OCDMA network. The detailed understanding and critical comparison between these approaches has been covered comprehensively in other chapters in this book, but, in order to illustrate the building blocks required for deployment of OCDMA systems, it is necessary to introduce the functionalities that need to be implemented directly in the optical domain for the range of alternatives. Hence, the integration technologies that best suit any module implementation will be highlighted.

In summary there are several generic coding strategies that can be adopted to create an OCDMA network:

- Incoherent encoding
- Coherent encoding
- Spectral encoding
- Hybrid techniques harnessing a combination of time, frequency, and spectral encoding

Spatial encoding has also been utilized (e.g., to transmit image data) but this area will not be considered.

8.3.2 ENCODING/DECODING ARCHITECTURES

The key function of any OCDMA system is that of encoding/decoding. Irrespective of the network architecture, e.g., star, bus, or mesh, a data signal is encoded at the transmitter and after distribution through the network, a decoder has to be capable of extracting the correct channel identified by a unique code—in the presence of other codes—in order for the user to recover the data. For full flexibility the encoder/decoder has to exhibit tunability in that it must be able to generate any of the optical code family used in the implementation. There are, of course, variations to full flexibility dependent on the connectivity requirements specified of the network, for example, broadcast and select may have a range of different transmitters (e.g., television channels) uniquely identified by a fixed code, but the receiver has to have the capability to decode all of the channels selectively.

8.3.2.1 Incoherent

With the incoherent approach, the encoding/decoding is based on the summation of powers. In its most basic implementation the approach uses long sparse time spreading codes that rely on the avoidance of pulses (chips); with these unipolar codes the phase, frequency, and spectral content are not important. Most simply, the encoder

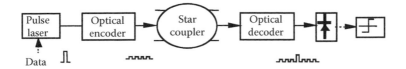

FIGURE 8.11 Schematic of an incoherent OCDMA network.

is a passive optical architecture that generates an impulse response that is a train of pulses (chips), which are delayed and scaled replicas of an input laser's pulses (modulated by the data). At the receiver, specific sequences are recognized by performing a correlation through a matched filter (Figure 8.11).

The most straightforward geometry is a 1:P power splitter/combiner (P is the number of "1"s in the code) with (fixed) optical delay lines (Figure 8.12), where the delays correspond to the time position of the chip within the code. Since each code requires a splitter/combiner pair, the main limitations of the geometry are power loss and the lack of code tunability. If the number of chips in a sequence of code weight P is large, the coding processes lead to a loss in total bit energy by a factor of P. The interrelationship between the data and chip rate will define the amount of delay that needs to be provided within the scheme, which in turn determines the best implementation technology.

An alternative is a ladder network [63], an n-stage network consisting of $n + 1$ fixed 3dB couplers connected in cascaded double fiber links (Figure 8.13). The total bit energy reduction is only a factor of two, regardless of P. These 2^n chips are produced from a single pulse passing through the encoder, although there is a restriction on the code families that can be generated.

An extension to the ladder is the tunable delay line architecture (TDL). Couplers are substituted by 2×2 optical switches that route the signal to the upper or lower

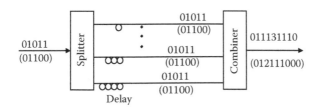

FIGURE 8.12 Incoherent encoder/decoder (matched filter).

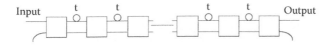

FIGURE 8.13 Ladder network encoder [63]. t is the chip delay.

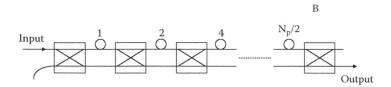

FIGURE 8.14 Tunable delay line encoder (TDL). This can be viewed as a base unit (B) that can be incorporated in other more complex coders to place a chip within a specific timeslot. N_p is the tuning range of the TDL.

branch (bar state or cross state; Figure 8.14) [64]. The switches are set to select the delay according to the position each pulse should occupy in the code word. The TDL is truly programmable and can generate any family of block codes. The extent of the architecture and, hence, the length of code is limited by the implementation technology, with the crosstalk performance of the switches being a fundamental issue due to the generation of interferometric noise.

8.3.2.2 Coherent

The matched filtering approach can be extended to phase coded sequences. The decoding operation relies on retrieving phase information akin to procedures commonplace in the electronics domain. Data bits at the transmitter are spread by bipolar sequences $(+1,-1)$, corresponding to phase $(0,\pi)$. Coherent decoding improves the auto-correlation peak while maintaining the cross-correlation at comparatively low levels, enhancing SNR and system performance [65]. These coded chips are summed up coherently; light sources with a coherence length greater than the chip length must be used.

The encoding/decoding geometry is very similar to those used with incoherent systems but each chip now is subject to independent phase modulation (Figure 8.15). The sequence is encoded by an optical-tapped delay line encoder with different delays and predetermined phase shifts on each branch, producing an optical bipolar code. Unlike time spreading, the delay is incremented linearly, mapping the sequence in adjacent timeslots; the phase provides the orthogonality required of the code family. The speed of the optical phase modulator is determined by the rate at which the codes need to be changed.

Ladder networks have also featured in coherent systems demonstrations. Figure 8.16 is an example of a ladder network comprising asymmetric delays in each arm, the

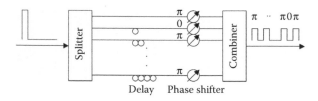

FIGURE 8.15 Coherent encoder/decoder (matched filter).

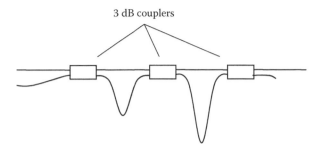

FIGURE 8.16 Ladder network encoder used in coherent matched filter systems [65].

delay increasing with the order of the ladder. The phases of the individual pulses in the encoder incur unique phase changes as they propagate through the ladder determined by the combination of delays. A matched structure is used as the decoder to reconstruct the original pulse through the summation of amplitudes; all the other combinations are summed up incoherently since their encoders are not matched to the particular decoder. The ladder encoder can be further enhanced by incorporating phase modulation in the asymmetric delay arms. In addition, the utilization of both ladder outputs — at the expense of using two (spatial) optical fiber links to effect a connection — improves system performance. This two-channel coherent decoding approach improves the autocorrelation further and the sidelobes disappear since the signals from the two spatial channels cancel each other in a balanced detector.

The main implementation obstacle in this and, indeed, with any coherent transmission is the requirement for phase stability and control. The many issues and requirements for coherent transmission have been tackled extensively in the early 1980s before the development of optical amplifiers removed the need to increase system power budget through coherent detection. The many conclusions reached then are becoming increasingly more relevant, and the evolution of the technology, especially narrow linewidth laser diode sources, has reduced the barriers to deployment of these types of concepts.

If the above conclusion is indeed valid, then a raft of other OCDMA coherent system concepts becomes potentially viable. The most flexible are systems utilizing a local oscillator (LO) in the decoding process (Figure 8.17) [66]. At the transmitter,

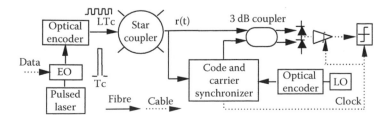

FIGURE 8.17 Schematic of a coherent detection OCDMA network.

the code sequence is generated as described above. At the receiver, another optical code sequence is locally generated via a pulsed LO followed by an optical encoder. Through proper synchronization of the code and optical carrier phase, the local code is multiplied by the received signal, chip by chip via a 3 dB coupler followed by a balanced detector. The output of the balanced detector is an electrical signal that comprises the autocorrelation and cross-correlation components. The signal is subsequently integrated over a bit duration and discriminated to recover the data bit. Receiver realization represents a demanding challenge; however, bringing together the optical functions with the electronic control and processing into an integrated module obviates some of the fundamental problems that faced the pioneering coherent community of the 1980s.

Techniques such as Differential Phase Shift Keying (DPSK) avoid the stringent phase requirements and the need for a local oscillator. Each stream is modulated by the data in a differential phase shift keying (DPSK) format and coded by two encoders to produce bipolar sequences [66]. By delaying the optical pulse stream in the upper branch by $2T_b$, double the bit duration, the streams are multiplexed alternatively in the time domain with T_b differential delay. At each receiver, the signal is split into two branches input to a pair of decoders that are matched to the encoders. One of the streams, delayed by $2T_b$, is multiplexed with the other stream chip-by-chip using a 3 dB coupler and a dual balanced detector. Thresholding is performed after integration of the detector output over $[0, 2T_b]$.

8.3.2.3 Hybrid

Encoding in both the time and frequency domain affords greater flexibility in the choice of optical codes and results in increased system capacity [67]. For example, time spreading codes can be integrated with wavelength-hopping algorithms producing chips within a code sequence at different wavelengths. If the number of wavelengths is equal to the pulses to be colored, the system is referred to as symmetric. A more general case is the asymmetric case where the number of spreading pulses and the number of the hopping patterns are different. An overcolored system means that there are more wavelengths than pulses while an undercolored system means that the wavelengths will be reused.

A number of hybrid (time and frequency) encoders are possible. Making use of advances in mainstream WDM networking, arrayed waveguide gratings (AWGs) in cascade with parallel optical delay lines [68] is one alternative (Figure 8.18). Coders consist of AWGs to multiplex the input range of wavelengths, which then access the appropriate fiber delays of different lengths according to the wavelength. The combination of wavelength and delay defines the wavelength-hopping and time-spreading code. Decoders are identical to the encoder with conjugate delay loops. A variation of the approach uses AWG and mirrored fiber delay lines [69].

Tunable structures can also be implemented based on AWG, switches, and fixed delay lines [70]. An AWG filters a broadband or multiwavelength input signal producing a number of chips at different wavelengths. The first set of switch and delay lines places these chips into the desired blocks while the second set positions the chips within the block.

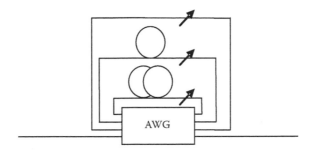

FIGURE 8.18 An encoder/decoder using AWG and delay lines in feedback mode [68].

An alternative family of coders has at its core Bragg gratings, at this time most readily implemented by optical fiber gratings, again making use of the advances brought about by the extensive deployment of WDM transmission. This coding device is a logical combination of two encoders: a set of integrated delay lines and diffraction gratings that provide spatial differentiation of the frequency spectrum into frequency "bins" (spectral amplitude encoding CDMA [71]). The Bragg gratings achieve the frequency spectral slicing and their positions in the fiber perform the same function as the fiber delay lines (Figure 8.19).

The coder can consist of a series of Bragg gratings all written at the same wavelength. However each grating can be tuned independently (e.g., for fiber devices using piezo-electric control) to adjust the Bragg wavelength to a given wavelength defined by the corresponding code placement operator. One advantage of this approach is that only one phase mask is needed to write any encoder or decoder; the tuning of each grating will select the code used. The gratings effectively spectrally and temporally slice an incoming broadband pulse into several components. In the encoder, the peak wavelengths would be placed in reverse order of the encoder to achieve the decoding function [72]. A number of different architectures based on

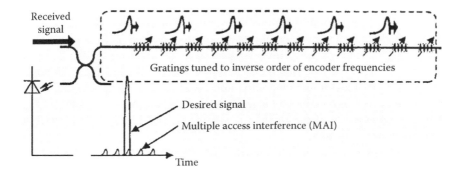

FIGURE 8.19 An example of a 2D encoder utilizing gratings (in this case fiber Bragg gratings) [71].

the same principles are possible [73–76]. The architecture also lends itself to implementing new algorithms such as those [77] including the time blank patterns between two adjacent symbols. A time blank represents the absence of frequency symbols in code sequences and does not interfere with frequency components. The group delay caused by the wavelength dispersion, even through short transmission distances, can be compensated by other grating devices.

A coder structure comprising a $1 \times \tilde{w}$ power divider/combiner, with each arm connected to a linear array of n equally spaced FBGs, generates two-dimensional (2D) spatial signature patterns (SSPs) [78]. In total, there are $n \times \tilde{w}$ FBGs designed at the same wavelength. One unique feature of 2D SSPs is that they are not restricted to a single pulse per row or per column; they may comprise multiple pulses per row (MPPR) or multiple pulses per column (MPPC).

There are also variations on the receiver configuration that improve system performance: optical threshold detection [79], time gating [63], and the use of dynamic optimal thresholding [80] are possible extensions to the basic configurations detailed above.

8.3.3 ENCODER/DECODER DEMONSTRATORS

Demonstrations of the encoding/decoding functions are central to bringing clarity and understanding of the system requirements and are crucial in determining the critical factors in any future module development.

A striking example is the technology demonstrator utilizing wavelength/time matrix codes [81–83]. The "central office" generates coded carriers—time-frequency comb of RZ pulses at a chip interval of 100 psec and repetition frequency of 1.25 GHz using eight wavelengths—distributed to users through a tree topology. This is a noteworthy example of utilizing commercial-off-the-shelf (COTS) components installed in a 19-in rack which in addition surfaces issues of control and management, facilitating further design upgrades. The decoders are based on discrete AWGs, 1×4 couplers and arrays of delay lines. The 48 (1×4) couplers are in four 2U enclosures and the 18 AWGs are in an 8U enclosure.

A further example is the world leading OCDMA transmission experiments utilizing state-of-the-art transversal filters [84–89]. The core unit, a monolithic silica-on-silicon platform, is multifunctional consisting of an array of programmable taps (Mach-Zehnder interferometer configuration controlled through (thermal) phase control), feeding a bank of delay lines (spaced at lengths corresponding to the chip interval of 5 psec) each of which can be phase (thermal) modulated (Figure 20). The module provides programmability, able to control out any variations in loss through the device and produce a range of phase coded optical sequences (see Figure 8.16). The complexity of the module is significant and indicates the possibilities using this material system.

Recent experimentation has indicated the requirements of an incoherent 2D OCDMA system. A pair of mux/demux selecting the wavelength, each input to a tunable delay line provide the necessary coder functions [90,91]. A more integrated module [92,93] implemented on a silica-on-silicon substrate integrates an AWG for wavelength mux/demux and 16 latticelike variable delay lines assigned to each

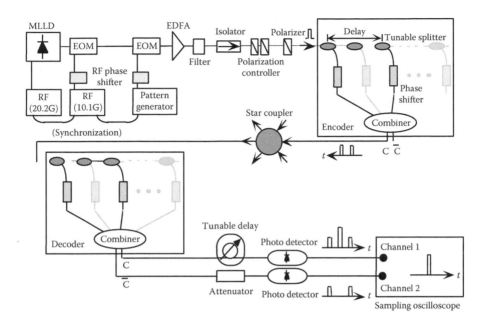

FIGURE 8.20 Although displayed in a more extensive system schematic, the coder/encoder implementation is on a monolithic silica-on-silicon platform and consists of an array of programmable taps feeding a bank of delay lines each of which can be phase modulated. (Courtesy of H. Sotobayashi [85].)

wavelength channel. Each variable delay line comprises a cascade of asymmetric Mach-Zehnder interferometers, with the length difference in each arm being doubled as the cascade progresses. The delay is controlled through thermo-optic switches. The module is capable of generating flexible codes at 10 Gbit/sec and is a very impressive example of a highly integrated module.

Most demonstrations of 2D system based on gratings have utilized fiber Bragg devices [74–76]. The OKI Corporation of Japan offers a commercial encoder/decoder based on FBGs [94].

8.4 INTEGRATED MODULES

8.4.1 GENERAL

The optimum approach to implementing the required module functionality is governed by the OCMDA strategy. Before proceeding to draw some conclusions, it is valuable to set a few basic design guidelines.

The interplay between the optical and electronic domains is core to the design, inherently linked to the maximum data rate per channel supported by the system, which in turn determines the chip rate (through the chosen code family). The optical domain provides very narrow (pico-second) temporal pulses that can be harnessed to enable high transmission capacity. There is great scope to utilize these narrow pulses to either create a larger family of codes or support higher base data rates. Indeed, the narrower the chip interval becomes, the shorter the optical delays required within the coders becomes, resulting in more compact modules amenable to integrated optical solutions.

However, electronics cannot, as yet, carry out functions on pico-second timescales. In many OCDMA implementations, data is discriminated by the requirement to detect within the chip duration (followed by thresholding). Therefore, approaches that rely on the integration of the signal over one-bit intervals (then discriminated by thresholding) are more amenable to the utilization of ultra-short (femto-second) pulses. In addition, if the system control and management operates at the bit interval, appropriate additional signal processing — such as techniques employed in the RF domain to overcome cochannel interference or electronic techniques to overcome dispersion — can be implemented and becomes one of the design options.

It should not be forgotten that other physical layer impairments that routinely have to be considered in the design of optical transmission are still vital to the network design. Power budget, dispersion, extinction ratio, and the effect of amplified spontaneous emission (ASE) are examples of the factors that overlay the particular design issues considered here [95].

8.4.2 BASIC FUNCTIONS

The basic functions that make up the pool needed to implement the range of encoders/decoders are:

- Sources of narrow optical pulses at many wavelengths
- Passive waveguides that distribute the signal, e.g., splitters/combiners
- Delay lines, realized through physical lengths of optical waveguide whether it be an optical fiber for longer delays (5 nsec per meter) or more likely integrated optical waveguides
- Filters, both classical gratings and arrayed waveguide gratings
- Optical switches and gates
- Amplitude and phase modulators—in some more advanced scenarios frequency (wavelength) modulation (conversion) might prove attractive
- Receivers; high speed photo-detectors, balanced detectors and coherent receivers

Combinations of the above functions yield the range of encoder/decoder geometries required. There are also advantages to utilizing advanced all-optical processing concepts to improve system performance, e.g., nonlinear loop mirrors incorporating a semiconductor optical amplifier to suppress cochannel terms or two-photon absorption to implement optical thresholding.

8.4.2.1 Sources

The requirement of the system's source is to produce a stream of narrow pulses at a range of wavelengths with the solutions being exclusively the domain of semiconductors, especially InP. There has been substantial progress in commercializing a range of laser diode sources. The DFB is the workhorse of many optical transmission systems; the integration of in-line, on-chip modulators has also occurred using, most significantly, electro-absorption intensity modulators at rates up to 40 Gbit/sec. For coherent implementations, the linewidth is a vital parameter.

A solution, therefore, might be to utilize a bank of these—either with or without an integrated modulator—at the appropriate wavelengths, merged by a simple combiner (monolithic or hybrid) to create a stream of modulated pulses at different wavelengths as input to the encoder. Producing narrow pulses (~pico-second), however, complicates the scenario and requires mode-locking approaches to be implemented. Here the electronic drive and control is provided at board level.

There also been significant progress made in tuneable laser products. Many of these are quite sophisticated examples of monolithic integration incorporating in-line modulators, especially given the complexity of the multistage processes used in their manufacture. Dependent on the application, a single tunable source capable of rapid tuning might provide the required solution. Alternatively, a limited number of tunable sources organized to provide specific bands of wavelengths from the total multiplex can be engineered. Again, issues of mode locking, drive, and limited output power are important considerations.

A more cost-effective route might be the array of gain blocks-AWG multiplexer tandem that provides, simultaneously, a number of wavelengths each able to be modulated independently. The wavelengths are automatically selected by the multiplexer pass-band. This can be achieved through both monolithic and hybrid integration and potentially allows a large number of wavelengths to be offered in one small module. Again issues of narrow pulse generation and power output limitations are important. It is undoubtedly the case, given the lossy encoder geometries, that some form of booster amplification would need to be incorporated (e.g., on-chip SOA or an external erbium doped fiber amplifier) prior to launch into the network.

An alternative geometry uses external cavity architectures realized through waveguide gratings in tandem with gain elements (with the output merged through a combiner). The planar grating architecture can be produced in silicon/SOS and hybrid integrated with the array of semiconductor gain elements.

It is worth noting that one powerful transmitter solution has already been demonstrated. A supercontinuum (SC) pulse source provides the ideal output; narrow pulses across an extremely wide bandwidth and high repetition rate [96]. However, this is a very complex geometry that is expensive; a clear economic case must be surfaced for its commercial deployment. However, for laboratory OCDMA experimentation, it is ideal.

8.4.2.2 Passive Waveguides

Passive waveguides used for signal distribution and the implementation of time delays are predominately the exclusive domain of silicon/SOS. Commercial products are and have been offered in these material systems. For some of the more advanced monolithic integration devices, passive waveguides are also required on InP.

Both of these waveguiding systems produce very low loss waveguides and are the only route to providing significant time delays that are commonplace in encoders/decoders.

8.4.2.3 Filters

In a planar format, filtering is most readily produced through gratings. Due to the fundamental principles of operation of the in-line grating, more advanced mux/demux are realised through relatively lossy architectures. A good alternative is the AWG, which can also be utilized in more advanced applications such as routing nodes. Both of these components are best produced in silicon/SOS, although examples in InP have been demonstrated.

8.4.2.4 Switches and Gates

For flexibility in code generation and decoding, switches and gates are essential. The rate at which the code needs to be selected will define the best technology solution. Thermally controlled switches, based on phase modulation in interferometric geometries, are possible in both silicon/SOS platforms at msec timescales. Recent progress in silicon has pushed down the timescales to the MHz range.

Faster operation necessitates the use of InP where Mach-Zehnder interferometer and directional coupler devices can yield fast (nsec) switches. Alternative switch fabrics can be achieved through the ON/OFF property of SOAs; in the ON state the device provides gain while in the OFF state it provides significant isolation (in excess of 50 dB). In this case, the switch fabric is constructed through gates within more complex splitting architectures like trees. The realization of large switching matrices is limited.

The tunable delay line is one of the more important basic units (Figure 8.14). It relies on switches and delay lines; therefore due to loss considerations, silicon/SOS implementations are the preferred solution. The size and, hence, the tuning range is dependent on the loss and isolation of the switches. With respect to the latter requirement, the SOA is powerful since it provides significant isolation in the OFF state. However, the accumulation of ASE noise then becomes the limiting factor.

Comprising an AWG and gating, the wavelength selector is a potentially valuable unit. It can be realized in both monolithic (SOS, silicon and InP) and hybrid (InP plus SOS) platforms. Such a unit provides the capability to select certain wavelengths from a large multiplex. The gates can be SOAs or Mach-Zehnder switches (InP, silicon/SOS) or variable optical attenuators (VOAs in silicon/SOS). In tandem with either static or, more importantly, tunable delay lines they represent important configurations, especially if they are extended to incorporate the array of multiwavelength sources described in Section 8.4.2.1.

It is worth noting that lithium niobate (LiNbO$_3$) directional coupler switches have also been integrated in hybrid platform with SOS.

8.4.2.5 Modulators

Intensity modulators are used to imprint the data onto the optical carrier. The monolithic integration of laser diodes plus in-line modulators is the ideal solution.

Phase modulation for producing coherent-like code sequences is usually embedded within the delay line coding/decoding structure (Figure 8.15). Thus, based on the arguments exercised in Section 8.4.2.2, the optimum approach is on silicon/SOS using thermal modulation.

Some approaches using coherent correlation detection might require faster (at the chip rate) modulation (Section 8.3.2.2). This necessitates the use of external modulators, the most mature of which are based on $LiNbO_3$.

8.4.2.6 Receivers

Receiver choice is driven by whether decisions are being made over a bit or chip duration. Forty Gbit/sec receivers are routinely available with promise of photodetectors approaching 100 Gbit/sec demonstrated within the laboratory. For some of the more advanced OCDMA approaches, balanced detection is required; 40 Gbit/sec devices in InP are available.

This module is the most amenable to near-term OEIC realisation. A photodetector with a transimpedance amplifier on a single material system (InP and SiGe) has been proven with enhanced performance. Hybrid implementation of a coherent receiver module (3 dB coupler plus balanced detector) has also been demonstrated with good performance.

8.5 CONCLUSIONS

Integration is absolutely necessary if advanced implementations of all-optical networking principles such as OCDMA are to be deployed.

In general, the integration technology roadmap is well understood and in itself does not provide any great insight. Adoption of a hybrid approach is the most obvious strategy, simply reflecting the maturity of the current technology. Harnessing elements needed for the production of a range of multifunctional modules is best achieved by using components optimized in particular material systems. Thus, at this stage:

- Signal distribution, filtering, and "slow" signal switching/gating and phase modulation are best achieved in SOS/silicon material systems. Most of these functions can be, and have been, integrated on one substrate.
- Signal generation, high-speed modulation, and detection are best achieved in InP. Advanced transmitters with in-line modulators on a single chip are available, some fully qualified; complex receivers at 40 Gbit/sec, are also available.

This conclusion is limited since it carries no indication regarding the scale of the implementation. In 2D OCDMA scenarios the size of the wavelength pool (and hence code pool) will have significant ramifications. For example, >10 wavelengths precludes the use of discrete sources due to cost; then more complex hybrid realizations become appropriate, e.g., multiwavelength sources based on AWGs. In this way parallel integration can be used to great effect.

Hybrid solutions will dominate in the near term and will make use of greater levels of monolithic integration when they become available. Even as InP manufacturing processes mature, bringing higher yields and lower costs, it is still unclear whether the performance of functions relying on low loss passive waveguides will be sufficiently compelling to surpass those produced in SOS/silicon.

Opto-electronic ICs are the ultimate goal, especially since the electronic control plane is just as important as the optical physical layer. InP is in pole position but silicon and its bandgap-engineered versions, will potentially provide a range — albeit more limited — of cost effective solutions. It is a fact that the experience the silicon sector has acquired in high volume, high yield, and highly complex IC manufacture should never be dismissed and the application of the same rigorous methodology to the creation of a new breed of devices should yield useful products.

The OCDMA community should take the opportunity to reexamine the possibilities that phase and frequency present given the advantageous correlation properties that result, using the many conclusions reached by the coherent communications community in the 1980s as the base. Integrated modules are very important in this respect providing solutions that can operate in controlled environments under complex electronic control.

ACKNOWLEDGMENTS

The author would like to acknowledge the following, whose advice has helped the production of this chapter. Thank you to Ian Armstrong, Taher Bazan, Graeme Maxwell, Mark Hesketh, Hideyuki Sotobayashi, Antonio Mendez, Varghese Baby, Lawrence Chen, Robert Runser, and Ted Sargent for providing input to the chapter.

REFERENCES

[1] Whitney, P. S. (2002). Hybrid integration of photonic systems. Proceedings Electronic Components and Technology Conference. 578–582.

[2] Kato, K., Tohmori, Y. (2000). PLC hybrid integration technology and its application to photonic components. *IEEE J. Sel. Areas Quant. Electron.* 6:4–13.

[3] Smit, M. K. (2002). InP photonic integrated circuits. Proceedings LEOS. 543–544.

[4] Himeno, A., Kato, K., Miya, T. (1998). Silica based planar lightwave circuits. *IEEE J. Sel. Topics Quant. Eletcron.* 13:2320–2326.

[5] Kaiser, R., Heidrich, H. (2002). Optoelectronic/photonic integrated circuits on InP between technological feasibility and commercial success. *IEICE Trans. Electron.* E85–C.

[6] Krauss, T. F., Wu, L., Wilson, R., Karle, T. (2002). Photonic crystal based integration. Proceedings LEOS. 572–573.

[7] www.axsun.com.

[8] Syms, R. R. A., Moore, D. F. (2002). Optical MEMS for telecoms. *Materials Today*:26–35.

[9] Gates, J., Meuhlner, D., Cappuzzo, M., Fishteyn, M., Gomez, L., Henein, G., Laskowski, E., Ryazansky, I., Shmulovich, J., Syvertsen, D., White, A. (1998). Hybrid integrated silicon optical bench planar lightwave circuits. Proceedings Electronic Components and Technology Conference. 551–559.

[10] Collins, J. V., Lealman, I. F., Fiddyment, P. J., Jones, C. A., Walker, R. G., Rivers, L. J., Cooper, K., Perrin, S. D., Nield, M. W., Harlow, M. J.(1995). Passive alignment of a tapered laser with more than 50% coupling efficiency. *Electron. Lett.* 31:730–731.

[11] Renaud, M., Keller, D., Sahri, N., Silvestre, S., Prieto, D., Dorgeuille, F., Pommereau, F., Emery, J. Y., Mayer, H. P. (2001). SOA-based optical network components. Proceedings Electronic Components and Technology Conference.

[12] Lemoff, B. E., Buckman, L.A., Schmit, A.J., Dolfi, D.W., (2000). A compact, low-cost WDM transceiver for the LAN. Proceedings Electronic Components and Technology Conference. 711–716.

[13] Hashimoto, T., Nakasuga, Y., Yamada, Y., Terui, M., Yanagisawa, M., Akahori, Y., Tohmori, Y., Kato, K., Suzuki, Y. (1998). Multichip optical hybrid integration technique with planar lightwave circuit platform. *IEEE J. Light. Technol.* 16:1249–1258.

[14] Yamada T., Hashimoto T., Ohyama T., Akahori Y., Kaneko A., Kato K., Kashahara R., Ito M. (2001). New planar lightwave circuit (PLC) platform eliminating Si terraces and its application to opto-electronic hybrid integrated modules. *IEICE Trans. Electron.* E84–C.

[15] Hunziker, W., Vogt, W., Melchior, H., Leclerc, D., Brosson, P., Pommereau, F., Ngo, R., Doussiere, P., Mallecot, F., Fillion, T., Wamsler, I., Laube, G. (1995). Self-aligned flip-chip packaging of tilted semiconductor optical amplifier arrays on Si motherboard. *Electron. Lett.* 31:488–490.

[16] Pabla, A. S., Ford, C. W., Collins, J. V., Schmidt, M. (1999). Hybrid integration of an optical processor on a LIGA motherboard. Proceedings LEOS. 2:501–502.

[17] Maxwell, G., Manning, B., Nield, M., Harlow, M., Ford, C., Clements, M., Lucas, S., Townley, P., McDougall, R., Oliver, S., Cecil, R., Johnston, L., Poustie, A., Webb, R., Lealman, I., Rivers, L., King, J., Perrin, S., Moore, R., Reid, I., Scrase D. (2002). Very low coupling loss, hybrid-integrated all-optical regenerator with passive assembly. Proceedings European Conference on Optical Communications. Paper PD3.5.

[18] Collins, J. V., Lealman, I. F., Waller, R., Ford C. W., Kelly, A. (1996). A generic technology for low cost semiconductor optoelectronic devices. Proceedings LEOS. 2:103–104.

[19] Lealman, I. F., Rivers, L. J., Harlow, M. J., Pirrin, S. D., Robertson, M. J. (1994) 1.56 μm InGaAsP/InP tapered active layer multiquantum well laser with improved coupling to cleaved single mode fibre. *Electron. Lett.* 30:857–859.

[20] Sato, R., Suzuki, Y., Yoshimoto, N., Ogawa, I., Hashimoto, T., Ito, T., Sugita, A., Tohmori Y., Toba, H. (1999). A 1.55 mm hybrid integrated wavelength-converter module using spot-size converter integrated semiconductor optical amplifiers on a planar-lightwave-circuit platform. *IEICE Trans. Commun.* E82–B.

[21] Mino, S., Yoshino, K., Yamada, Y., Terui, T., Yasu, M., Moriwaki, K. (1995). Planar lightwave circuit platform with coplanar waveguide for opto–electronic hybrid integration. *J. Light. Technol.* 13:2320–2326.

[22] Mino, S., Ohyama, T., Hashimoto, T., Akahori, Y., Yoshino, K., Yamada, Y., Kato, K., Yasu, M., Moriwaki, K. (1996). High frequency electrical circuits on a planar lightwave circuit platform. *J. Light. Technol.* 14:806–811.

[23] Hashimoto, T., Nakasuga, Y., Yamada, Y., Terui, H., Yanagisawa, M., Moriwaki, K., Suzuki, Y., Tohmori, Y., Sakai, Y., Okamoto, H. (1996). Hybrid integration of spot-size converted laser diode on a planar lightwave circuit platform by passive alignment technique. *Photon. Technol. Lett.* 8:1504–1506.

[24] Jackson, K. P., Flint, E. B., Cina, M. F., Lacey, D., Kwark, Y., Trewhella, J. M., Caulfield, T., Buchmann, P., Harder, C., Vettiger, P. (1994). A high-density, four-channel, OEIC transceiver module utilizing planar-processed optical waveguides and flip-chip, solder-bump technology. *J. Light. Technol.* 12:1185–1191.

[25] Tajima, K., Nakamura, S., Ueno, Y., Sasaki, J., Sugimoto, T., Kato, T., Shimoda, T., Itoh, M., Hatakeyama, H., Tamanuki, T., Sasaki, T. (1999). Hybrid integrated symmetric Mach-Zehnder all-optical switch with ultrafast, high extinction switching. *Electron. Lett.* 35:2030–2031.

[26]   Jones, C. A., Cooper, K., Nield, M. W., Rush, J. D., Walker, R. G., Collins, J. V.,
        Fiddyment, P. J. (1994). Hybrid integration of a laser diode with a planar silica
        waveguide. *Electron. Lett.* 30:215–216.

[27]   Mino, S., Yoshino, K., Yamada, Y., Yasu, M., Moriwaki, K. (1994). Optoelectronic
        hybrid integrated laser diode module using planar lightwave circuit platform. *Electron.
        Lett.* 35:1888–1890.

[28]   Mino, S., Ohyama, T.., Akahori, Y., Yanagisawa, M., Hashimoto, T., Yamada, Y.,
        Tsunetsugu, H., Togashi, M., Itaya, Y., Shibata, Y. (1998). High-speed optoelectronic
        hybrid-integrated transmitter module using a planar lightwave circuit (PLC) platform.
        Photon. Technol. Lett. 10:875–877.

[29]   Mino, S., Ohyama, T., Akahori, Y., Hashimoto, T., Yamada, Y., Yanagisawa, M.,
        Muramoto, Y. (199). A 10Gb/s hybrid-integrated receiver array module using a planar
        lightwave circuit (PLC) platform including a novel assembly region structure. *J. Light.
        Technol.* 14:2475–2482.

[30]   Jones, C. A., Cooper, K., Nield, M. W., Rush, J. D., Thurlow, A. R., Walker, R. G.,
        Ayliffe, P. J., Harrison, P. M. (1995). 2.4Gbit/s transceiver using silica waveguides on
        a silicon optical motherboard. *Electron. Lett.* 31:2208–2210.

[31]   Inoue, Y., Oguchi, T., Hibino, Y., Suzuki, S., Yanagisawa, M., Moriwaki, K., Yamada, Y.
        (1996). Fiber-embedded wavelength-division multiplexer for hybrid-integrated trans-
        ceiver based on silica-based PLC. *Electron. Lett.* 32:847–848.

[32]   Maxwell, G. D., Kashyap, R., Sherlock, G., Collins, J. V., Ainslie, B. J. (1994). Dem-
        onstration of a semiconductor external cavity laser using a UV written grating in a
        planar silica waveguide. *Electron. Lett.* 30:1486–1487.

[33]   Tanaka, T., Hibino, Y., Hashimoto, T., Kasahara, R., Inoue, Y., Himeno, A., Itoh, M.,
        Abe, M., Oohashi, H., Tohmori, Y. (2001). PLC-type hybrid external cavity laser
        integrated with front-monitor photodiode on Si platform. *Electron. Lett.* 37:95–96.

[34]   Akahori, Y., Ohyama, T., Oguma, M., Kato, K., Yamada, Y. (1998). Hybrid integrated
        differential photoreceiver module for photonic packet switching systems using a planar
        lightwave circuit platform. *Photon. Technol. Lett.* 10:869–871.

[35]   Sasaki, J., Hatakeyama, H., Tamanuki, T., Kitamura, S., Yamaguchi, M., Kitamura, N.,
        Shimoda, T., Mitamura, M., Kato, T., Itoh, M. (1998). Hybrid integrated 4 × 4 optical
        matrix switch using self-aligned semiconductor optical amplifier gate arrays and silics
        planar waveguide circuit. *Electron. Lett.* 34:986–987.

[36]   Akahori, Y., Obyama, T., Yamada, T., Kamei, S., Ishii, M., Kasahara, R., Nakamura,
        M., Okayasu, M., Oohashi, H., Yamakoshi, K. (2001). WDM interconnection using
        PLC hybrid technology for 5-Tbit/s electrical switching system. Proceedings European
        Conference on Optical Communications. Paper We.A.2.5.

[37]   Kasahara, R., Yamagisawa, M., Sugita, A., Ogawa, I., Hashimoto, T., Suzuki, Y.,
        Magari, K. (2000). A compact optical wavelength selector composed of arrayed-
        waveguide gratings and an optical gate array integrated on a single PLC platform.
        Photon. Technol. Lett. 12:34–37.

[38]   Ebisawa, F., Ogawa, I., Akahori, Y., Takiguchi, K., Tamura, Y., Hashimoto, T., Sugita,
        A., Yamada, Y., Suzaki, Y., Yoshimoto, N., Tohmori, Y., Mino, S., Ito, T., Magari, K.,
        Kawaguchi, Y., Himeno, A., Kato, K. (1999). High speed 32 channel optical wave-
        length selector using PLC hybrid integration. Technical Digest Conference Optical
        Fiber Communications. 3:18–20.

[39]   Ito, T., Ogawa, I., Yoshimoto, N., Magari, K., Ebisawa, F., Yamada, Y., Yoshikuni, Y.,
        Hasumi, Y. (1998). Dynamic response of high-speed wavelength selector using hybrid
        integrated four-channel SS-SOA gate array on PLC platform. *Electron. Lett.*
        34:494–496.

[40] Sato, R., Ito, T., Magari, K., Ogawa, I., Inoue, Y., Kasahara, R., Okamoto, M., Tohmori, Y., Suzuki, Y. (2004). 10-Gb/s low-input-power SOA-PLC hybrid integrated wavelength converter and its 8-slot unit. *J. Light. Technol.* 22:1331–1337.

[41] Soole, J. B. D., Poguntke, K., Scherer, A., LeBlanc, H. P., Chang-Hasnain, C., Hayes, J. R., Caneau, C., Bhat, R., Koza, M. A. (1992). multistripe array grating integrated cavity (MAGIC) laser: a new semiconductor laser for WDM applications. *Electron. Lett.* 28:1805–1807.

[42] Armstrong, I., Andonovic, I., Bebbington, J., Michie, C., Tombling, C., Fasham, S., Kelly, A. E., Chai, Y. J., Penty, R. V., White, I. H. (2004). Hybridisation platform demonstrating all optical wavelength conversion at 10 and 20Gbit/s. Proceedings Conference Optical Fiber Communications. Paper ThS3.

[43] Lealman, I. (2000). Hybrid integration technologies for semiconductor lasers and optical amplifiers. Proceedings LEOS. 2:702–703.

[44] Maxwell, G., Manning, B., Nield, M., Harlow, M., Ford, C., Clements, M., Lucas, S., Townley, P., McDougall, R., Oliver, S., Cecil, R., Johnston, L., Poustie, A., Webb, R., Lealman, I., Rivers, L., King, J., Perrin, S., Moore, R., Reid, I., Scrase, D. (2002). Very low coupling loss, hybrid-integrated all-optical regenerator with passive assembly. Proceedings European Conference Optical Communications. Post deadline paper PD3.5.

[45] Janz, C. (2000). Integrated SOA-based interferometers for all-optical signal processing. Proceedings European Conference on Optical Communications. 2:782–783.

[46] Mestric, R., Porcheron, C., Martin, B., Pommereau, F., Guillemot, I., Gaborit, F., Fortin, C., Rotte, J., Renaud, M. (2000). Sixteen channel wavelength selector monolithically integrated on InP. Proceedings Conference Optical Fiber Communications. 1:81–83.

[47] Ratovelomanana, F., Vodjdani, N., Enard, A., Glastre, G., Rondi, D., Blondeau, R., Joergensen, C., Durhuus, T., Mikkelsen, B., Stubkjaer, K. E., Jourdan, A., Soulage, G. (1995). An all-optical wavelength converter with semiconductor optical amplifiers monolithically integrated in an asymmetric passive Mach-Zehnder interferometer. *Photon. Technol. Lett.* 7:992–994.

[48] Spiekman, L. H. (1997). All-optical Mach-Zehnder wavelength converter monolithically integrated with an /4-shifted DFB source. Proceedings Conference Optical Fiber Communications. Paper PD10.

[49] Janz, C., Dagens, B., Bisson, A., Poingt, F., Pommereau, F., Gaborit, F., Guillemot, I., Renaud, M. (1999). Integrated all-active Mach-Zehnder wavelength converter with record signal sensitivity and large dynamic range at 10Gb/s. Proceedings Conference Optical Fiber Communications. 4:30–32.

[50] Pommereau, F., Mestric, R., Martin, B., Rao, E. V. K., Gaborit, F., Leclerc, D., Porcheron, C., Renaud, M. (1999). Optimisation of butt-coupling between deep-ridge and buried ridge waveguides for the realisation of monolithically integrated wavelength selector. Proceedings Conference InP and Related Materials. 401–404. Paper We.A.1-7-1.

[51] Kikuchi, N., Shibata, Y., Okamoto, H., Kawaguchi, Y., Oku, S., Ishii, H., Yoshikuni, Y., Tohmori, Y. (2002). Monolithically integrated 64-channel WDM channel selector with novel configuration. *Electron. Lett.* 38:331–332.

[52] Doerr, C. R., Joyner, C. H., Stulz, L. W. (1999). 40-wavelength rapidly digitally tunable laser. *Photon. Technol. Lett.* 11:1348–1350.

[53] den Besten, J. H., Broeke, R. G., van Geemert, M., Binsma, J. J. M., Heinrichsdorff, F., van Dongen, T., de Vries, T., Bente, E. A. J. M., Leijtens, X. J. M., Smit, M. K. (2002). A compact digitally tunable seven-channel ring laser. *Photon. Technol. Lett.* 14:753–755.

[54]   Herben, C. G. P., Maat, D. H. P., Leijtens, X. J. M., Leys, M. R., Oei, Y. S., Smit, M. K. (1999). Polarization independent dilated WDM cross-connect on InP. *Photon. Technol. Lett.* 11:1599–1601.

[55]   Ahn, J. H., Oh, K. R., Kim, J. S., Lee S. W., Kim, H. M., Pyun, K. E., Park, H. M. (1999). Uniform and high coupling efficiency between InGaAsP-InP buried hetero-structure optical amplifier and monolithically butt-coupled waveguide using reactive ion etching. *Photon. Technol. Lett.* 8:200–202.

[56]   Marsh, J. H., (1993). Quantum well intermixing. *Semicon. Sci. Technol.* 8:1136–115.

[57]   Lockwood, D. J., Pavesi, L., eds. (2004). *Silicon Photonics.* Topics in Applied Physics, vol. 94. Heidleberg: Springer Verlag.

[58]   Lockwood, D. J., Pavesi, L. (Eds) (2004). *Silicon photonics: Monolithic silicon microphotonics.* Topics in Applied Physics. 94:89–120.

[59]   Kawachi, M. (1996) Recent progress in silica-based planar lightwave circuits on silicon. *IEEE Proceedings-Optoelectronics* 143: 257–262.

[60]   Murata, K., Enoki, T., Sugahara, H., Tokumitsu, M. (2003). ICs for 100Gbit/s data transmission. Proceedings of 11th GAAS Symposium. 457:460.

[61]   Peiyi, C. (2001). Development of SiGe materials and devices. Proceedings Solid-State and Integrated-Circuit Technology. 1:570–574.

[62]   Wong, H. (2002). Recent developments in silicon optoelectronic devices. Proceedings International Conference on Microelectronics. 1:285–292.

[63]   Marhic, M. E. (1993). Coherent optical CDMA networks. *J. Light. Technol.* 11:854–863.

[64]   Hobnes, A. S., Syms, R. R. A. (1991). Switchable all-optical encoding and decoding using optical fiber lattices. *Optics Commun.* 86:24–28.

[65]   Chang, Y. L., Marhic, M. E. (1992). Fiber-optic ladder networks for inverse decoding coherent CDMA. *J. Light. Technol.* 10:1952–1962.

[66]   Huang, W., Andonovic, I. (1999). Coherent optical pulse CDMA systems based on coherent correlation detection. *IEEE Trans. Commun.* 47:261–271.

[67]   Tancevski, L., Andonovic, I. (1996). Hybrid wavelength hopping/time spreading schemes for use in massive optical networks with increased security. *J. Light. Technol.* 14:2636–2647.

[68]   Yegnanarayanan, S., Bhushan, A. S., Jalil, B. (2000). Fast frequency-hopping time spreading encoding/decoding for optical CDMA. *Photon. Technol. Lett.* 12:573–575.

[69]   Yu, K., Shin, J., Park, N. (2000). Wavelength-time spreading optical CDMA system using wavelength multiplexers and mirrored fiber delay lines. *Photon. Technol. Lett.* 12:1278–1280.

[70]   Min, S., Yoo, H., Won, Y. H. (2003). Time-wavelength hybrid optical CDMA system with tunable encoder/decoder using switch and fixed delay line. *Optics Commun.* 216:335–342.

[71]   Zaccarin, D., Kavehrad, M. (1993). An optical CDMA system based on spectral encoding of LED. *Photon. Technol. Lett.* 4:479–482.

[72]   Slavik, R., La Rochlle, S. (2002). Multiwavelength single-mode Erbium doped laser for FFH-CDMA testing. Proceedings Conference Optical Fiber Communications. Paper WJ3.

[73]   Fathallah, H., Cortès, P. Y., Rusch, L. A., LaRochelle, S., Pujol, L. (1999). Experimental demonstration of optical fast frequency hopping-CDMA communications. Proceedings European Conference on Optical Communications. Paper Tu.B2.

[74]   Ben Jaâfar, H., LaRochelle, S., Cortès, P. Y., Fathallah, H. (2001). 1.25 Gbit/s transmission of optical FFH-CDMA signals over 80 km with 16 users. Proceedings Conference Optical Fiber Communications. Paper Tu.V3.

[75] Wada, N., Sotobayashi, H., Kitayama, K. (2000). 2.5 Gbit/s time-spread/wavelength-hop optical code division multiplexing using fiber Bragg gratings with supercontinuum light source. *Electron. Lett.* 36:815–817.

[76] Chen, L. R., Smith, P. W. E. (2000). Demonstration of incoherent wavelength-encoding /time-spreading optical CDMA using chirped Moiré grating. *Photon. Technol. Lett.* 12:1281–1283.

[77] Bin, L. (1997). One coincident sequences with specific distance between adjacent symbols for frequency-hopping multiple access. *IEEE Trans. Commun.* 45:408–410.

[78] Chen, L. R. (2001). Flexible fiber Bragg grating encoder/decoder for hybrid wavelength-time optical CDMA. *Photon. Technol. Lett.* 13:1233–1235.

[79] Inaty, H., Shalaby, H. M. H., Fortier, P. (2002). A new transmitter-receiver architecture for non-coherent multirate optical fast frequency CDMA system with fixed optimal detection threshold. *J. Light. Technol.* 20:1885–1894.

[80] Ng, E. K. H., Sargent, E. H. (2002). Optimum threshold detection in real-time scalable high-speed multiwavelength optical code division multiple access LANs. *IEEE Trans. Commun.* 50:778–784.

[81] Mendez, A. J., Hernandez, V. J., Bennett, C. V., Lennon, W. J., Gagliardi, R. M. (2003). Optical CDMA (O-CDMA) technology demonstrator (TD) for 2D codes. Proceedings LEOS. 2:1044–1045.

[82] Mendez, A. J., Gagliardi, R. M., Feng, H. X. C., Heritage, J. P., Morookian, J-M. (2000). Strategies for realizing optical CDMA for dense, high-speed, long span, optical network applications. *J. Light. Technol.* 18:1685–1696.

[83] Mendez, A. J., Gagliardi, R. M., Hernandez, V. J., Bennett, C. V., Lennon, W. J. (2003). Design and performance analysis of wavelength/time (W/T) matrix codes for optical CDMA. *J. Light. Technol.* 21:2524–2533.

[84] Sotobayashi, H., Kitayama, K-I. (1999). Broadcast-and-select OCDMA/WDM network using 10Gbit/s spectrum-spliced Supercontinuum BPSK pulse code sequences. *Electron. Lett.* 35:1966–1967.

[85] Sotobayashi, H., Chujo, W., Kitayama, K. -I. (2001). 1.52Tbit/s OCDMA/WDM (4 OCDMA × 19 WDM × 20Gbit/s) transmission experiment. *Electron. Lett.* 37:700–701.

[86] Sotobayashi, H., Chujo, W., Kitayama, K. -I. (2002). 1.6-b/s/Hz 6.4-Tb/s QPSK-OCDMA/WDM (4 OCDMA × 40 WDM × 40 Gb/s) transmission experiment using optical hard thresholding. *Photon. Technol. Lett.* 14:555–557.

[87] Sotobayashi, H., Chujo, W., Kitayama, K. -I. (2002). Transparent virtual optical code/wavelength path network. *IEEE J. Sel. Areas Quant. Electron.* 8:699–704.

[88] Sotobayashi, H., Chujo, W., Kitayama, K. -I. (2004). Highly spectral efficient optical code division multiplexing transmission system. *IEEE J. Sel. Areas Quant. Electron.* 10:250–258.

[89] Saeki, I., Nishi, S., Murakami, K. (1999). All-optical code division multiplexing switching network based on self-routing principle. *IEICE Trans. Electron.* E82-C:187–193.

[90] Glesk, I., Baby. V., Brès, C. -S., Xu, L., Rand, D., Prucnal, P. R. (2004). Experimental demonstration of a 2.5Gbps incoherent 2D OCDMA system. Proceedings of CLEO.

[91] Baby. V., Bres, C. -S., Glesk, I., Xu, L., Prucnal, P. Wavelength aware receiver for enhanced 2D OCDMA system performance. *Electron. Lett.* 40:385–387.

[92] Takiguchi, K., Shibata, T., Itoh, M. (2002). Encoder/decoder on planar lightwave circuit for time-spreading/wavelength-hopping optical CDMA. *Electron. Lett.* 38:469–470.

[93] Takiguchi, K., Itoh, M. (2003). Transmission experiment using silica waveguide based encoder/decoders for time-spreading/wavelength-hopping CDMA. *Electron. Lett.* 39:1813–1814.

[94]   Kutsuzawa, S., Minato, N., Oshiba, S., Nishiki, A., Kitayama, K. (2003). 10Gb/s/spl times/2ch signal unrepeated transmission over 100 km of data rate enhanced time-spread/wavelength-hopping OCDM using 2.5Gb/s-FBG en/decoder. *Photon. Technol. Lett.* 15:317–319.

[95]   Feng, H., Mendez, A., Heritage, J., Lennon, W. (2000). Effects of physical layer impairments on 2.5G/s optical CDMA transmission. *Optics Express* 7:2–9.

[96]   Sotobayashi, H., Kitayama, K-I. (1998). 325 nm bandwidth supercontinuum generation at 10Gbit/s using dispersion-flattened and non-decreasing normal dispersion fibre with pulse compression technique. *Electron. Lett.* 34:1336–1337.

9 Optical CDMA Network Security

Peter A. Schulz and Thomas H. Shake

CONTENTS

9.1 INTRODUCTION

Many papers have cited enhanced security as an advantage of building an OCDMA network [1–5]. The idea is that if multiple codes operate simultaneously, it would be nearly impossible to get any meaningful information from the data signal. In contrast, intercepting data from a WDM (wavelength division multiplexed) network would be straightforward. The data for any particular channel in a WDM network could be obtained with use of the appropriate wavelength filter. The analogous attempt of using an optical CDMA filter is not effective because the number of optical codes can greatly exceed the number of wavelengths being used by many orders of magnitude.

An optical spectrum can identify each and every wavelength being used for communication. In a WDM system, this would allow detection and demodulation of each data stream. This need not be true for an OCDMA system. For an OCDMA link, the spectral composition can be essentially independent of the number of codes being used in the communication link. On the other hand, the optical power might give some indication of the number of users transmitting at any time.

Until recently [6,7], most arguments advocating OCDMA for secure communication in the research literature have been qualitative and vague, and the few more mathematically rigorous analyses have been incomplete. For example, some analyses of OCDMA security have focused on showing that a very large number of codes can be made available through particular forms of OCDMA coding. This can be highly misleading from a security perspective, because having a very large number of available codes is a necessary but not sufficient condition for good security. Furthermore, evidence that many of the qualitative arguments found in the literature may not be correct can be found in published analyses of radio frequency (RF) CDMA technology (e.g., for cell phones). CDMA technology used for cell phones can use a huge number of codes but is not secure unless explicit encryption algorithms are employed. [8]

While confidentiality is the best-known aspect of security, there are many other types of security as well (see Figure 9.1). Little work on OCDMA security outside the area of confidentiality has been published so far, and the potential of OCDMA to provide security functions such as jamming resistance, covertness, and authentication

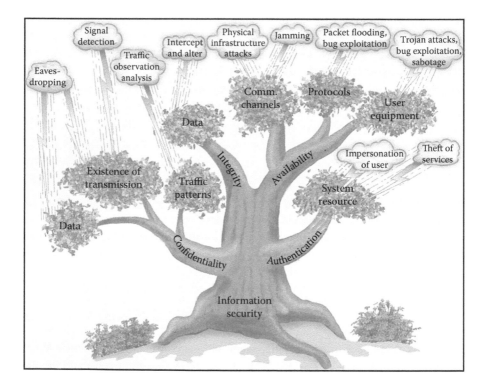

FIGURE 9.1 (Color Figure 9.1 follows page 240.) The security tree. Security is often divided into branches of confidentiality, integrity, availability, and authentication. Each of these branches has assets to be protected (in foliage). Some threats for the various assets are identified (in ominous clouds).

remains an open area of research. Nevertheless, confidentiality is one of the most important and widely needed types of security, and this chapter will focus primarily on summarizing results of research on OCDMA confidentiality.

This chapter first introduces the various classes of security for those readers who may not be familiar with the field of security and encryption. The most obvious issues associated with the use of OCDMA for network security are mentioned. The next section discusses how the degree of security in an OCDMA network can be quantified. The conclusion summarizes our present understanding of security for an OCDMA network.

9.2 SECURITY

9.2.1 GENERAL SECURITY ISSUES

Security and obscurity are often confused [9]. Obscurity refers to information or data that may be difficult or near impossible to find. For example, data may be hidden within large files (e.g., movie or picture files) without significantly affecting the output. But for anyone who knows where to look, the data can be easily read. Such techniques are often useful for passing relatively small amounts of information. This data is traditionally not considered secure.

Security deals with data that is encoded in such a manner that decoding is difficult or impossible without some secret information, even if the coded or encrypted form of the data is easily read. This secret information is typically passed in a separate channel between users. Optical CDMA is better at obscuring data than at securing it. This chapter will focus on security.

Within the field of security there are a number of branches, which are listed in Figure 9.1. Issues arising in the different branches are distinct though overlapping. Confidentiality assures that only the intended receivers of information actually receive the information. Integrity means that the information received is the same as that transmitted. Availability means that information transmitted is not "lost" or destroyed en route. Authentication allows the receiver to ascertain the identity of the transmitter and vice versa.

Confidentiality can be compromised to various degrees. In the worst case, an eavesdropper can directly read the data. Even if an eavesdropper cannot read the data, the knowledge that two particular people (or computers) are communicating may compromise confidentiality. Some information is available to potential adversaries even if only the traffic patterns can be measured.

The integrity of information assures that the data is not intercepted and altered. This particular area is usually concerned with information at higher layers of the network stack (e.g., application layer) and not with physical or data link layer. It is difficult to imagine how the integrity would be affected (positively or negatively) by the use of OCDMA. Consequently, we will not consider this subject further.

The availability of the information has to do with whether information transmission can be blocked or disrupted in some way. This disruption can occur anywhere in the networking stack. Disruption at the physical layer can occur simply by cutting the communication channel. In the world of optical networking this can happen because a backhoe mistakenly cuts an optical fiber, because a fiber is disconnected inadvertently when a telecommunications company alters its fiber optic connectivity, because of hardware malfunction, etc.

At the physical layer, information can be disrupted by addition of an unauthorized signal.

OCDMA may be advantageous for increasing network availability in the presence of interfering signals, because the signal is spread over many wavelengths. Thus jamming a given OCDMA signal may require significantly more power than jamming a WDM signal. To delve into this issue requires knowledge of the detailed architecture of a given OCDMA network. To date no detailed architecture has been devised that would significantly increase availability over a WDM network. However, OCDMA technology is still in its infancy, and such architectures may well appear as more research is done in this area.

Authentication is an embodiment of the need to trust. Authentication is especially problematic because it requires many steps: (1) an agreed method of coding and decoding; (2) an agreed format for a digital "certificate," which associates a user's identity with an electronic code; (3) a method of generating the certificate; and (4) a general set-up whereby we can rely upon and trust the certificate. Any of these steps could be attacked to compromise authentication. At the physical and data link layer, the method of coding and decoding could be provided by an OCDMA transmission format. However, no such scheme has been proposed to date. At the end of this chapter we outline how OCDMA may be applied to this problem.

9.2.2 OPTICAL CDMA NETWORK SECURITY ISSUES

Imagine that you are trying to design a secure network. As a secure-network designer, you will want to consider the strategies of an eavesdropper who is trying to compromise network security. To provide security, you will devise strategies to thwart the eavesdropper. We will refer to such strategies as countermeasures. In this section, eavesdropper strategies and some of your possible countermeasures will be mentioned. It is only by repeated efforts at alternately optimizing strategies for the eavesdropper and devising your countermeasures that secure networking can be achieved. Since the field of OCDMA networking is still quite young, this discussion represents only a starting point. Nevertheless, it provides a broad foundation for analyzing the security properties of OCDMA systems, even as OCDMA technology matures and new techniques emerge.

The most commonly used architecture for OCDMA networks is the star network (see Figure 9.2). In the star network, data from all users are collected at some central point and then distributed to all users. Such an architecture is typical of a LAN (local area network). Indeed the Ethernet protocol was designed with this architecture in mind. This architecture is not particularly efficient for resource allocation, but is simple and cheap. Unfortunately, this architecture also makes life easy for the eavesdropper.

An eavesdropper in an OCDMA network may tap signals from various locations within the network. He may commandeer an authorized user terminal or may tap signals from network fibers. For the purposes of code interception and to attack the confidentiality of the data, it is advantageous to tap isolated user signals, avoiding

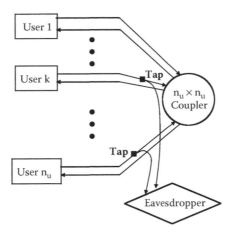

FIGURE 9.2 Schematic of a star network, a commonly used OCDMA network architecture. In this architecture, users receive signals from all other users. The links are totally passive. An eavesdropper tapping into the optical fiber can isolate an individual user by tapping into the fiber from that user to the coupler.

the multiple user interference (MUI) that is characteristic of CDMA systems and giving the eavesdropper access to an isolated user signal. A typical broadcast star LAN carries individual user signals over approximately 50% of its total fiber length (the user-to-star coupler links). This gives an eavesdropper much opportunity to tap into individual user signals.

The eavesdropper may tap off the signal from one of the users and read the optically coded signal of a targeted user. If that user is using the most commonly used optical transmission format of on-off keying, then the eavesdropper's job is very simple. The eavesdropper does not need to decode the signal; she can just read the ones and zeros directly. It does not even matter how many codes there are; there is no security! Much of the Optical CDMA literature, when speaking of security, has missed this argument.

Figure 9.3 illustrates one of many possible OCDMA schemes. In this case each code is a particular arrangement of different colors at different time slots (known as chips) within a single data bit [10]. It should be clear that if only one data stream is present, the total energy in each bit time is all the information needed to obtain the data bits. Deciphering the data bits when multiple users are present becomes more difficult—especially, as in the case illustrated, when the data streams are asynchronous.

The network designer can design some countermeasures to make the eavesdropper's job a bit more difficult. The network designer might assume the eavesdropper could not listen on the transmit side of any user. But this breaks a cardinal rule of secure design—a good security design must consider *all* possible attack scenarios.

So, the network designer could ask the users to reduce their power level to put their transmissions at a level where the eavesdropper does not have enough

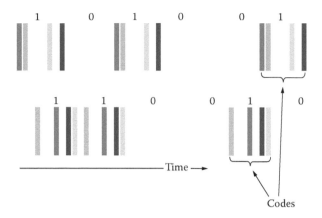

FIGURE 9.3 **(Color Figure 9.3 follows page 240.)** Schematic of two, asynchronous incoherent OCDMA channels. On–offkey modulation is being used. Note that no decoding is needed if only one channel is present.

signal-to-noise ratio to cleanly measure the data. However, this strategy has several problems: (1) the user receivers are already receiving only a small fraction of the transmitted power due to star coupler power splitting, and (2) the user receivers must filter out other optically coded signals from the desired coded signal. Both of these problems reduce the receiver's signal-to-noise ratio. So reducing the transmitter power may not be an option. In addition, even if the transmitter power is reduced, the eavesdropper could tap a little more power to maintain their data detection capability. For security in general, relying on signal-to-noise ratio advantage for the receiver relative to the eavesdropper is a bad idea. Too many factors beyond the network designer's control (e.g., malfunctioning hardware, new developments in receiver technology) can cause this relative advantage to evaporate.

Another bad idea for the network designer is to depend on the secrecy of network parameters for data confidentiality. When doing a security analysis, assume that an adversary knows or may know the communication network parameters. These same principles are applied in the analysis of cryptographic systems, and are often stated in the form of Kerckhoffs's principle, which essentially states that one should assume that the eavesdropper knows everything about the cryptographic algorithm except for the key that each user employs [10].

Clearly, the network designer must think harder! One option for the network designer is to use a different transmission format (other than on-off keying). The user could send out a particular optical code for a "1" and an orthogonal optical code for a "0." We will refer to this technique as *2-code keying*. Now a power detector (no matter how good the signal-to-noise ratio is) could not measure the data. The eavesdropper is now forced to figure out the coding in order to decode the signal and obtain the data, which is a far more complicated and difficult task than tapping the user's signal and detecting the power transmitted during each bit period.

Note that even this simple (yet clever) change has its costs:

- Twice as many codes are needed.
- The increase in codes will increase multiuser interference, which may reduce the number of users.
- The increase in codes adds complexity to the network, which may increase the cost of network management.

This is a common issue with security: increased security has a cost. Reduced performance and increased complexity are common side effects of increased security.

9.2.3 CODE SPACE SIZE

Let us briefly review the reasons that lead to the expectation of OCDMA providing some degree of data confidentiality. Each OCDMA transmitter/receiver pair is assumed to use a specific code. The receiver uses the exact knowledge of the code to separate the transmission from other users transmitting on different codes and from random channel and receiver noise. It is difficult for an eavesdropper to correctly demodulate the OCDMA signal without knowing the code being used, especially if there are multiple users transmitting simultaneously on different codes. If an OCDMA coding scheme has a very large number of possible codes and the transmitted data does not appear to offer any information on the code, then an eavesdropper would have to perform a brute force search through half of the codes, on average, before finding the proper code to demodulate a given user's data. Thus the first measure of the degree of security potentially available from OCDMA encoding is the size of its code space (the number of different codes that might be used by an individual user). This can vary greatly depending on the type of OCDMA code and its parameters.

The assumptions used in a security analysis can strongly affect the results. We often assume that the user has some advantage because the eavesdropper can measure only the transmitted signal and is not aware of the decoded data. This assumption is not always valid. One of the most important code-breaking examples of the twentieth century was the Allies breaking the German Enigma code in World War II. In that case, the Allies used both the plain text (i.e., decoded) message and the coded message in order to figure out how to break the code [9]. This type of code breaking is called a *known-plaintext* attack (as opposed to a *ciphertext-only* attack, which is the assumption of this chapter).

A commonly used assumption in the cryptography community is that potential eavesdroppers are technologically sophisticated, have significant resources, and know a great deal about the signals being transmitted [9]. For an OCDMA network, the eavesdropper would know the data rates, the type of encoding, and the structure of the codes—but not the particular code that an individual user employs. The eavesdropper might gain this kind of knowledge in a number of ways—by bribing the manufacturers of the transmission equipment, for example, or by obtaining and reverse engineering a user terminal. For an OCDMA waveform to provide effective security, an eavesdropper in possession of both detailed knowledge of the OCDMA

TABLE 9.1
Code Space Size for Four Categories of OCDMA Coding

Code Type	Representative Code Type Examples	Parameters for Code Space Calculation	Potential Code Space Size
Time-spreading	OOCs Prime codes EQC codes	Prime code 10,000 time slots	~10^2 Codes [11–13]
Time-spreading/ wavelength- hopping	Prime-hop codes Multi-λ OOCs	Prime-hop code 32 wavelengths 1024 time slots	~10^{30} Codes [3,14]
Spectral amplitude	Hadamard codes m-Sequences	m-Sequence code 511 mask elements	~10^4 Codes [15]
Spectral Phase	Hadamard codes Random codes	Random code 511 mask elements	~10^{150} Codes [16]

waveform and a user terminal must still not be able to decipher a user's data without also knowing the exact code in use by that particular user. This can only be accomplished if the total number of possible codes (known as the *code space* size) of the OCDMA waveform is large enough that the eavesdropper cannot perform an exhaustive search of the code space and obtain a given user's code in any reasonable amount of time.

Table 9.1 compares the code space sizes of four common categories of OCDMA techniques. The code sizes are approximate with the exact number dependent on a number of code parameters.

The first category, time-spreading codes (using a single wavelength), includes optical orthogonal codes [11], prime codes [12], and EQC codes [13]. These codes all have small code spaces of the same order as the square root of the time-spreading factor. These code spaces are not large enough to deter brute-force searching techniques. However, they do tend to work well in an asynchronous network.

Kitayama and coworkers [17] have implemented a larger code space through coherent processing and the use of Gold codes. This larger code space is the size of the time-spreading factor, which they have demonstrated as large as a factor of 127. Even this larger code space is not large enough to deter brute-force searching techniques. In addition, Gold codes do not work well in an asynchronous network.

The second category, time-spreading/wavelength-hopping codes, can be viewed as an extension of time-spreading codes into two dimensions (time and wavelength), and can also be viewed as an analog to RF frequency hopping [14]. These codes can be designed to have a very much larger code space size than the one-dimensional time-spreading code. The resulting code space sizes can be large enough to prevent a brute force code space search from being successful in any reasonable amount of time [3].

For example, consider time spreading with 961 time slots using an optical prime code with $p = 31$ (note 31 is prime and 961 is 31^2). In the second dimension, we consider a transmitter with $N_W = 31$ wavelengths. The transmitter would fill 31 of the 961 time slots with light at one of the 31 wavelengths with the restriction that no wavelength gets used twice. The first of the time slots with a one can be at any of the 31 wavelengths. The second time slot can use any of the remaining 30 wavelengths; the third slot can use any of 29 wavelengths, etc. Then, there are $N_W!/(N_W - p)! = 31!/(31-31)!$ different wavelength combinations that can be used. The large number of codes does not imply that any more users can be added to the network compared with time spreading only. Nor are these codes independent. That is, many of the different wavelength combinations will have very poor cross correlation properties. So, if one particular time spreading/wavelength code is used, the other time-spreading/wavelength-hopping codes must be carefully selected to have good cross correlation properties. As long as the code assignment is secure, then the eaves-dropper will have little information to know a priori what the assigned code is for any particular user.

The codes described in this example have the security advantage that all users have identical optical spectra. So, the optical spectrum reveals no information about the user (either in number or identity). That the optical spectra are identical does not prevent an eavesdropper from learning about the user's code (and thereby to obtain the transmitted data); it only makes the eavesdropper's task more difficult, which is good from the network designer perspective.

The third and fourth categories in Table 9.1 represent spectral encoding tech-niques — amplitude encoding and phase encoding, respectively. Spectral amplitude encoding [15] relies on code sequences with particular properties to maintain a reasonable degree of orthogonality among different users' coded signals. The spec-tral amplitude codes in [15] require either Hadamard sequences or maximal length sequences (m-sequences) as their basis, and these codes are still fairly limited in code space size. And while time-spreading codes may be implemented with code lengths of the thousands or even tens of thousands, depending on the data rate, implementation constraints for spectral coding masks limit feasible codes to lengths of a few hundred or so. For a code length of 511 amplitude mask elements, one can calculate that there are 48 different m-sequences that could be used as codes [18], and each of these sequences can be shifted by one or more code elements to produce a distinct code. This produces a maximum of about 25,000 (48×511) possible codes. This is a considerably larger code space than that produced by most time spreading codes, but still quite small compared with time-spreading/wavelength-hopping codes.

Spectral phase encoding has similar code mask implementation constraints to spectral amplitude encoding. However, analysis has shown that spectral phase encoding may be able to support a reasonably large number of simultaneous users at low bit error rates (BERs) by employing random codes [16]. These performance calculations apply to average performance, though, and many possible code sets chosen randomly might have well below average BER performance. Still, a network management function could select a set of random codes for some desired number of users, and could then pseudo-randomly refine the set of codes until the overall performance of the code set met the desired BER specifications. The resulting set of codes could still appear random to an eavesdropper trying to guess which individual codes had been selected, and he would thus have to search a large fraction of the code space before being successful. Random code choice allows the code space to be very large indeed. For example, a 511-element phase mask can generate 2^{511} ($\sim 10^{150}$) possible codes. Time-spreading/wavelength-hopping codes and spectral phase codes appear to be two of the most promising code types for generating OCDMA code spaces that are large enough to prevent successful brute force code search attacks. A very large code space is necessary (though not sufficient) for security. In addition, the optical spectrum for both of these codes reveals nothing about the user, which also is an important prerequisite for security.

9.2.4 CODE PROPERTIES RELEVANT TO SECURITY

Two particularly important mathematical properties of OCDMA codes for both communication performance and security are their auto-correlation and cross-correlation functions. Let a code be represented by a vector of elements, a, where each element a_i consists of a scalar value. These values are usually restricted to a small number of possibilities (e.g., 0 and 1). Each element specifies the value of a coding parameter that determines the properties of the coded, transmitted waveforms in time and wavelength. For the example of a time spreading prime code with 961 time slots, the code would consist of 961 elements, each of which is a 1 or a 0. A 1 indicates the presence of a code pulse (of which there are 31) in a given time slot and a 0 indicates no code pulse in the time slot.

If we consider two codes, a and b, of length N, the correlation functions can be written as:

$$C_j = \sum_{i=1}^{N} a_{j-i} b_i \qquad (9.1)$$

where the auto-correlation is obtained if a and b are the same codes and the cross correlation is obtained if a and b are different codes.

Ideally, the autocorrelation function should be a delta function $C_j = \Delta(j)$. If the auto-correlation differs from this ideal, the receiver may have a harder time maintaining timing synchronism. The effect on the eavesdropper of a poor auto-correlation has not been investigated.

The cross-correlation function is important both for the user and the eavesdropper. Ideally, the cross correlation should be zero (i.e., codes should be perfectly orthogonal). For the user, the cross correlation is directly related to the amount of cross talk between channels. If the cross correlation between two codes that are simultaneously being used in the network is large, then those users see a large "coherent" background associated with the wrong user, causing interference between the two channels. For the eavesdropper, any cross correlation between a guessed code and the actual code allows the eavesdropper to (1) partially recover the decoded data (although suffering a poorer signal to noise ratio compared with the authorized user) and/or (2) hone in on the correct code. Clearly the choice of codes and, specifically, their cross-correlation properties, can have a strong impact on network security.

For an asynchronous network the requirements are even more stringent; partial cross-correlations should be small. That means not only the sum in Equation 9.1 should be small, but any partial sums should be small as well. This is most important for the users because large partial sums increase the noise level. The eavesdropper may be able to take advantage of large partial cross correlations as well.

Other properties of OCDMA codes that affect security will depend upon the particular type of code. For example, the numbers of times slots and wavelengths used in a time-spreading/wavelength-hopping code will affect not only the code space size, but will also affect an eavesdropper's ability to correctly obtain the code through specialized signal tapping techniques described below. The number of code elements in spectral amplitude and phase codes similarly affects an eavesdropper's ability to detect such codes.

9.2.5 Optical CDMA Secure-Network Designer Strategy

Brute force searching for an individual user's code is a very inefficient attack strategy for the eavesdropper whenever the code space is large. Intelligent eavesdroppers will seek other forms of attack if they are available. For the OCDMA techniques with large code spaces described above, there are more efficient forms of attack. One form of attack considered here is based on the observation that many OCDMA transmitter designs broadcast the very thing that is the key to keeping the user's data confidential: the code word itself. If an eavesdropper can detect the transmitted signals accurately enough, he may be able to solve for the code word, which he can then use to detect subsequently transmitted information until the code word is changed.

You can try to make it difficult for an eavesdropper to accurately detect the signal in the channel. If the signal-to-noise ratio is low, the eavesdropper will not be able to accurately solve for the code. You can notify users to send low power signals. The eavesdropper will then have a low signal-to-noise ratio, which makes accurate channel measurements difficult. Solving for the code becomes essentially impossible without accurate channel measurements. The eavesdropper's ability to solve for the code can also be decreased by increasing the code complexity, which can decrease the eavesdropper's signal-to-noise ratio per code element. With this overall approach, the eavesdropper's ability to solve for the code can be determined by classical detection theory [19]. The degree of confidentiality produced by this approach will depend on the signal-to-noise ratio that an eavesdropper can attain when attempting to detect the user's coded signals.

In addition, the users can change codes frequently — more frequently than an eavesdropper could detect the channel waveform and solve for the code. The required rate of code reconfiguration depends on the time required for an eavesdropper to accurately detect the channel waveform and solve for the code. This time depends, in turn, on the signal-to-noise ratio that the eavesdropper is able to obtain, and on the code complexity. The effectiveness of code reconfiguration thus depends on how difficult the transmitter can make it for the eavesdropper to detect codes by observing the channel.

Unfortunately, code reconfiguration comes at an enormous cost in network management overhead. If code reconfiguration happens often, the network manager must make sure the new codes are sent securely and that all parties on the network are able to synchronously change codes. Transmitters and receivers that do not synchronously change codes lose data or worse, their ability to communicate at all.

9.3 QUANTIFYING OCDMA CONFIDENTIALITY

9.3.1 LESSONS FROM THE RADAR DETECTION PROBLEM

We will now assume that the network designer can produce a design that avoids the pitfalls discussed in previous sections of this chapter. This will mean avoiding reliance on obscurity rather than security techniques, avoiding on-off-keying, and designing an OCDMA code that produces a very large code space — large enough that brute force searching for an individual user's code word takes a very long time. In this case, an eavesdropper must rely on techniques such as code detection. We now consider the effectiveness of this eavesdropping strategy.

The exact techniques required for code detection depend on the type of code being transmitted; therefore, it is necessary to choose a particular type of code to quantify the effectiveness of this type of attack. The first part of this section examines the confidentiality performance of time-spreading/wavelength-hopping encoding [3,14] against a code detection attack. Time spreading encoding on a single wavelength [10,11] can be considered as a special case of time-spreading/wavelength-hopping. The analysis presented here treats the eavesdropper's code interception problem as a problem in classical detection theory [19]. The eavesdropper taps a coded transmission of a particular user and performs the necessary calculations to derive the transmitter's code word from these transmissions. The resulting code will have some probability of error, which will depend strongly on the signal-to-noise ratio at the eavesdropper's receiver.

In theory, an eavesdropper can use a receiver that is highly similar to a radar receiver to intercept this type of signal and determine the code [19,20]. In a classic radar receiver, a time interval that is expected to contain a radar return pulse is divided into a number of shorter intervals, and each short interval is examined to see if it contains a signal pulse or not. For OCDMA code detection, the eavesdropper can divide each data bit duration into n_c time intervals, or "bins" (Figure 9.3), and determine whether an energy pulse is present or not in each one. This can be done by implementing a filter that is matched to an individual code pulse and sampling the output of the filter once per time bin.

The performance of this type of receiver can be determined using the mathematics of classical radar detection theory. The optimum implementation of this type of receiver would be a coherent detection receiver and an exact matched filter. However, a more likely implementation would be an optical amplifier, followed by an optical filter that is approximately matched to the code pulses, with a photodiode used to detect the output of the optical matched filter [21,22]. The output of the

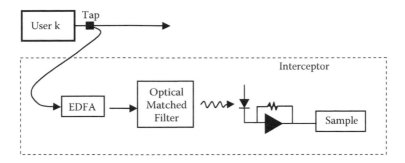

FIGURE 9.4 Incoherent receiver for code intercepting detector.

electronic detector is then time sampled. Such a code interceptor is shown in Figure 9.4.

This code interception strategy generalizes to time-spreading/wavelength-hopping coding in a straightforward way. Given a code using n_c time chips and n_λ wavelengths, the receiver structure in Figure 9.4 can be replicated n_1 times. If n_1 is too large for this to be practical, a reasonable number of wavelength channels can be implemented and scanned sequentially over the different wavelength bands covered by the coded signal. This would produce a trade-off between the number of wavelength channels implemented in the code intercepting receiver and the time required to detect the code with a given degree of statistical reliability.

The implementability of the required detector structures for the eavesdropper is also a significant issue. Reasonable approximations to the required optical matched filters are currently available, and should not pose great difficulty for modest numbers of different encoding wavelengths. But time sampling the envelope-detected outputs of these filters quickly and accurately enough is quite difficult for high data rate transmissions. For example, if sampling were done in real time for a user signal at 1 Gbit/sec. and $n_c = 961$, each wavelength channel would have to be sampled at a rate of nearly 1 THz. Real-time sampling technology is currently available at rates of 20 GHz (for 8-bit samples) in commercial, off-the-shelf oscilloscopes. This is well short of the required THz rate for the above example.

A number of possibilities exist for increasing the effective sampling rate, however. The technique of *equivalent time sampling* is currently used in high-bandwidth sampling oscilloscopes, and allows very high effective sampling rates, up to 800 Gb/sec in commercial instruments. Optical means for equivalent time sampling have also been demonstrated [23]. These techniques require good time synchronization and moderately large numbers of input sampling passes (each sample would be taken from a different data bit, in the code interception context). This would significantly increase the time required to process and detect a given code word. A second possibility for implementing high effective rate time-sampling of time-spreading/wavelength-hopping signals would be to capture one or more user data bit transmissions in an optical recirculating loop or use a stretched time sampler. Either of these techniques would allow the actual sampling to be done at a reasonably low rate. This could produce the required effective sampling rates without requiring large numbers of data bits to be processed.

The detector for the eavesdropper is assumed to consist of a set of optical filters, followed by photodetectors with timing measurement on the output of the photodetectors. The optical filters could be implemented with a grating in a spectrometer. A fast array with as many as 40 photodetectors is now commercially available [24]. So, a technically sophisticated eavesdropper could listen to the optical code with a relatively small number of components.

The figure of merit used for code interception performance calculations is the probability that the eavesdropper can detect the user's entire code word with no errors, denoted by $P_{correct}$. This probability will depend on the type of detection processing and on the amount of time the eavesdropper observes the user's signal for each detection; it can be calculated from two quantities that are staples of classical detection analysis — the probability of missing a transmitted pulse in a given time bin, P_M, and the probability of falsely detecting a pulse in a bin where none was transmitted, P_{FA}. If the code interceptor makes a code word decision based on observing the transmitted signal for a single data bit interval, the overall probability of error-free code word detection is given by

$$P_{correct} = (1 - P_M)^W (1 - P_{FA})^{(n_C n_\lambda - W)} \qquad (9.2)$$

The first term represents the probability of not missing any of the W pulses that are transmitted during a data bit. The second term is the probability of detecting pulses where there are none in any of the $n_c\, n_\lambda - W$ time bins where pulses are not transmitted during a data bit. This equation is identical in form to the detection equation used in radar analysis.

While this analysis is fairly straightforward, the eavesdropper may have additional information that makes the probability of obtaining the correct code word even more likely. For instance, Equation 9.2 does not assume the eavesdropper knows that there are exactly W pulses in the code. If the eavesdropper knew there were exactly W pulses, they could drop or add pulses for those bins that had been just above or below the detection threshold, respectively. This would increase the probability of obtaining the correct code.

The eavesdropper's ability to correctly detect user code words is strongly dependent on the SNR at the intercepting receiver, and it follows that the degree of confidentiality provided is also a strong function of this SNR. Since the eavesdropper's SNR is a function of a number of system design and operation parameters, this means that the degree of confidentiality provided by OCDMA techniques will also be a function of these system design and operation parameters. Since the degree of confidentiality of user data is dependent on the SNR at the eavesdropper, it is important to quantify how low this SNR could be made through intelligent system design. This design is not completely straightforward, though, because it must involve a trade-off between communication performance and confidentiality for the authorized users. A detailed analysis of this trade-off is presented in [6]. We summarize the results here.

Confidentiality is *decreased* by an increase in the eavesdropper's tapping efficiency, by an increase in the number of taps in the star coupler (which reduces the fraction of transmitted power that reaches each authorized user and requires each user to transmit more power), or by a decrease in the eavesdropper's receiver noise level relative to the authorized user's receiver noise level. Confidentiality is *increased* by an increase in the combining efficiency of the user receivers (allowing an overall decrease in transmitted user power to maintain an acceptable BER); or by an increase in the weight of the code words (which divides the energy per bit into more, hence lower energy, code pulses). Equation 9.2 also implies a further increase in confidentiality if the length of the code, $n_c\, n_\lambda$, is increased.

9.3.2 An Example

Figure 9.5 shows an example of confidentiality performance vs. system capacity for a straw man set of system design parameters. The solid curve plots the eavesdropper's approximate probability of error-free code detection (integrating signal energy over one data bit period) vs. the fraction of theoretical system capacity that can be attained for a specified maximum BER. The straw man design specifies 100 potential users connected to a broadcast star network with $n_u = 100$ taps. The users each employ

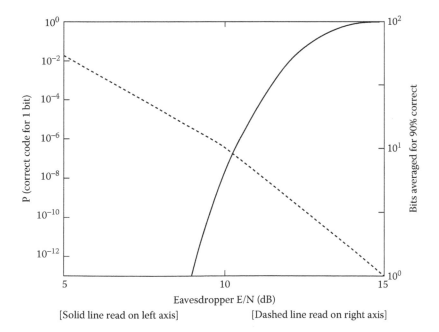

[Solid line read on left axis]   [Dashed line read on right axis]

FIGURE 9.5 Eavesdropper's detection performance for 2-code keying. As the eavesdropper's signal to noise improves, the probability of estimating the correct code improves (solid). Even with poor signal to noise, the eavesdropper can estimate the correct code if the eavesdropper uses data from multiple bits (dashed). Even with 5 dB bit energy-to-noise ratio, the eavesdropper can estimate the correct code within 100 bits. With more energy the eavesdropper needs fewer bits.

time-spreading/wavelength-hopping codes with $n_c = 961$ timeslots, $n_\lambda = 31$ wavelengths, and $W = 31$ code pulses per data bit. These parameters would be produced by a 31,31 Prime Hop Code as specified in [25], although they may apply to other types of codes as well.

Most encoding schemes for OCDMA use code words that are relatively far apart in Hamming distance; this allows relatively good orthogonality properties among multiple users transmitting simultaneously. However, if the eavesdropper knows the structure of the code (e.g., that Prime Hop Codes are being used), then an intercepted code word — which may contain detection errors — can be compared with the set of allowable code words. The allowable code word nearest in Hamming distance to the intercepted code word would then be chosen. In this case, the coding structure — designed primarily for good cross correlation properties — will function much like an error correcting code for the eavesdropper, possibly allowing the eavesdropper to take an intercepted code word with errors and correct the errors. If a completely random coding scheme were employed, where any possible combination of code chips could represent a user's code word, then an eavesdropper could not improve its interception performance in this way. In time-spreading/wavelength-hopping coding schemes, however, such random coding would lead to variable weight codes, which is problematic for maintaining the cross-correlation properties among multiple user codes that are required for good communication performance with multiple simultaneous users.*

A further, and more dramatic, improvement in the eavesdropper's code interception performance can be obtained by processing and combining code transmissions from multiple data bits. The eavesdropper can use exactly the same detector structure but can accumulate samples in each of n_c bins (Figure 9.3) over multiple data bits. The eavesdropper must maintain chip synchronization so that the same bins can be sampled repeatedly on multiple data bits. The case of greatest interest for multiple bit combining is where the transmitter uses 2-code keying (remembering that on-off-keying provides no security). Accumulating multiple bits from a 2-code keyed OCDMA data stream using time-spreading/wavelength-hopping encoding produces the superposition of the two code words, C1 and C2, in the eavesdropper's detector. Since the codes are designed to be as mutually orthogonal as possible, it is almost certainly possible to separate the two individual code words from their superposition. For example, it is quite simple to examine the superposition of two prime codes and determine the individual code words, especially if the two code words are synchronized in time [12]. Note that the eavesdropper's detected superposition of C1 and C2 will always be synchronized if they are from a single transmitter using 2-code keying and the eavesdropper has attained bit synchronization. The eavesdropper's detection performance for combining multiple bits in this manner is shown by the red curve in Figure 9.5, which plots the number of bits that must be combined to achieve a 90% probability of correct code word detection at various signal-to-noise ratios. Note that even at low signal-to-noise ratios where the

---

* In spectral phase encoding schemes, random coding may be feasible for a reasonable number of simultaneous users. [6,16] The confidentiality properties of spectral phase encoded OCDMA are similar to those of time-spreading/wavelength-hopping encoding, and are analyzed in detail in [7].

probability of correct detection is negligibly small for single-bit detections (the black curve in Figure 9.5), a 90% probability of correct detection can be obtained by combining fewer than 100 bits.

9.3.3 What can the Secure-Network Designer do?

The example in the preceding section is representative of what can be achieved with currently proposed OCDMA technology [6,7]. Changing the codes frequently can increase the confidentiality somewhat. But to achieve strong confidentiality against code detection attacks, code reconfiguration rates probably need to approach the data rate, since the eavesdropper's advantage from combining multiple data bits increases quite rapidly.

The confidentiality of OCDMA encoding can be broken convincingly given either a high enough SNR or sufficient time to accumulate and combine signal energy from encoded signal transmissions. The fact that a potential eavesdropper may have to expend a great deal of resources to successfully implement a code detector capable of breaking this confidentiality is significant, though. It implies that OCDMA may offer confidentiality protection against adversaries who lack resources to develop, procure, or use such technology. This degree of confidentiality may be suitable for protecting information that is not of high enough value for a potential adversary to justify the expense and difficulty of implementing the required detector.

The standard technique for providing data confidentiality today is source cryptography. It is instructive to compare the confidentiality characteristics of OCDMA encoding with those of cryptography to assess the usefulness of OCDMA for this type of security. The confidentiality of time-spreading/wavelength-hopping OCDMA can be broken, in theory, in a very short amount of time (for example, 100 bits) even at relatively low SNRs and for reasonably complex codes. By comparison, the amount of time necessary to break the confidentiality of a state-of-the-art encrypted signal is usually measured in tens of years or more. Furthermore, the degree of confidentiality from OCDMA encoding is dependent on the many design parameters of the communication system that affect the amount of signal power available to the eavesdropper. In contrast, the confidentiality of an encrypted signal depends in no way on the design of the communication system.

Somewhat greater confidentiality can be obtained through OCDMA encoding if the eavesdropper can be forced to detect multiple signals simultaneously. However, this advantage is lost if a single user ever transmits without other users also transmitting, or if the eavesdropper is able to isolate individual user transmissions in a particular fiber. Cryptography has no such limitations. A variety of different OCDMA encoding techniques not specifically considered in this chapter display these same features [6,7] and are not a substitute for cryptography.

9.4 OCDMA FOR AUTHENTICATION

While OCDMA appears to provide security that is weak compared with encryption there are cases when even weak security can be important. Consider an example where an Unmanned Air Vehicle (UAV) were designed to provide a communications hub capable of providing a high-bandwidth optical link to ground vehicles. In order to provide a relatively inexpensive optical terminal, a modulatable retro-reflector might be used on the ground. In this case, the users of the ground vehicle would want some assurance that the light illuminating the vehicle is from an authenticated source, not from an adversary seeking to locate and destroy it. If the signal were in the form of an optical code, the code could be used for secure enough authentication that the ground terminal would be willing to uncover its retro-reflector.

Once the retro-reflector was uncovered a link could be established between the ground terminal and the UAV using higher level authentication tools. The higher-level authentication would provide strong security but this cannot happen until the link is established. There indeed may be more examples where optical CDMA provides a security role for optical networks; however, our work has shown that the security provided by optical CDMA is not as strong as previously thought.

9.5 CONCLUSIONS

OCDMA provides security at the physical layer of a communication network. Some of the types of security shown in Figure 9.1 are more appropriately placed at the physical layer than others. For example, cryptography (usually done at the application, transport, or network layers) provides much greater confidentiality than any OCDMA techniques that have been proposed so far in the literature. On the other hand, most OCDMA waveforms in the literature are inherently spectrally spread, and may be able to provide some of the protections classically associated with spread-spectrum waveforms, such as jamming protection and covertness [8]. As noted above, OCDMA might also be used to provide a form of authentication at the physical layer. The suitability of OCDMA for providing these types of security remains an open topic for research.

Even though cryptography provides superior confidentiality performance, OCDMA can provide some confidentiality advantages compared with standard optical communication technologies such as WDM, where a commercial off-the-shelf detector can be purchased to read the data. However, this is only true if the modulation format is *not* on-off-keying. Whether or not this degree of confidentiality is sufficient for a given purpose depends largely on the value of the information being protected, and the likelihood that an adversary will be willing to expend the resources necessary to read the information.

It is clear that source cryptography provides a much greater degree of confidentiality than does OCDMA encoding. In principle, this conclusion applies to any form of encoding that can be represented by a linear, time-invariant transfer function. On the other hand, an intelligently encoded OCDMA signal can force a potential eavesdropper to implement a sophisticated and possibly expensive detector in order to be able to break the user's confidentiality. Rapid reconfiguration of codes can increase the difficulty of interception but also makes the network management more difficult.

REFERENCES

[1] Karafolas, N., Uttamcandani, D. (1996). Optical fiber code division multiple access networks: a review. *Optical Fiber Technology* 2:149–168.

[2] Iverson, K., Hampicke, D. (1995). Comparison and classification of all-optical CDMA systems for future telecommunication networks. *Proc. of the SPIE* 2614:110–121.

[3] Tancevski, L., Andonovic, I., Budin, J. (1995). Secure optical network architectures utilizing wavelength hopping/time spreading codes. *IEEE Photon. Tech. Lett.* 7(5): 573–575.

[4] Torres, P., Valente, L. C. G., Carvalho, M.C. R. (2002). Security system for optical communication signals with fiber Bragg gratings. *IEEE Trans. Micro. Theory and Tech.* 50(1):13–16.

[5] Sampson, D. D., Pendock, G. J., Griffin, R. A. (1997). Photonic code-division multiple-access communications. *Fiber and Integrated Optics* 16:129–157.

[6] Shake, T. H. (2005). Security performance of optical CDMA against eavesdropping. *J. Lightwave Tech.*, 23(2): 655–670.

[7] Shake, T. H. (2005). Confidentiallity performance of spectral phase encoded optical CDMA. *J. Lightwave Tech.*, 23(4): 1652–1663.

[8] Simon, M. K., Omura, J. K., Scholtz, R. A., Levitt, B. K. (1985). *Spread Spectrum Communications*. Rockville, MD: Computer Science Press.

[9] Schneier, B. *Applied Cryptography*, 2nd ed. New York: John Wiley & Sons.

[10] Kwong, W. C., Perrier, P. A., Prucnal, P. R. (1991). Performance comparison of asynchronous and synchronous code-division multiple-access techniques for fiber-optic local area networks. *IEEE Trans. Commun.* 39(11): 1625–1634.

[11] Salehi, J. A. (1989). Code division multiple-access techniques in optical fiber networks: Part I: Fundamental principles. *IEEE Trans. Commun.* 37(8):824–833.

[12] Yang, G. -C., Kwong, W. C. *Prime Codes*. Boston, MA: Artech House.

[13] Marhic, S. V., Kostic, Z. I., Titlebaum, E. L. (1993). A new family of optical code sequences for use in spread spectrum fiber-optic local area networks. *IEEE Trans. Commun.* 41(8):1217–1221.

[14] Fathallah, H., Rusch, L. A., LaRochelle, S. (1999). Passive optical fast frequency-hop CDMA communications system. *IEEE Journal of Lightwave Tech.* 17(3): 397–405.

[15] Kavehrad, M., Zaccarin, D. (1995). Optical code-division-multiplexed systems based on spectral encoding of noncoherent sources. *IEEE Journal of Lightwave Tech.* 13(3):534–545.

[16] Salehi, J. A., Weiner, A. M., Heritage, J. P. (1990). Coherent ultrashort pulse code-division multiple access communication systems. *IEEE Journ. Lightwave Tech.* 8(3): 478– 491.

[17] Matsushima, K., Wang, X., Kutsuzawa, S., Nishiki, A., Oshiba, S., Wada, N., Kitayama, K. (2004). Experimental Demonstration of Performance Improvement of 127-Chip SSFBG En/Decoder Using Apodization Technique. *IEEE Phot. Tech. Lett.* 16(9): 2192–2194.

[18] Ojanpera, T., Prasad, R., eds. (1998). *Wideband CDMA for Third Generation Mobile Communications*. Boston: Artech House, see p. 110.

[19] Helstrom, C. W. (1968). *Statistical Theory of Signal Detection*, 2nd ed. New York: Pergamon Press.

[20] Skolnick, M. I. (2001). *Introduction to Radar Systems*, 3rd ed. Boston: McGraw Hill.

[21]   Humblet, P. A. (1991). Design of optical matched filters. IEEE GLOBECOM '91 2:1246–1250, 2–5.

[22]   Humblet, P. A., Azizoglu, M. (1991). On the bit error rate of lightwave systems with optical amplifiers, *IEEE J. Lightwave Tech.* 9(11): 1576–1582.

[23]   Han, Y., Jalali, B. (2003). Photonic time-stretched analog-to-digital converter: Fundamental concepts and practical considerations. *IEEE J. Lightwave Tech.* (21):3085–3103.

[24]   http://usa.hamamatsu.com/en/products/solid-state-division/ingaas-pin-photo-diodes/image-sensor-array/g8909-01.php?&group=1# (December 2004).

[25]   Tancevski, L., Andonovic, I. (1994). Wavelength hopping/time spreading code division multiple access systems. *Elect. Lett.* 30(17): 1388–1390.

10 Optical CDMA Network Architectures and Applications

Robert J. Runser

CONTENTS

10.1 INTRODUCTION

The flexibility and diversity of optical CDMA approaches are reflected in the variety of applications that have been proposed since the first optical CDMA experiments were published in the mid-1980s. This chapter reviews many of the proposed and demonstrated applications of optical CDMA techniques that have been studied over the past two decades. Although this review is not comprehensive, it paints a broad landscape of potential applications of optical CDMA in commercial networks, computer systems, and military environments. Since optical CDMA is currently an active area of research, it would be premature to single out just one "killer application" for the technology. As a result, this chapter covers a wide spectrum of possible applications that have been discussed in the research literature. Since full chapters of this book are already devoted to the use of optical CDMA in network security (Chapter 9) and hybrid OCDMA/WDM overlay networks (Chapter 7), emphasis in this chapter has been placed on other applications areas to illustrate the broad applicability of optical CDMA to novel areas in communications. Previous chapters have also introduced various technologies for realizing optical CDMA networks. The mapping of a particular optical CDMA technique to a given application can depend on many factors including network scalability, device integration, system cost, and environmental robustness. Although many application demonstrations have focused on a particular type of optical CDMA, most optical CDMA technologies can be applied to many of the applications presented here.

In as much as the applications motivate the continued pursuit of optical CDMA, the network architectures provide a foundation on which real optical CDMA networks can be constructed in a variety of different environments. This chapter provides some of the novel architectures that have been proposed for interconnecting nodes using optical CDMA and their advantages for transmission in both fiber and free space.

10.2 LOCAL AREA NETWORKS

Many proposed and demonstrated optical CDMA systems have targeted high bandwidth, multiuser local area networks (LANs) as a key driving application and use for the technology. Indeed, applications in the LAN environment have motivated many networking technologies including the first practical implementation of Ethernet [1], which is the most widely deployed LAN protocol today. The carrier sense multiple access and collision detect (CSMA/CD) methods developed for Ethernet provide a means for individual users to statistically share data capacity on a single broadcast infrastructure on demand without synchronization or coordination among the users. The simple protocols for Ethernet have evolved over time to accommodate the need for increased bandwidth. The low cost and practical implementation of optical fiber and integrated optoelectronics have recently enabled LAN data rates to accelerate from 10 and 100 Mb/sec for legacy Ethernet to 1 and 10 Gb/sec per user supported by Gigabit Ethernet. To address the challenges of new high-speed services, Ethernet has migrated to a switched architecture where the arbitration on a given path is provided by electronic switches connected via optical fiber to Gigabit Ethernet clients.

While Ethernet and Gigabit Ethernet continue to play a dominant role in LAN environments, the scalability of these networks is presently limited by the electronic technology in the hubs and switching nodes used to actively resolve contention on the LAN. Upgrading Gigabit Ethernet networks to individual user data rates beyond 10 Gb/sec will prove to be challenging in the near term and not necessarily competitive with the general trends followed by previous generations of Ethernet products. This limitation is especially important given the many terabits per second of fiber bandwidth theoretically available to individual network users. Ultimately, the electronic bottleneck at switching nodes presently limits the scalability, flexibility, and cost of the overall LAN network and places important constraints on the adoption of new, higher data rate interfaces that will be developed within the next decade.

Optical CDMA networking techniques possess many of the same advantages that the original designers of Ethernet had in mind during the formulation of the first CSMA/CD protocols for shared media [1]. Many Optical CDMA implementations and coding techniques are designed to function over broadcast network architectures, without requiring synchronization among the individual users. Low latency, on-demand connections can be established through the use of unique codes that provide the physical address of the desired destination. Optical CDMA has also been shown to be a flexible technology that can accommodate various data rates and classes of service over the same network infrastructure. The interconnection among the users via an optical star network provides fully transparent links to all destinations without requiring bandwidth limiting electronics for routing or switching at the central node. Optical CDMA has the potential to scale and adapt gracefully to dynamic service needs and unpredictable network loading conditions.

While optical CDMA has the potential to provide scalability beyond the limits of today's LAN environments that are still dominated by electronic switching technologies, it is also an attractive approach when compared to other multiple access optical networking technologies that have been proposed over the last two decades. In many optical layer multiple access schemes, the underlying technology is initially designed to support the maximum number of potential users, which requires building the network interfaces to the most stringent design tolerances. For instance, in an optical time division multiple access (OTDMA) network, the maximum number of users is determined by the minimum optical time slot width that can be accommodated in a fixed time frame, which is related to the period of the maximum user data rate. An OTDMA network that can support 25-psec time slots can accommodate a maximum of 40 simultaneous users operating at 1 Gb/sec. Similarly, wavelength division multiple access (WDMA) networks require stringent control over the optical wavelength spacing and filter bandwidth to accommodate individual users on uniquely assigned optical channels. A WDMA LAN operating in the C-band region of the infrared communications spectrum (wavelengths in the range of 1530 and 1565 nm) could accommodate about 44 users using a wavelength spacing of approximately 0.8 nm (or 100 GHz in the frequency domain). In both cases, any node added to the network must conform to the design tolerance for the maximum network capacity regardless of how many users are active on the network at initial deployment. This increases the cost of the optical network interfaces and the complexity of the technology required to implement optical multiple access LANs.

In flexible networks of the future, it will be desirable for networks to change their quality of service (QoS) and data rate provided to the individual users on the fly according to their current application requirements. Optical CDMA provides a more flexible and robust bandwidth sharing technique for adaptable network interfaces. The cost and complexity of each optical CDMA node can be designed for the data rate and service desired at a given node. Additionally when users do not require the maximum data rate or the highest QoS, these codes can be assigned to other users to free available bandwidth for high priority applications. The optical network interfaces can also be simplified and potentially lower cost than equivalent implementations of optical LANs using OTDMA and WDMA. Using two-dimensional incoherent wavelength-time coding techniques such as those described in Chapter 2, the weight of the code, determined by the number of wavelengths, establishes the quality of service (QoS) for a particular session. It is not necessary for nodes that are assigned to lower QoS levels to access all wavelengths available on the optical CDMA LAN since encoding and decoding the full weight of the code is not required. Through trade-offs in system performance, cost, network size, and QoS, optical CDMA provides an inherently flexible optical multiple access environment that retains many of the same benefits of traditional electronic LANs but with the scalability of optics.

This section reviews the architectures and deployment environments for optical CDMA in various LAN environments. Within the context of LAN applications, optical CDMA is treated as a multiple access network where the bandwidth of the optical medium is physically shared among many users with network addresses assigned to unique codes in the optical CDMA code space. The architectural choices for optical CDMA generally fall into six main categories: star, tree, ring, bus, mesh, and free-space networks. The fiber architectures are reviewed in Figure 10.1 and Figure 10.2. The optical CDMA transmitters (Tx) and receivers (Rx) on the networks described in Figure 10.1 use passive optical couplers to access the fiber broadcast media and decode the data using optical CDMA correlation techniques. Ring and mesh architectures shown in Figure 10.2 require more complex access nodes that utilize active code selection to interface to the fiber media. Add-drop multiplexers (ADMs) specifically designed for optical CDMA can be used to extract and route codes on ring and mesh networks. The combination of architectural considerations, technology complexity, cost, and applications all contribute to determining the appropriateness of an optical CDMA technique for a given network application.

10.2.1 PASSIVE STAR NETWORKS

A star network is a general class of interconnectivity among many nodes with a centralized node that broadcasts all network transmissions to all receivers. Each user is assigned a unique address on the network that enables it to distinguish traffic destined for its receiver from the traffic on the rest of the network. The implementation of star networks in the optical domain is particularly attractive for LAN environments and short distance (< 10 km) interconnection networks. Star networks can provide transparent, bidirectional, full-duplex communications among many optical nodes. Optical star couplers are simple, passive devices that enable optical traffic from individual input optical fibers to be optically combined and broadcast to many output fibers. By delegating the channel selection to the individual node receivers, a star network can perform the same functions as a cross-bar switch, such as broadcasting and multicasting, without requiring a complex active central switching node [2]. Furthermore, individual nodes can be upgraded gracefully to support new services or higher data rates without replacing the fiber interconnection network. The star network is especially important for applications that require high bandwidth communications among many users, such as a distributed computer interconnect, where each destination node has approximately the same expected network demand.

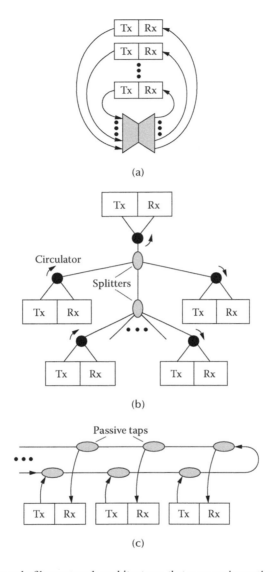

(a)

(b)

(c)

FIGURE 10.1 Example fiber network architectures that use passive optical taps, splitters, or circulators for optical distribution and node access for optical CDMA transmitters (Tx) and receivers (Rx). The fiber network topologies shown are: (a) star, (b) tree, and (c) folded bus.

Several classes of star networks relevant for optical CDMA implementations have been proposed and demonstrated, each having their own particular advantages.

10.2.1.1 Simple Star Networks

In an $N \times N$ star coupler, the optical power, P_{in}, from a given input fiber is broadcast to all N output fibers with a power splitting ratio of P_{in}/N per output port. Singlemode fiber star couplers as large as 32×32 are commercially available. By adding $M \times 1$

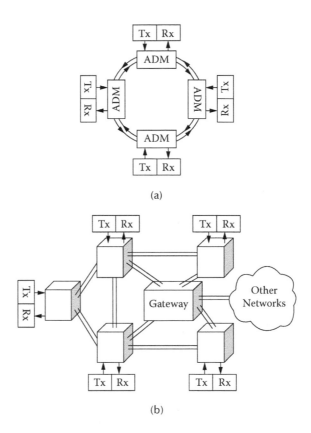

(a)

(b)

FIGURE 10.2 Fiber network architectures using active nodes to select optical CDMA codes: (a) 2-fiber ring using add-drop multiplexers (ADMs) to insert and remove codes from the fiber network, (b) an irregular mesh network that performs routing and switching of optical CDMA transmissions with a Gateway node translates optical CDMA addresses and formats between interconnected networks.

optical combiners and $1 \times M$ optical splitters to the input and output ports, respectively, of an $N \times N$ star coupler, a star network as large as $NM \times NM$ can be constructed as shown in Figure 10.3. Simple commercial components comprising of 32-port devices where $M = N = 32$ enable the interconnection of more than 1000 optical nodes using off the shelf components.

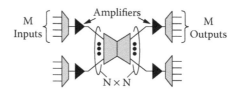

FIGURE 10.3 A general architecture for an amplified star network that can support (MN) nodes.

Scaling star networks to support large-scale optical CDMA networks has been studied in detail. Of principle concern for large-scale star networks are the optical losses associated with splitting the optical signal to many users. Fortunately, optical amplification can be applied where losses exceed the practical limits of laser transmission power and receiver sensitivity. An extensive power budget and signal to noise ratio analysis using photon counting statistics was performed for a 32×32 star network by Razavi and colleagues. [3,4]. By incorporating optical amplifiers to compensate for the optical CDMA encoder/decoder losses as well as the splitting loss expected in a 32×32 star coupler, a time-spreading, direct sequence optical CDMA system can be deployed to support 32 simultaneous users at a bit error rate (BER) < 1E-08.

Several novel optical CDMA approaches have been applied to demonstrate star networks. In the simplest form, the encoder output is multiplexed onto a single port of the fiber optic star network. Likewise, each output of the star is connected to a single decoder. Star networks are easily implemented to support asynchronous optical CDMA schemes where timing alignment of the individual users utilizing the network is not required. The fiber lengths from the individual users to the star coupler can vary depending on their geographic location from the star. For synchronous schemes, however, the propagation time from the individual encoders to the star must be the same to properly multiplex the codes into their appropriate time positions in chip or frame synchronous optical CDMA networks. For either system, the tolerances on the outgoing fiber lengths from the star to the decoders are not strictly important in so far as fiber propagation effects and protocol latency constraints are satisfied.

While appropriate for LAN environments, star networks are not necessarily practically scalable over large geographic distances. Long distance fiber routes from a central location to the individual endpoints, for instance, may not be readily available or could cause extensive propagation delay relative to the geographic distance between adjacent users. Without additional redundant transmission networks, a failure at the star coupler could adversely impact the entire network. The ease of implementation of star networks, however, has made this simple architecture the dominant fiber topology considered in optical CDMA LAN networks.

10.2.1.2 Spatial Encoding on Star Networks

Although the majority of time-spreading codes, wavelength-hopping codes, and their two-dimensional coding combinations assume that the encoder output and decoder input are coupled onto a single port of the star interconnection network, other star architectures have been proposed for specialized types of optical CDMA. As previously discussed, the losses in a large-scale star network can be compensated by adding optical amplifiers to the individual inputs and outputs of the star coupler. However for LAN environments, optical amplifiers are costly components and may not be competitive with legacy electronic switching technologies available in the LAN today. It has also been shown in previous chapters that multidimensional codes can support larger numbers of simultaneous users compared to one-dimensional codes for a given level of network performance. A novel approach for multidimensional encoding

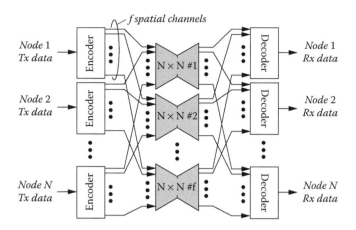

FIGURE 10.4 Temporal/spatial optical CDMA star network adapted from [6]. f star couplers provide spatial channels as a dimension of the optical CDMA code to interconnect N nodes.

involves the use of multiple fiber paths in the architecture to provide a spatial dimension to the code space. It was originally proposed as a two-dimensional coding technique for multifiber ring networks [5]. The general architecture for temporal/ spatial optical CDMA configured for a star network is shown in Figure 10.4. In the simplest case for an N-node network which uses single pulse per row codes (SPR) [6], each pulse from a direct sequence, time spreading code of length, L, is distributed to a unique fiber spatial channel from a total of f such channels, and broadcast to all users through separate $N \times N$ star couplers. More generally, the spatial dimension can be added to any code family including wavelength-time codes for additional scalability and redundancy in system design and code weight. This design enables lower loss multiplexing and distribution of the codes by eliminating the need for multiplexers and combiners at the encoder outputs and demultiplexers and splitters at the decoder inputs. For simple single wavelength time spreading SPR codes where $L = f$, the network has f^2 times lower loss compared to the single star architecture. Additionally, new classes of codes have been developed to take advantage of the architecture. The implementation of SPR codes is energy efficient since the architec- ture has the unique advantage of exhibiting no autocorrelation side lobes. Shorter temporal code lengths can be used on this network with the same performance as longer temporal codes deployed over single star architectures [7].

Despite these advantages, there are several trade-offs for implementing this architecture. The interconnection network is relatively complex since it requires f star couplers. Additionally, time-spreading codes require that the relative path delays to and from the star coupler array for a given user are properly managed to produce the correct timing alignment for transmitted codes and the proper decoder correlation alignment for received codes. Although in asynchronous time spreading optical CDMA it is not necessary for the absolute time delay of all users to be globally synchronized, strict fiber path length management for each user to and from the star coupler array is required which can be difficult if the network is geographically dispersed over a large area or supports high data rates. Nonetheless, short distance interconnection networks can benefit from the low loss architecture inherent in this implementation of a star network. The redundancy of the multiple paths of the interconnection network could be added to other two-dimensional coding schemes for added network robustness and resiliency.

10.2.2 TREE NETWORKS

In telecommunications networks today, the multipoint-to-multipoint communications provided by star networks are not necessarily efficient for the traffic demand patterns in access networks. In many access environments, the traffic patterns represent point-to-multipoint architectures. An example of an emerging fiber access technology that utilizes a tree structure to aggregate and distribute information is shown in Figure 10.5. Known generally as a passive optical network (PON), the signals between the optical line terminal (OLT) and the end user optical network unit (ONU) remain entirely in the optical domain. The structure is designed to facilitate low cost deployment of fiber to the home (FTTH) and fiber to the business (FTTB) services for the distribution of voice, video, and data over a single infrastructure. Although PONs require carriers to invest in new deployments of single-mode optical fiber, the architecture ultimately provides a low cost, easily managed, broadband access network that can supply each user with more than 100 Mb/sec capacity. The link layer protocols for PONs vary, but the basic concept is generally the same. Many different approaches can be used to separate traffic in the upstream and downstream directions on the PON. In the general tree architecture previously shown in Figure 10.1(b), optical circulators are used to separate the two directions of the network to allow for the same wavelength to be used in both directions. It is also possible to use separate fibers. However, the most common approach is to use two

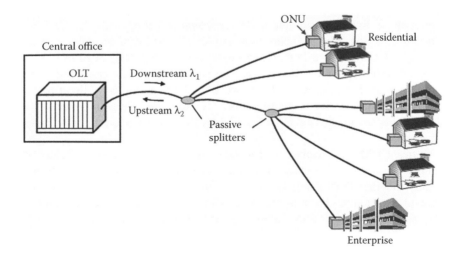

FIGURE 10.5 Basic fiber optic tree architecture for implementation of a passive optical network (PON).

different wavelengths as described in Figure 10.5. A downstream wavelength, λ_1, from the OLT is passively split and broadcast to all nodes on the network. The ONUs select the traffic appropriate for each terminal by matching a simple binary address associated with the media access control (MAC) protocol. In the upstream direction, a separate wavelength, λ_2, is used to carry data from each ONU to the OLT through the same fiber splitter network in the reverse direction.

Since all users transmit in the upstream direction toward the OLT on the same wavelength, an optical multiple access technique is required. Both WDMA and TDMA have been proposed for PONs. Unfortunately, both schemes suffer from different drawbacks associated with the technology required for achieving multiple access. Although WDMA techniques can leverage off the shelf components including tunable filters, lasers, and wavelength demultiplexers, the cost of the wavelength tuned optical components today tends to be too high for a PON network to be deployed cost effectively. TDMA schemes that rely on segmenting the bandwidth on the upstream wavelength into time slots have thus far dominated the protocols for PON access leveraging existing Ethernet and ATM hardware and standards. However, these techniques rely on periodic resynchronization of each ONU with the OLT and time slot resource allocation based on the network demand. The distance between the ONU and OLT can be as large as 20 km, resulting in additional latency for coordinating media access. Resource reservation techniques and fairness complicate the TDMA PON protocols and can limit the scalability and ultimate data rate achievable on PONs.

An alternative approach to support multiple access to the OLT in PON networks is the use of optical CDMA in the *upstream* direction. In this architecture, the OLT is responsible for decoding more than one code simultaneously to support multiple users on the network. Optical CDMA implementations that use electronic post-processing for decoding the upstream information can also be used in this environment given that the individual user data rates are generally modest. The large potential code space offered by most CDMA techniques allows each ONU encoder to be fixed to a unique code for communicating with the OLT in the upstream direction. The asynchronous arrival of data from the ONUs is a key advantage of CDMA as a PON technology. The active ONUs need only be synchronized at the session level when they are permitted to use a given code. During session periods, burst access to the network media is permitted without requiring multiuser synchronization or coordination. By providing a large number of potential parallel upstream channels, optical CDMA PONs can perform gracefully under loading [8].

Optical CDMA as a technology for access networks and more specifically PONs has only just begun to be explored in the research literature [8]. An analysis of a 32-node Ethernet PON using synchronous time-spreading optical CDMA in the upstream direction has been proposed by Wu and colleagues [9]. In this implementation, perfect difference codes are used by each ONU to communicate with the OLT. Performance estimates indicate that all 32 nodes can be supported with a BER < 1E-09. Although the system requires synchronization, it is a first step toward the practical use of optical CDMA in access networks and demonstrates the compatibility of optical CDMA techniques with today's PON fiber architecture.

Despite these advantages, the deployment of the optical or electronic processing required to implement CDMA access in PONs and other access network architectures is primarily determined by the relative cost of the components needed in the encoders at the ONU and the decoders in the OLT compared to current implementations that leverage widely available Ethernet and FSAN (Full Service Access Network, [10]) standards-based approaches. Optical CDMA may yet play an important role in the future scalability of optical access networks since optical CDMA technologies are designed to provide multiple access communications in shared fiber networks.

10.2.3 RING AND BUS NETWORKS

Basic fiber optic ring structures provide the foundation for collector and aggregation networks in metropolitan areas. These networks are easy to manage, provide redundancy, are efficient to deploy over a wide geographic area, and have many simple protocols established for protection switching and restoration in the event of a node failure or fiber cut. With the addition of a WDM overlay on a fiber optic ring network, diverse logical network topologies such as mesh and hub networks can be defined at the wavelength layer while retaining many of the advantages offered by the physical layer ring that underlies the network. Although most operational rings today still rely on electronic switching at each node to perform add/drop functions, reconfigurable optical add/drop multiplexers (R-OADMs) are emerging to provide optical layer reconfigurability at the wavelength level. Each R-OADM uses fixed or tunable optical bandpass filters to extract one or more desired wavelengths from the fiber ring. Likewise, fixed or tunable lasers at the transmitter can be used to add traffic to the ring. In optically transparent ring architectures, careful design methods including dynamic channel equalization, connection management, and filtering are needed to prevent the build-up of amplified spontaneous emission (ASE) noise from optical amplifiers and circulating optical cross-talk which can lead to chaotic lasing effects on the ring [11,12].

Given the wide deployment and high reliability inherent in ring networks, optical CDMA techniques that are compatible with the installed ring infrastructure are important for practical carrier environments. Rings provide many of the same features of the star network, namely a broadcast medium where each node can transmit data to any other node on the ring. Optical CDMA systems, however, present several unique design challenges for transparent rings. Unlike the star networks, which do not have closed lightpaths, codes from individual users added to a transparent optical ring can complete a round-trip path leading to the build up of noise, interference, and chaotic lasing effects. A closed optical path on an optical CDMA network may significantly increase the bit error rate (BER) of the system and ultimately lead to unavailability of the network or particular codes.

While multiplexing optical CDMA codes onto the fiber ring can be easily facilitated by passive fiber couplers, dropping or removing the energy associated with a given code depends on the system implementation and coding technique adopted on the network. An active "code drop" filter is needed to extract and remove the energy from a correlated code at its destination and to forward the codes destined for downstream nodes on the ring undisturbed. Active add/drop optical nodes specifically

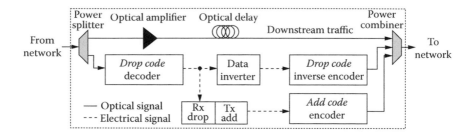

FIGURE 10.6 Proposed architecture for a fiber optic code division add/drop multiplexer (ADM) for incoherent spectrally encoded optical CDMA adapted from [13]. The received (Rx) data at the node is recovered by the *Drop Code* Decoder and transmitted (Tx) data is inserted onto the network via the *Add Code* Encoder. By combining the output of the *Drop Code* Inverse Encoder with the downstream network traffic, the original data pattern of the dropped code is cancelled out of the aggregate optical signals transmitted to the network.

for optical CDMA have only recently received attention from the research community. A reconfigurable optical add/drop multiplexer (ADM) was proposed and analyzed by Wu and colleagues [13] for incoherent spectrally encoded complementary optical CDMA [14]. The functional architecture and detailed subsystems for the ADM node are shown in Figure 10.6 and Figure 10.7, respectively. Optical signals from the network are split into two paths at the node. The upper path represents the traffic that is transmitted to the downstream nodes while the lower path carries the signals that are processed by the node. A drop code decoder is programmed with the desired code to correlate the data destined for the node using complementary decoding and balanced detection. The node recovers the data from the correlated code through a standard digital data recovery circuit. To eliminate the data pattern of the dropped code from the downstream traffic, the data pattern is inverted and encoded with the dropped code. The inverted dropped code signal is passively combined with a delayed copy of the aggregate downstream traffic that emerges from the optical delay line. Since this encoding scheme is based on balanced detection of the direct and complementary coding sequences, this has the effect of canceling the dropped code data from subsequent downstream receivers. The downstream signals are also combined with the transmitted data at the node from the add code encoder and sent to the next node on the network.

The add/drop node was analyzed for both ring and bus network architectures using $N-1$ rows from an $N \times N$ Hadamard matrix as the basis for the optical CDMA coding technique [13]. For equivalent encoder output powers, the bus network was shown to support more than twice as many optical nodes as the ring network for the same system BER performance [13]. The reduction in performance of the ring network resulted from the continuous circulation of noise on the closed fiber ring. Although the ADM effectively cancelled out the dropped data from the network, it did not remove the energy associated with the dropped code. Indeed, additional noise was added to the network at each node from the inverse encoder, which limited the scalability of this ADM in ring topologies.

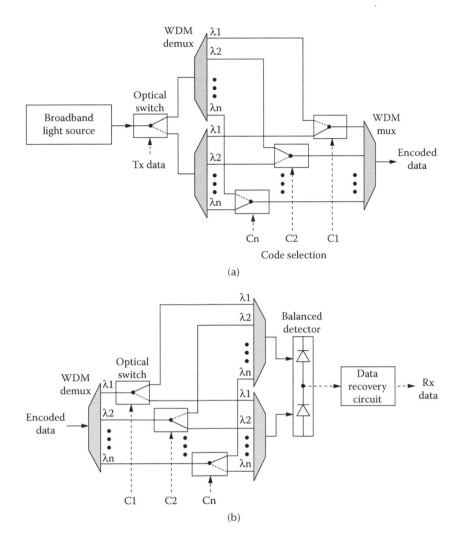

FIGURE 10.7 Subsystem architectures for the incoherent spectrally encoded optical CDMA ADM: (a) encoder and (b) decoder detail adapted from [14]. Data is encoded by assigning the direct and complementary codes to 1 and 0 binary symbols, respectively, through the control of a 1 × 2 optical switch in the encoder. Decoded data is correlated by a matched decoder with balanced detection. Different codes are selected by setting the appropriate states of the code selector switch control bits {C1, C2, . . . , Cn}.

To increase the scalability of optical CDMA on ring networks, an ADM is needed that removes the energy from the dropped code to prevent circulation of noise and other signals. The node architecture for an all-optical add/drop multiplexer for two-dimensional wavelength-time incoherent CDMA is shown in Figure 10.8. For data encoded on this network, the presence and absence of a particular code represent binary "1"s and "0"s, respectively. Incoming network traffic passes into the drop

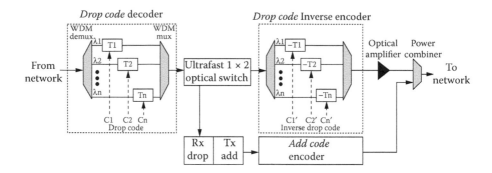

FIGURE 10.8 Add/drop node for incoherent two-dimensional wavelength-time optical CDMA systems. The ultrafast optical switch directs the energy from the autocorrelation peak output of the drop code decoder to the node receiver for data recovery. The cross correlation signals (downstream network traffic) that fall outside the time window of the switch are directed to the drop code inverse encoder where they are realigned with the inverse of the drop code time delays and transmitted to the next node on the network.

code decoder, which produces an autocorrelation peak when 1s from the desired data channel are present. By synchronizing an ultrafast 1×2 optical switch with the autocorrelation peak, only the optical pulse energy that falls within the autocorrelation time slot duration is switched to the node data receiver labeled "Rx drop" in Figure 10.8. Cross-correlation signals that fall outside of the autocorrelation switching window are directed toward the drop code inverse encoder. Since the cross-correlation signals represent live network traffic for downstream nodes, the time delays imposed by the drop code decoder will cause this traffic to deviate from its originally intended sequence. Therefore, the drop code inverse encoder is used to cancel out these time delays to reform the original sequence for the downstream traffic. The downstream signals are passively combined with data transmitted and encoded via the add code encoder and sent to the next network node. This ADM architecture more closely resembles the properties of R-OADM nodes used on transparent DWDM rings. Energy from dropped signals is switched out of the ring and not permitted to circulate indefinitely.

The ADM architecture for incoherent two-dimensional wavelength-time optical CDMA shown in Figure 10.8 was experimentally demonstrated by C.-S. Bres and colleagues using a 4-node network scenario [15]. The system used an incoherent 4-wavelength, 101-time slot carrier hopping prime code sequence for encoding the data on the network. In this experimental node, an ultrafast all-optical switch known as the Terahertz Optical Asymmetric Demultiplexer (TOAD) [16] was used to open a narrow temporal optical sampling window (< 10 ps) triggered by an optical control pulse that was synchronized to the user data rate of 2.5 Gb/sec on the network. The all-optical switch directed the autocorrelation peak at the output of the drop code decoder to the data receiver and the cross-correlation signals to the drop code inverse decoder. The dropped code exhibited a BER < 1E-09 in the presence of multiuser access interference from three network nodes.

In much the same way that connections can be provisioned and reconfigured on DWDM rings using wavelength selective filtering, add/drop nodes for optical CDMA can be used to establish dynamic logical connections among nodes physically inter-connected on a transparent optical ring. By eliminating the data signal carried by the dropped code from the downstream traffic, the network can re-use these codes to enhance the system scalability. Active code selection is also useful for reducing the amount of multiuser access interference noise when the energy from the dropped code can be optically switched out of the network. Although most of these techniques have been studied for deployment on fiber ring networks, they can also be applied to folded bus networks. In most cases, bus networks exhibit greater scalability in terms of the number of nodes since amplifier noise and crosstalk are not allowed to circulate around the network indefinitely. However, careful design practices similar to those used on DWDM ring networks can be used to reduce optical noise impairments on fiber ring networks.

The development and implementation of optical CDMA add/drop multiplexers are important for deployment of the technology over existing telecommunication fiber networks especially in metropolitan areas. Logical topologies such as point-to-point, mesh, and hub-and-spoke communications architectures can be defined through the appropriate assignment of codes to individual nodes on a physical fiber ring network. This approach is similar to logical overlay networks on a DWDM ring where point-to-point communications are defined by assigning individual users a unique wavelength or set of wavelengths that are multiplexed with the optical channels from other nodes and transmitted over a physical fiber ring. Add/drop multiplexers for optical codes enable fiber rings to use the scalability and high cardinality of optical CDMA to support a large number of users, diverse topologies, and provide statistical access to the shared fiber infrastructure. These new subsystems for managing codes on optical rings can easily be extended to other multihop networks such as bus and mesh architectures.

10.2.4 MESH NETWORKS

Although there are many classes of mesh networks, a general characteristic of a node on a mesh architecture is the availability of more than one physical input and output connection. The wide area optical backbone in large telecommunication provider networks tends to be organized as an irregular physical mesh topology that follows the traffic patterns and geography of the service area. Core nodes in large cities, for instance, may require five distinct DWDM fiber optic connections to adjacent cities. An irregular mesh network was previously shown in Figure 10.2(b). In this network, each node is responsible for both adding and dropping traffic destined for its location and forwarding the rest of the traffic to the appropriate output ports.

Although the implicit addressing capability inherent in optical CDMA transmissions in principle can support multipoint optical routing on mesh networks, it is difficult to implement in a practical networking environment without adding a large amount of complexity to the optical node. If an arbitrary optical transmission enters a node containing an unknown subset of optical destinations encoded using CDMA, a routing node must provide optical correlators for all possible code sequences that may be present on that particular link. Much like a multihop packet network, the correlated transmissions are routed to their appropriate destinations based on the assignment of code address classes to destination ports including those dropped locally at the routing node. Hierarchical code design, much like addressing on packet networks, is needed to simplify the routing node design for practical implementation. For instance, an optical CDMA code supporting up to 1000 codes each mapped to distinct users may be subdivided into 10 classes of 100 codes where each class is mapped to a physical destination port at a routing node. Only common aspects of the 10 code classes are correlated by the optical CDMA address recognition hardware to route the optical CDMA packets or circuits to their destinations.

Finally, gateways between networks are also important for enhancing the scalability of optical mesh networks. Gateways can provide a number of functions including address translation, format conversion, and optical signal regeneration. For optical CDMA, optical code translation and format conversion may be required to bridge optical subnetworks or to provide interoperability with traditional DWDM systems. As with the development of ADM technology, these functions tend to be specific to a particular type of optical CDMA deployed on a given network.

Optical code conversion between LANs has been demonstrated by Gurken and colleagues [17]. In this code-conversion gateway node, incoming two-dimensional wavelength-time codes were optically mapped onto a new code by performing both wavelength translation and time slot reassignment on the incoming code. All-optical wavelength conversion was performed by an integrated periodically poled lithium niobate (PPLN) waveguide device that simultaneously converted four wavelengths from the original code to a new set of four wavelengths represented in the new code. The time chips for each wavelength were aligned to new time slots by using a serial cascade of fiber Bragg gratings tuned to the center channels of the new wavelengths and temporally separated on the fiber to produce the time sequence for the new code.

Only recently have researchers begun addressing the networking aspects of optical CDMA node architectures essential for interconnecting optical LANs across the wider area network. More work is needed to realize the full optical routing capabilities and gateway functions required to support large scale transparent optical mesh networks.

10.2.5 Free Space and Wireless Infrared Communications

CDMA techniques have established a strong presence in commercial RF wireless systems due to the many unique advantages of spread spectrum in the RF domain, including asynchronous access, multiple access interference rejection, and efficient use of available RF spectral bands. Many of these advantages have led to the wide deployment of CDMA in mobile cellular networks over wide areas, but the overall RF access data bandwidth in large networks is still limited due to the availability of frequency bands in the RF spectrum and the requirements on electronic processing technology. As a result, delivery of high speed data services directly to residential areas using wireless techniques are still in an early stage of development and have not yet demonstrated the ability to provide the quality of service and bandwidth required to be competitive with wireline broadband access technologies.

Free space optics (FSO) is an emerging technology to provide high-speed access services to residential and business users. Using existing RF towers or line of site positions on rooftops, point-to-point optical links can be established in a metropolitan area. Free space optics offers many advantages over RF communications, in general, including a largely unregulated spectrum and services that can support data bandwidths approaching gigabits per second. Free space optics services can be established quickly and cost effectively compared to time-consuming deployment of wireline infrastructure such as optical fiber or coaxial cable networks [18]. Additionally, high capacity FSO networks can be established quickly in emergency situations to backup damaged wireline infrastructure or for mobile ad hoc networks in military operations on the battlefield.

The majority of FSO approaches have focused on point-to-point connectivity. However, these networks could be further enhanced if multiple access techniques could be applied to the architecture. An approach to provide multiuser free space optical networks was first studied by Gagliardi [19] using optical CDMA communications among a constellation of satellites. The architecture, however, could generally be applied to any type of FSO network including ground-based rooftop access networks. The concept of a multimultiuser FSO network is shown in Figure 10.9. multiuser communications are provided by highly diverging optical beams that optically broadcast signals to nodes that intersect the beam cone at a distance from the transmitter. Since the transmissions from all nodes are received at each node, optical CDMA can be used as an addressing and selection mechanism for the communications that are intended for the receiving node.

The particular type of CDMA required for enabling robust, free space networks is important so that many of the advantages of the FSO network are retained. Asynchronous, incoherent CDMA approaches are simple to implement, potentially lower in cost, and allow the FSO to retain many of the advantages of deployment including ad hoc network establishment. Both on-off-keyed (OOK) and pulse-position

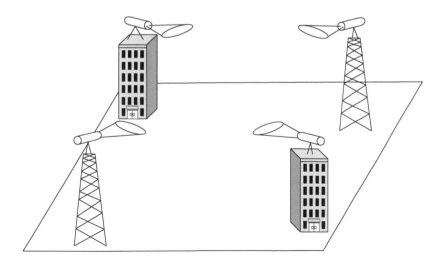

FIGURE 10.9 Architecture for a multiuser free space optical network. Highly diverging optical beams enable the optical output from the transmitters to be received by multiple nodes simultaneously. Optical CDMA can be used as the multiple access protocol in the network.

modulated (PPM) optical CDMA schemes have been proposed and analyzed for FSO networks [19,20]. In the PPM scheme, M-ary optical orthogonal codes (OOCs) with length F and weight K are used in the design for the optical CDMA FSO system. Each timeframe is divided into M slots where each slot was further subdivided into F chips where F determines the length of the sequence. This approach was analyzed for its robustness to atmospheric effects such as scintillation. Compared to OOK intensity modulation, PPM is particularly well suited for FSO applications since it is more robust to power fading caused by scintillation [20]. OOK techniques, on the other hand, require an adaptive threshold for determining ones and zeros in the presence of scintillation. Figure 10.10 shows the theoretical BER of the FSO M-ary PPM optical CDMA system for an $N = 3$ user network. The atmospheric conditions, described by the logarithmic variance of the scintillation parameter σ_s^2, were varied to simulate the turbulence strength for plane and spherical waves at a path length of 250 m [20]. The data rate per user was set to 156 Mb/sec using a wavelength of 830 nm and an assumed background light level, P_b, of −45 dBm. As shown in the analysis, changes in atmospheric conditions can dramatically impact the FSO system performance. However, forward error correction (FEC) can be applied to the system for scintillation values $\sigma_s^2 > 0.1$ to provide reliable data communications.

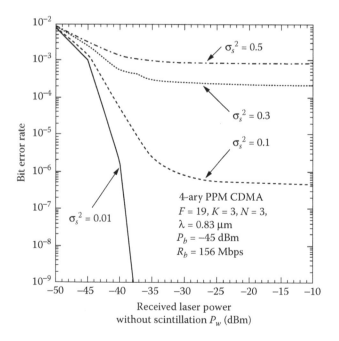

FIGURE 10.10 Theoretical BER versus the received laser power without scintillation, P_W, for a 4-ary PPM CDMA FSO system. The code parameters were set to $F = 19$ and $K = 3$ for an $N = 3$ user network with a single user data rate of $R_b = 156$ Mb/sec at a wavelength of $\lambda = 0.83$ μm. The atmospheric conditions, σ_s, were varied as a parameter of the model which included an assumed background light level of $P_b = −45$ dBm. (After [20], © 2003 IEEE.)

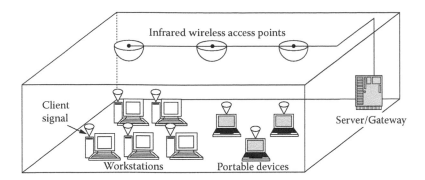

FIGURE 10.11 Architecture for a free space optical LAN within a building. Portable devices and computers access the network through diffuse infrared optical nodes located in the ceiling.

Another important area for optical CDMA techniques in free space environments is in wireless infrared communications for local area networks (LANs). The concept of wireless infrared communications within a LAN environment is shown in Figure 10.11. One of the earliest studies of optical CDMA in free space optical communications was performed by Elmirghani and colleagues [21] for indoor diffuse infrared communications in computer networks. This scheme uses a simple PPM CDMA technique to provide connectivity within a small LAN located in a single room. By using CDMA as the multiple access scheme for the individual terminals, the network does not require synchronization or coordination of the transmitters at the local stations. Wireless infrared access points located in the ceiling enable users to access a fixed wireline network with portable devices. Uplink communications between the portable stations and the access points can be achieved by using low cost LED transmitters at each portable station. Likewise the access point broadcasts data in the downlink direction and the portable nodes detect the address of the broadcast to select data destined for the local node.

The scalability of large infrared LANs can be limited by the shared bandwidth in the downlink direction especially in large areas where many access points may be required to cover an area. Many techniques for sharing the bandwidth and coordinating the transmission of multiple users to the access points are possible including TDMA, WDMA, and space-division multiplexing [22]. CDMA techniques using intensity modulated direct sequence codes, however, can provide multiple access communications among many users without requiring complex synchronization while still leveraging low cost electronics for performing code detection and transmission.

Figure 10.12 shows a comparison of various multiplexing techniques for a diffuse wireless infrared network divided into adjacent hexagonal cells where each cell has a maximum throughput of 10 Mb/sec. The system optical average power efficiency is evaluated as a function of the cell size. The figure of merit on the vertical axis, γ, is a measure of the required signal to noise ratio for unit optical path gain to achieve a system BER of 1E-09. It is proportional to the square of the transmitted optical power. An increase in optical transmitter power of 5 dB, for example, corresponds

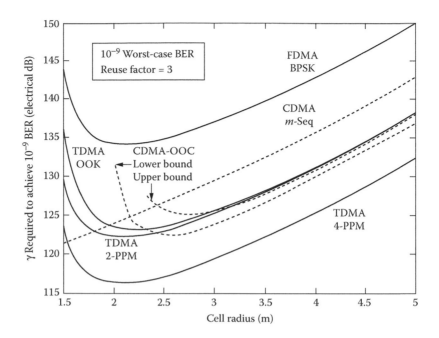

FIGURE 10.12 Comparison of the power efficiency for different multiple access modulation schemes for infrared wireless LANs. For small cell radii, *m*-sequence CDMA is the only technique that is not limited by cochannel interference from adjacent cells. (After [22], © 1997 IEEE.)

to a 10 dB change in γ [22]. Two optical CDMA techniques that use optical orthogonal codes (OOC) and *m*-sequences are compared to infrared TDMA PPM and FDMA using Bipolar Phase Shift Keying (BPSK) modulation. It is desirable in wireless infrared networks, especially for portable stations, to provide a given BER performance (e.g., BER < 1E-09) with the lowest transmitted optical power, which is represented by smaller values of γ. All techniques require increasing the transmitter power level as the cell radius increases, an indication that they are approaching noise-limited operation. On the other hand, as the cell radius approaches relatively small values (<2 m), *m*-sequence optical CDMA is the only technique that is not limited by the increased amount of cochannel interference from the adjacent cells for small cell radii. This performance advantage for extremely small cell radii is attributable to the higher co-channel noise rejection inherent in the *m*-sequence coding approach.

Optical CDMA techniques have a wide range of many promising applications for FSO networks from optical access networks to small wireless infrared LAN environments. CDMA techniques enable multiple users to access the bandwidth of the network without complex system coordination. The modulation formats for optical CDMA also enhance the communications in the presence of atmospheric effects that can degrade optical signals in FSO networks. For specialized micro-cell optical wireless LANs, CDMA can provide power efficient communications compared to other multiple access technologies. By developing cost effective CDMA encoding and decoding techniques, CDMA approaches may become an important innovation for next generation FSO networks and mobile ad hoc high bandwidth optical communications platforms. Future advancements in optical CDMA–based FSO networks that include all-optical correlation technology and spectrally efficient coding techniques may eventually achieve many of the same advantages that have made RF CDMA networks successful today.

10.2.6 Network Implementation Considerations

While the previous discussion of optical CDMA LAN applications demonstrates the usefulness of multiple-access capability at the physical layer, other networking issues at higher layers in the network must also be addressed to practically implement a viable, reliable, and scalable network. Optical CDMA provides an inherent ability to distinguish the origin or destination of a message through correlation of the optical code imposed upon the data transmission. Optical codes are a direct mechanism for assigning a unique link layer address to each node on the network. The assignment, distribution, and use of optical codes within the link or network layers, however, requires the implementation of a protocol for managing access to the codes active on a network at any given time. These protocols and management policies are essential for enforcing fairness, providing contention resolution, and ensuring quality of service guarantees. They may be implemented in a centralized management system, negotiated through multiple access protocols distributed throughout the network, or some combination of both techniques. Coordination and synchronization of code assignment throughout the network is also required to ensure network configuration consistency. This can be a challenging problem if the number of nodes active on the network is dynamically changing. Network synchronization is an especially important issue for coding techniques that depend on chip or frame synchronization among network users. Finally, monitoring the overall performance of the network through a management system is also required to identify degradation and fault conditions that may occur during operation.

The networking aspects of optical CDMA are an important area for future research for the architectures presented here. Since optical CDMA is still in the early stages of development, standards and protocols for implementing, managing, and deploying these networks have not yet been developed. A few studies have begun to analyze the performance of multiple access and link layer protocols for optical CDMA [23–28] but more work is needed. The flexibility of optical CDMA, however, enables it to adapt to many different network and management models that have already been used in both telecommunications and data networks. By combining standard media access protocols with different optical CDMA approaches, practical multiple access optical networks can be realized.

10.3 APPLICATIONS DEMONSTRATIONS

Although data networks are the most important driver for multiple access approaches, many different technologies can be used to provide shared access to the network as evidenced today by the proliferation of Ethernet and RF wireless technologies. However, optical CDMA has several key advantages over traditional approaches in other specialized applications and environments. This section reviews several novel demonstrated and proposed applications that highlight the unique advantages of optical CDMA.

10.3.1 MULTIMEDIA NETWORKS USING OPTICAL CDMA

The flexible design and variety of choices of CDMA systems allow for many types of data transmission to be supported over optical networks. The first section of this chapter discussed LAN applications over distributed network architectures as a main feature of CDMA. This section presents the advantages CDMA offers to connection-oriented and broadcast-based data sources such as video and multimedia transmission. By building flexibility into the code space and networking architecture, CDMA networks can support many multimedia services, varied data rates, and multiple quality of service levels in circuit-oriented networks.

10.3.1.1 Video Distribution

Cable networks have relied for many years on fiber infrastructure for carrying RF broadcast channels to residential and business subscribers. In analog cable systems, these hybrid-fiber coaxial (HFC) networks use a single optical wavelength to support multiple analog RF subcarriers. Each subcarrier transports the RF modulation of a unique television channel to a distribution point where the carriers are detected and redistributed over the coaxial infrastructure.

While subcarrier modulation is a relatively simple means for carrying RF analog channels, it has several shortcomings. Since all of the subcarrier channels share the same optical carrier, the power per channel is reduced as additional channels are added. Nonlinearities and intermodulation distortion also occur when multiple sub-carriers are mixed and modulated onto the optical carrier. TDM and WDM based multiplexing techniques offer greater scalability than subcarrier modulation but are more complex and expensive to deploy. In addition, TDM schemes require synchronization among the individual media channels, which is difficult to achieve without electronic processing.

Code division techniques for asynchronously multiplexing video streams onto an access network can be less complex than TDM or WDM but more scalable than subcarrier multiplexing. An incoherent, asynchronous optical CDMA technique using pseudo-orthogonal codes for encoding pulse position modulated (PPM) video channels has been analyzed and demonstrated in an asynchronous video distribution network [29]. The encoders and decoders are constructed from simple parallel fiber delay lines. By setting the appropriate delays to a given code, the correlation for that code reproduces the original video PPM modulation for video reception. With the increased speed of electronic processing available today, decoding the desired optical waveform for each channel could also be performed in the electronic domain using signal processing techniques similar to those used in RF wireless CDMA systems.

The scalability of the video distribution network using optical CDMA was calculated for different PPM frame sizes [29]. As shown in Figure 10.13(a), up to 50 channels carrying compressed video data in the 1 to 20 Mb/sec range can be supported for PPM frames with up to 256 time slots per frame. The bit error rates

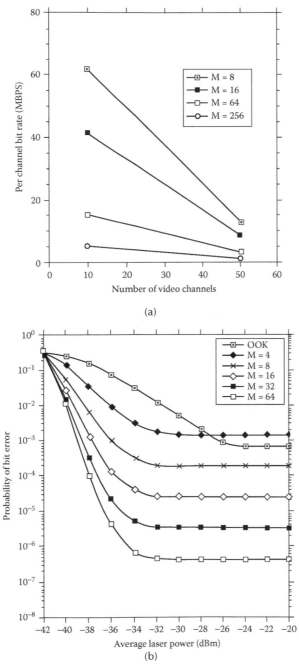

FIGURE 10.13 The scalability and performance of a PPM-based optical CDMA network for video distribution parameterized for different PPM frames given by the number of time slots, M, per frame: (a) the scalability of the network in terms of number of channels and the supported data rate, and (b) the bit error rate for a 50 channel network with code weight 4 compared to a non-PPM formatted signal using on-off keying (OOK). (After [29], © 1993 IEEE.)

for a 50 channel network with various values for the PPM frame were calculated and compared to coding with simple on-off keying (OOK) techniques. As shown in Figure 10.13(b), the PPM format outperforms simple OOK modulation for the same class of pseudo-orthogonal codes. This is a direct result of the additional spreading in time inherent in the PPM format to reduce overall crosstalk in the system.

Broadcast transmission, such as video, can utilize these optical CDMA techniques for providing network scalability while supporting the simplicity of asynchronous channel transmission and coding techniques that only use a single optical wavelength.

10.3.1.2 Image Transmission Using Spatial Spread
Spectrum Techniques

The majority of the optical spread spectrum techniques that have been discussed thus far have included codes that utilize the time, optical frequency, phase domains, and their combinations for encoding data transmissions. Architectures were also discussed in Section 10.2.1.2 for using parallel transmission fibers for spatially spreading the optical signal across multiple independent channels. This technique can also be used to encode and multiplex images carried by multicore optical fiber over a common fiber path. This approach, which combines the features of optical image processing with multiple access inherent in optical CDMA, has been proposed and demonstrated for encoding two dimensional pixel arrays by Kitayama and colleagues [30,31]. The architecture is shown in Figure 10.14. A magnified two-dimensional (2D) $N \times N$ bit plane, comprising individual pixels of an image (or binary data from a planar interconnect), is encoded using a 2D optical orthogonal signature pattern (OOSP) of dimension $M \times M$ where M^2 is the CDMA spreading factor. Each magnified pixel of the original image is encoded by the signature pattern such that the encoded image has $MN \times MN$ pixels. The encoded image is constructed by taking the Hadamard product of the overlapping matrix elements of the magnified bit plane and OOSP. Each node on the network uses a unique OOSP to encode the planar data from its input image or 2D data sequence. The encoded 2D images from each user are superimposed and focused onto a multicore imaging fiber. At a minimum, the multicore fiber must have $MN \times MN$ cores to support the pixel resolution required to transmit the encoded images. After propagation through the multicore fiber, the images are broadcast to all nodes and decoded using the corresponding OOSPs at the receivers. After establishing a pixel threshold for processing the decoded image, the original input image is recovered at the desired receiver.

The architecture shown in Figure 10.14 is based on the first demonstration of optical spatial CDMA for encoding and transmitting images through a multicore optical fiber [31]. A white light source illuminates a liquid crystal spatial light modulator (SLM) programmed with a unique 8×8 OOSP for each node. This pattern is transmitted onto a magnified 2×2 input bit-plane image (2D planar digital data) using the second SLM so that each pixel of the image is encoded by the 8×8 OOSP. As a result, any portion of the input image or OOSP that is set to a pixel value 0 produces a 0 value in the encoded image. In this example, the encoded image becomes a 16×16 pixel matrix. The encoded images from each user are multiplexed

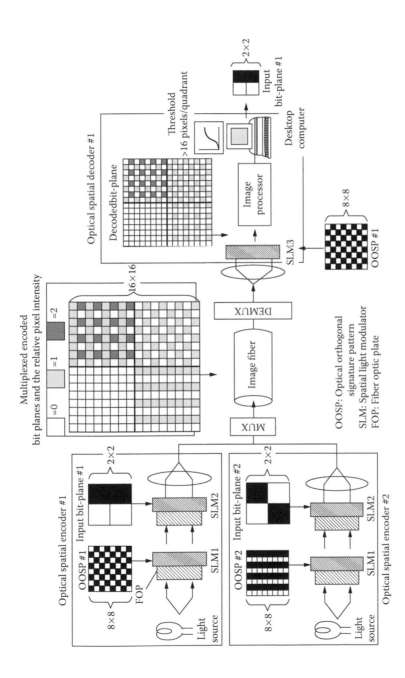

FIGURE 10.14 Example of spatial optical CDMA for multiplexing two-dimensional images into a shared multicore optical fiber. The encoded image from each transmitter is formed by the multiplication of the OOSP with each pixel of the input bit plane. The multiplexed image transmitted over the multicore fiber is a three-value matrix that represents the superposition of the encoded image intensities from the individual sources. (After [31], © 1997 IEEE.)

to form a 16×16 composite image representing a three-valued matrix of data from all nodes. Increasing intensity level in a given pixel is represented by negative shading in Figure 10.14 such that the darkest pixel has the highest intensity level denoted by a value of 2. In the demonstration, the multiplexed images were focused onto a portion of a 30,000 pixel multicore fiber and transmitted over 16 m. At the output of the multicore fiber, an SLM decoder is set to the desired OOSP. Similar to the encoding processing, each pixel of the composite image or OOSP that is set to 0 will result in a 0 pixel value at the image processor. A CCD camera and computer are used to analyze the decoded image and perform optical thresholding of each pixel quadrant to recover the original binary value of the 2×2 image bit plane. Similar to optical CDMA correlation techniques, a threshold is set so that images encoded with other OOSPs are not detected. Since the OOSPs must satisfy orthogonality relations, the appropriate threshold setting will yield images encoded only with a given OOSP. In the example described in Figure 10.14, the image will be properly decoded for a threshold set between 16 and 32 pixels of intensity 1 or greater for the four 8×8 quadrants of the 16×16 decoded matrix. The experiment was also performed using 16 pixel, 4×4 bit planes encoded with 8×8 OOSPs [31]. The images were successfully extracted at the output of the transmission fiber without pixel errors.

The construction of the optical orthogonal signature patterns for two-dimensional data encoding follow many of the same principles typical of optical CDMA codes including code orthogonality and large cardinality. In addition, each signature pattern is distinguishable from space-shifted versions of themselves on a 2D plane and any two different signature patterns in a set are distinguishable from each other, even with the existence of vertical or horizontal space shifts in the plane [30,32]. A detailed algorithm for the generating OOSPs suitable for 2D image encoding has been developed [30]. By analyzing different spatial spreading factors, M^2, for the OOSPs in this particular family, the bit error rate (BER) as a function of the number of multiplexed users was theoretically calculated. As shown in Figure 10.15, OOSPs of 11×11 are sufficient for encoding 2D images from up to 10 simultaneous users with a BER $\leq$ 1E-07 [30]. The size of the 2D images transmitted by each user is only limited by the practical constraints of manufacturing large dimensional multicore optical fiber.

Two-dimensional image multiplexing can be used in a variety of applications including medical imaging. This architecture can also be used as a multiplexing technique in high performance computing applications where parallel optical interconnections between processors and memory are imaged onto the encoded planar arrays and broadcast to other processing nodes. Future development of longer and larger multicore fibers coupled with high density, fast SLMs are critical for advancing the scalability and data rate of two-dimensional optical spatial CDMA.

10.3.2 Radio Over Fiber Networks

The use of CDMA techniques in RF networks has been well established commercially. The popularity and ubiquity of RF networks have led to a high density of RF devices especially within metropolitan areas. To increase the scalability of the RF

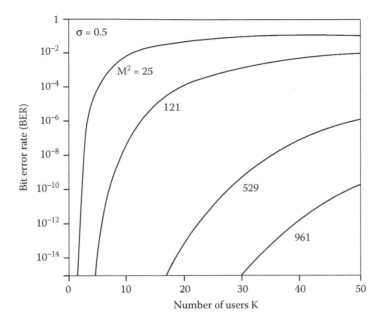

FIGURE 10.15 Calculated BER vs. number of simultaneous users for the spatial optical spread spectrum system using OOSPs generated by the prime sequence with spreading factor M^2. (After [30], © 1994 IEEE.)

network, micro or pico cellular areas are being defined to accommodate the large number of potential users in a metropolitan area. By dividing single cells into multiple micro cells, more users can share the same RF spectrum. Since the number of micro cellular areas can be many times the number of radio base stations (RBSs) managed in today's mobile networks, the practical use of these networks depends on the cost and ease of deployment of the RBS nodes that serve the microcells. It has also become increasingly important for each RBS to support multiple signal formats for digital voice, data, and legacy RF analog services for multiple service providers. The rapid changes in standards and RF technology make upgrading the large number of RBS receivers in a microcell network a difficult problem especially as providers continue to differentiate themselves by adopting new technologies.

One potential solution to simplify the deployment of microcellular networks that facilitates sharing among multiple providers is a converged support network called a Radio over Fiber (RoF) network [33]. As shown in Figure 10.16, each RBS is connected directly to a passive optical fiber network. The architecture of the RBS is designed to be low cost and adaptable to multiple RF signal formats without the need for hardware upgrades at the remote station. The analog RF signals from the antenna are directly modulated onto the envelope of an optical carrier and transmitted to the nearest radio control station (RCS) where they are demodulated by the appropriate service provider.

As with most optical access networks, many different approaches for distinguishing the modulated signals from each RBS on the fiber network have been

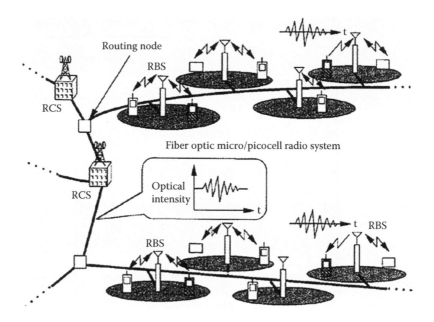

FIGURE 10.16 Radio over fiber (RoF) network concept. (RBS: radio base station; RCS: radio control station). (After [33], © 2003 IEEE.)

proposed including subcarrier and time division multiplexing. Optical CDMA, however, is also an attractive candidate since it is often difficult to maintain global time synchronization among the RBS nodes and the available code space can be significantly larger than the number of subcarriers that can coexist on the same wavelength. A direct sequence time-coded optical CDMA encoder/decoder architecture has been recently proposed and demonstrated for use in RoF networks [33]. The encoder architecture is shown in Figure 10.17. The received RF signal from the antenna is directly modulated onto the envelope of a laser diode. The envelope modulated optical signal is sampled by a high-speed optical switch or modulator. Under normal sampling conditions, the modulator is driven by a pulse sequence with a constant repetition frequency. However, to facilitate multiple access to the fiber medium for each RBS, a CDMA time code is used to impose a unique code signature on the sampled signal that can be distinguished from the other RBS nodes at a radio control station (RCS). The decoder architecture shown in Figure 10.18 is used to recover the desired RF signal from the composite optical waveform containing multiplexed signals from all RBS nodes on that branch of the network. The sampled output of the modulator is detected by a photodetector and passed through a bandpass filter to reconstruct the original RF signal. The optical carrier provides remote backhaul of the microcellular RF signal to a centralized facility for processing without requiring technology specific to a given RF format to be fielded at the remote station.

To faithfully sample and reconstruct the original RF signal using this optical CDMA approach, the bandwidth of the components and the code sequence frame rate must be compared to the desired spectrum from the RF antenna. For a naturally

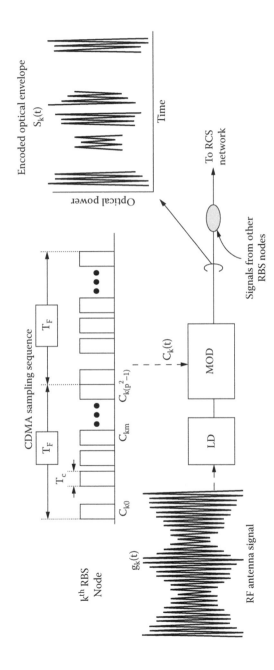

FIGURE 10.17 Architecture for an RBS encoder using a CDMA sequence, $c_k(t)$, to sample and optically encode the received RF signal, $g_k(t)$ from the antenna. Optically encoded RF signals from other radio base stations (RBS) are added to the network via an optical coupler and transmitted to the RCS fiber network. LD: laser diode; MOD: optical modulator. (Adapted from [33], © 2003 IEEE.)

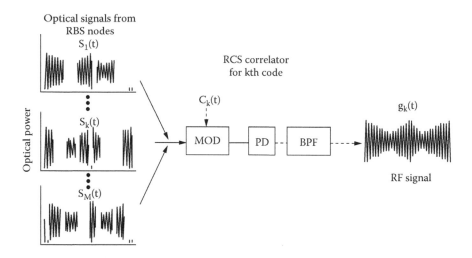

FIGURE 10.18 Architecture for a decoder at an RCS node for the kth RBS code. The optical modulator is driven by the original code sequence, $c_k(t)$, to select the desired signal from the composite optical transmissions from all RBS nodes. A photodetector (PD) followed by a bandpass filter (BPF) is used to convert the optically correlated signal to the original RF waveform. (Adapted from [33], © 2003 IEEE.)

sampled bandwidth limited signal, the sampling frequency must be at least twice the bandwidth of the RF signal so that $f_s \geq 2\Delta f_{RF}$ where Δf_{RF} is the RF signal bandwidth. For a given sampling code, $c_k(t)$, used to drive the optical switch, the code frame rate, $1/T_F$, should therefore satisfy the relation $1/T_F \geq 2\Delta f_{RF}$. For a direct code sequence based on the prime codes (see Chapter 2) with length p^2, for instance, the minimum chip width, T_c, for this system is given by $T_c = T_F/p^2$. The system scalability of the RoF network based on the code sampling approach has been analyzed using a variety of different codes and architectures for the decoder [33]. The analysis indicates that more than 50 RBS nodes with a 900 kHz RF bandwidth centered on a carrier frequency of 1.93 GHz can be supported using 10 GHz optical modulators or switches and a balanced detection scheme [33].

Multiple access optical networks can improve the scalability of the wireless system by providing RF backhaul transport to a central control station. This architecture supports low-cost microcellular networks by alleviating the remote node from performing modulation and demodulation functions for processing the RF signals. The flexibility in code design and ease of deployment of optical CDMA is an excellent candidate for providing optical layer multiple access for radio over fiber networks.

10.3.3 ROBUST LANS AND INTERCONNECTS FOR HOSTILE ENVIRONMENTS

The primary emphasis in the discussion of optical CDMA in LAN systems focused on architectures that provide multiple access and high capacity simultaneously. The maximum capacity of most optical CDMA LANs is analyzed as a function of the number of users since multiple access interference (MAI) ultimately limits the network performance. However, these analyses assume that the LAN and its subsystems are operating within their design specifications and tolerances. In a hostile environment such as a mobile military platform (e.g., aircraft), mechanical stresses and large thermal gradients across the system can dramatically impact the performance of optical components in a LAN or interconnect. In traditional WDMA or TDMA optical access architectures, strict temperature control is required to maintain wavelength or time slot registration to prevent collisions and outages on the network. This added complexity increases the cost of high-speed optical systems in hostile environments.

The design flexibility inherent in optical CDMA enables capacity in the network to be traded for robustness to environment effects. Codes with sufficient length and weight can be used to provide additional redundancy in the transmitted optical data to tolerate dynamic changes in system components. Indeed, one of the first optical CDMA schemes studied in the late 1980s was designed for its robustness to laser frequency drift — a common problem in optical communications and laser source design at the time [34]. Optical spread spectrum was used in a local area network to tolerate the random drift inherent in the laser center frequencies in the network. The same advantages hold for hostile environments where it may not be possible to guarantee frequency stabilization under dramatic temperature shifts in the platform or in access networks where temperature control adds increased cost to the system design.

More recently, fast-frequency hopping optical CDMA has been analyzed in the presence of large temperature variations [35]. By using broadband optical sources whose spectra are largely insensitive to temperature variations, encoders based on a serial cascade of fiber Bragg gratings (FBGs) are used to encode a wavelength hopping pattern on the transmitted data. The correlating receiver uses a tunable cascaded section of FBGs to select the desired transmission from the network. (Note that FBG encoders and decoders have been previously described in Chapter 4.) Although the center wavelengths of FBGs are sensitive to temperature changes, it can be shown that the temperature shift is nonfrequency selective [35]. The center frequencies of the FBGs are all shifted by the same amount in the presence of a temperature change as long as a temperature gradient does not also exist across the short serial array of FBGs. These temperature variations can be easily accommodated by making incremental adjustments to the tunable decoder. The decoder initially tunes to the known frequency-hopping pattern at a nominal temperature. It then performs a hunting step where it makes small incremental shifts to the center frequencies of all FBGs in the encoder simultaneously until it finds the maximum autocorrelation signal [35]. The model of the system showed that ambient temperature fluctuations of $\pm 100°$ C could be tolerated by this optical CDMA scheme. The results hold for any FFH encoding device using components that exhibit non–frequency selective temperature drift such as fiber Bragg gratings and distributed feedback lasers [35].

In addition to providing robustness to temperature changes at each of the encoders, the non–frequency selective drift allowed in the system has the counterintuitive advantage of supporting more users than the temperature stabilized version of the same system for a given system BER. In the case where there is a uniform distribution of temperatures between each of the encoders, the magnitude of the multiple access interference (MAI) is reduced since few of the encoders actually share frequency alignment among any of the chips used in their codes. In an analysis of the system, the number of simultaneous users at an average system BER of 1E-09 increased from 45 in the perfectly temperature stabilized system to 75 in the unstabilized system in the presence of a uniform distribution of temperatures across all encoders [35]. In a hostile environment, where nonuniform temperature variations are expected, the number of simultaneous users lies somewhere between these values.

While it is certainly possible to trade system capacity for robustness and resiliency using optical CDMA, certain FFH encoder architectures can actually outperform temperature-stabilized networks in hostile environments. By using optical CDMA approaches that do not require time synchronization, high bandwidth data networks can be constructed to meet the challenging environments of military platforms. Additionally, these CDMA approaches provide a low-cost solution for access networks and LANs where temperature stabilized components would not be cost effective.

10.3.4 OPTICAL SENSOR MULTIPLEXING

Due to the sensitivity of optical components to changes in temperature and length, optical sensors have become increasingly important for many commercial and military applications. By interconnecting multiple fiber optic sensors in a network, measurements of temperature gradients, stress, and mechanical vibration across the network are possible. Arrays of passive optical components such as fiber Bragg gratings or coils of fiber distributed over long passive folded buses enable many sensors to be multiplexed onto a single return path and received at the head-end of the bus. This architecture provides colocation of the active elements such as lasers and processing elements at a single head-end node without requiring the distribution of electrical power or processing electronics in the sensor array. A multiplexing technique for combining the optical sensor outputs onto a single fiber return path and a mechanism to uniquely distinguish the signal from each sensor is required at the head-end to read the output from each element.

Similar to the multiplexing techniques in multiple access networks, time division, wave division, and code division have all been proposed and demonstrated for use in fiber optic sensor arrays. Techniques for using code division multiplexing to uniquely address individual sensors along an array were initially investigated in the late 1980s and early 1990s [36–38]. Although there are several approaches for using code division in a fiber optic sensor, the principle of operation is illustrated in Figure 10.19 [38]. A single laser source at the head-end of the sensor array is modulated with a pseudo-random sequence (PRBS) that is transmitted along the upper fiber path common to each sensor element. A fiber optic coupler joins one arm of each sensor to the transmit path. The phase or intensity of the optical signal is modulated by the time-dependent transfer function of the sensor. (Note that phase modulation can be transformed into intensity modulation using interferometric techniques.) The output from each sensor is multiplexed onto the common fiber return path to form a composite signal that is analyzed by the head-end receiver.

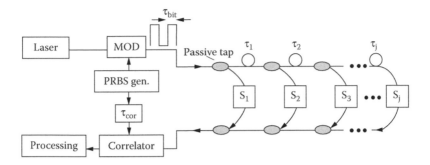

FIGURE 10.19 General architecture of fiber sensor array that uses code correlation to isolate the response of each sensor stage, S_j. The correlator uses a time delayed version of the original sequence to isolate the sensor response from an individual element by sweeping the delay, τ_{cor}, through all possible values.

In a large array it may be difficult to precisely control the time separation between the individual sensor stages during manufacturing and while in operation. On a long sensor array that is undergoing mechanical stress and strain, the time separation between the sensor stages can change dynamically. These effects can limit the ultimate scalability and accuracy of the sensor. The code correlation approach, however, does not require that the individual stages have uniform or precisely known delay. Each stage can have a unique delay as long as the condition $\tau_j > \tau_{bit}$ is satisfied where j represents a specific sensor along the array. As a result, the delay between the sensors does not have to be fixed or controlled accurately. It is sufficient that when $\tau_{cor} = \tau_j$, the response of the jth sensor stage is correlated from the composite optical signal.

Figure 10.20 shows an example of a fiber optic sensor array that uses code division techniques for distinguishing the response of individual hydrophones used to measure acoustic vibrations in the marine environment [39]. The sensor network architecture is formed from an array of N Sagnac interferometers where each interferometer

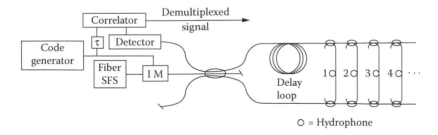

FIGURE 10.20 Example of a Sagnac sensor array using PRBS code correlation for separating the responses of adjacent hydrophone sensor stages. The multistage Sagnac interferometer converts differential phase shifts experienced by optical waveforms that pass through the same hydrophone sensor at different times into intensity variations detected by correlating the time shifted PRBS code at the receiver. (After [39], © 1999 IEEE.)

has a successively longer total propagation delay than the previous stage. To illustrate its operation as an acoustic sensor, consider the limiting case of a single Sagnac interferometer where only the first hydrophone stage is attached to the array. A broadband optical signal generated by a short coherence length source (such as a super-fluorescent source [SFS]) is split by a 3×3 coupler into identical copies that propagate in the clockwise (CW) and counterclockwise (CCW) directions around the fiber loop. A fiber delay line located on one side of the hydrophone provides a long path delay of at least 50 μsec (>10 km fiber) such that the CCW propagating light will pass through the sensor before the CW propagating signal. The hydrophone is constructed by tightly winding 10 to 100 m of fiber around an acoustic sensitive mandrel. The mechanical vibrations of the mandrel impart small variations on the relative optical phase between the CW and CCW signals through the mechanical expansion and contraction of the mandrel. These phase differences are converted into intensity variations in the Sagnac interferometer output at the head-end. Because the counter propagating signals will pass through the mandrel at different times, the relative phase difference, and hence maximum acoustic frequency response of the interferometer, is inversely proportional to the time delay provided by the fiber optic delay line. By low pass filtering (<10 kHz) and processing the output of an optical receiver, the acoustic vibrations of the hydrophone can be measured carried on the optical intensity envelope.

By incorporating additional fiber optic splitters in the loop of the interferometer as shown in Figure 10.20, multiple Sagnac interferometers and, therefore, multiple hydrophone sensors can be accommodated by a single head-end transmitter and receiver. The time varying acoustic response of multiple hydrophones can be measured since each successive sensor has a longer total path delay than the previous stage. To distinguish the individual response of each sensor, code correlation of the PRBS signature sequence is used. A fiber-based super fluorescent source (SFS) generates a broad spectral continuum that is intensity modulated with a pseudorandom PRBS code sequence. The bit duration of the modulated code, τ_{bit}, is shorter than the propagation delay, τ_j, between adjacent sensor stages. To detect the modulated signal from a particular sensor stage, a time delay sweeps the reference PRBS sequence through all possible delay values, τ_{cor}, to perform code correlation at the receiver. Electronic or optical correlation techniques can be used to distinguish the desired sensor modulated code pattern from the composite optical signal. Differential detection can also be applied to enhance the correlator contrast [39].

Many of the asynchronous optical CDMA correlation techniques discussed in previous chapters can be used in serial fiber sensor arrays to improve the correlator sensitivity, scalability, and signal to noise ratio. In addition to distinguishing the response of each sensor using a modulated PRBS sequence, passive optical CDMA encoders could be added to each sensor stage to impart a unique optical CDMA code that is distinguishable using a tunable optical decorrelator at the head-end receiver. This approach could leverage the processing gain inherent in an all-optical implementation without requiring electronic code correlation.

The robustness of code division techniques has the potential to enable practical implementations of simple passive multisensor fiber arrays. These systems, which are designed to measure the analog perturbations of optical sensors, differ substantially from the large body of research and analysis that has been devoted to optical CDMA data communications networks. Further study is required to quantify the sensitivity of code division approaches for various sensor applications. Nevertheless, fiber optical sensing is a promising area that can leverage many of the concepts and subsystems developed for optical CDMA and illustrates the wide range of possible applications of the technology to other fields of study.

10.4 SUMMARY

The variety of applications that have been pursued since the first optical CDMA experimental systems were initially proposed is a reflection of the flexibility inherent in the implementation of the coding techniques and network architectures. Optical CDMA LANs based on optical star and bus architectures have been analyzed in detail while fiber ring and mesh networks which require active code selection and forwarding are just beginning to be explored by the research community. Further work is needed to develop multiple access protocols, code distribution techniques, and performance management systems to implement optical CDMA in practical commercial and enterprise LANs and access networks. Although high bandwidth, multiuser optical LANs have largely motivated the development of optical CDMA, progress has also been made in applying the technology to many unique and specialized areas including spatial image multiplexing, radio over fiber networks, interconnects for hostile environments, and optical sensor networks. Through further integration of the subsystem components and continued progress in optical processing, lower cost modules for implementing optical CDMA systems may make many of these applications practical in future systems.

REFERENCES

[1] Metcalfe, R. M., Boggs, D. R. (1976). Ethernet: Distributed packet switching for local computer networks. *Communications of the ACM* 19:395–404.

[2] Nowatzyk, A., Prucnal, P. R. (1995). Are crossbars really dead? The case for optical multiprocessor interconnect systems. 22nd International Symposium on Computer Architecture, Santa Margherita, Ligure, Italy.

[3] Razavi, M., Salehi, J. A. (2002). Statistical analysis of fiber-optic CDMA communication systems: Part I: Device modeling. *J. Lightwav. Technol.* 20:1304–1316.

[4] Razavi, M., Salehi, J.A. (2002). Statistical analysis of fiber-optic CDMA communication systems: Part II: Incorporating multiple optical amplifiers. *J. Lightwav. Technol.* 20:1317–1328.

[5] Hui, J. Y. (1985). Pattern code modulation and optical decoding — A novel code-division multiplexing technique for multifiber networks. *IEEE J. Select. Areas Commun.* SAC-3:916–927.

[6] Park, E. Mendez, A. J., Garmire, E. M. (1992). Temporal/spatial optical CDMA networks-design, demonstration, and comparison with temporal networks. *IEEE Photon. Technol. Lett.* 4:1160–1162.

[7] Shivaleela, E. S., Sivarajan, K. N., Selvarajan, A. (1998). Design of a new family of two-dimensional codes for fiber-optic CDMA networks. *J. Lightwav. Technol.* 4:501–508.

[8]   Stok, A., Sargent, E. H. (2002). The role of optical CDMA in access networks. *IEEE Commun. Mag.* 14:83–87.

[9]   Wu, J., Gu, F. -R., Tsao, H. -W. (2004). Jitter performance analysis of SOCDMA-based EPON using perfect difference codes. *J. Lightwav. Technol.* 22:1309–1319.

[10] See the Passive Optical Networks Forum: www.ponforum.org.

[11] Xin, W., Chang, G. K., Meagher, B. W., Yoo, S. J. B., Jackel, J. L., Young, J. C., Dal, H., Ellinas, G. (1999). Chaotic lasing effect in a closed cycle in transparent wavelength division multiplexed networks. Optical Fiber Communications Conf. and the Intl. Conf. on Integrated Optics and Optical Fiber Communication, San Diego, CA, 1:246–248.

[12] Saleheen, H. I. (2001). Closed cycle lasing of ASE noise in a WDM ring network. The 4th Pacific Rim Conf. on Lasers and Electro-Optics 2:II-558–II-559.

[13] Wu, J., Lin, C. -L. (2000). Fiber-optic code division add-drop multiplexers. *J. Lightwav. Technol.* 18:819–824.

[14] Lam, C. F., Tong, D. T. K., Wu, M. C., Yablonovitch, E. (1998). Experimental demonstration of bipolar optical CDMA using a balanced transmitter and complementary spectral encoding. *IEEE Photon. Technol. Lett.* 10:1504–1506.

[15] Brès, C. -S., Glesk, I., Runser, R. J., Prucnal, P. R. (2004). All-optical OCDMA code drop unit for transparent ring networks. The 17th Annual Meeting of the IEEE Lasers and Electro-Optics Society, Nov. 7–11, Rio Grande, Puerto Rico, 2:501–502.

[16] Sokoloff, J. P., Prucnal, P. R., Glesk, I., Kane, M. (1993). A terahertz optical asymmetric demultiplexer (TOAD). *IEEE Photon. Technol. Lett.* 5:787–790.

[17] Gurkan, D., Kumar, S., Sabin, A., Willner, A., Parameswaran, K., Fejer, M., Starodubov, D., Bannister, J., Kamath, P., Touch, J. (2003). All-Optical wavelength and time 2-D code converter for dynamically-reconfigurable O-CDMA networks using a PPLN waveguide. Optical Fiber Communications Conf., Los Angeles, CA, 2:654–656.

[18] Willebrand, H. A., Ghuman, B. S. (2001). Fiber optics without fiber. *IEEE Spectrum* 38(8):41–45.

[19] Gagliardi, R. M. (1995). Pulse-coded multiple access in space optical communications. *IEEE J. Sel. Areas Commun.* 13:603–608.

[20] Ohtsuki, T. (2003). Performance analysis of atmospheric optical PPM CDMA systems. *J. Lightwav. Technol.* 21:406–411.

[21] Elmirghani, J. M. H., Cryan, R. A. (1994). New PPM-CDMA hybrid for indoor diffuse infrared channels. *Electron. Lett.* 30:1646–1647.

[22] Kahn, J. M., Barry, J. R. (1997). Wireless infrared communications. *Proc. of the IEEE* 85:265–298.

[23] Hsu, C. -S., Li, V. O. K. (1997). Performance analysis of slotted fiber-optic code-division multiple-access (CDMA) packet networks. *IEEE Trans. Commun.* 45:819–828.

[24] Hsu, C. -S., Li, V. O. K. (1997). Performance analysis of unslotted fiber-optic code-division multiple-access (CDMA) packet networks. *IEEE Trans. Commun.* 45:978–987.

[25] Muckenheim, J., Hampicke, D. (1997). Protocols for optical CDMA local area networks. Proc. Eur. Conf. Networks Optical Commun. (NOC '97), Antwerp, June, pp. 255–262.

[26] Stok, A., Sargen, E. H. (2002). System performance comparison of optical CDMA and WDMA in a broadcast local area network. *IEEE Commun. Lett.* 6:409–411.

[27] Shalaby, H. M. H. (2003). Optical CDMA random access protocols with and without pretransmission coordination. *J. Lightwav. Technol.* 21:2455:2462.

[28] Shalaby, H. M. H. (2004) Performance analysis of an optical CDMA random access protocol. *J. Lightwav. Technol.* 22:1233:1241.

[29] Gagliardi, R. M., Mendez, A. J., Dale, M. R., Park, E. (1993). Fiber-optic digital video multiplexing using optical CDMA. *J. Lightwav. Technol.* 11:20–26.

[30] Kitayama, K. -I. (1994). Novel spatial spread spectrum based fiber optical CDMA networks for image transmission. *IEEE J. Select. Areas Commun.* 12:762–772.

[31] Kitayama, K. -I., Nakamura, M., Igasaki, Y., Kaneda, K. (1997). Image fiber-optic two-dimensional parallel links based upon optical space-CDMA: Experiment. *J. Lightwav. Technol.* 15:202–212.

[32] Salehi, J. A. (1989). Code division multiple-access techniques in optical fiber networks: Part I: Fundamental principles. *IEEE Trans. Commun.* 37:824–833.

[33] Tsukamoto, K., Higashino, T., Nakanishi, T., Komaki, S. (2003). Direct optical switching code-division multiple-access system for fiber-optic radio highway networks. *J. Lightwav. Technol.* 12:3209–3220.

[34] Foschini, G. J., Vannucci, G. (1988). Using spread spectrum in a high capacity fiber-optical local network. *J. Lightwav. Technol.* 6:370–379.

[35] Fathallah, H., Rusch, L. A. (1999). Robust optical FFH-CDMA communications: Coding in place of frequency and temperature controls. *J. of Lighwav. Technol.* 17:1284–1293.

[36] Everard, J. K. A. (1987). Novel signal processing techniques for enhanced OTDR sensors. Proc. Fiber Optic Sensors II, The Hague, SPIE 798:42.

[37] Al–Raweshidy, H. S., Uttamchandani, D. (1990). Spread spectrum technique for passive multiplexing of interferometric fiber optic sensor. Proc. Fiber Optics '90, London, SPIE 1314:342.

[38] Kersey, A. D., Dandridge, A., Davis, M. A. (1992). Low-crosstalk code-division multiplexed interferometric array. *Electon. Lett.* 28:351–352.

[39] Vakoc, B. J., Digonnet, M. J. F., Kino, G. S. (1999). A novel fiber-optic sensor array based on the sagnac interferometer. *J. Lightwav. Technol.* 11:2316–2326.

Index

T - #0351 - 071024 - C404 - 234/156/18 - PB - 9780367391478 - Gloss Lamination